Organism and Environment

AF677048

Organism and Environment

Ecological Development, Niche Construction, and Adaptation

Sonia E. Sultan
Department of Biology and Program in Environmental Studies
Wesleyan University

OXFORD
UNIVERSITY PRESS

OXFORD
UNIVERSITY PRESS

Great Clarendon Street, Oxford, OX2 6DP,
United Kingdom

Oxford University Press is a department of the University of Oxford.
It furthers the University's objective of excellence in research, scholarship,
and education by publishing worldwide. Oxford is a registered trade mark of
Oxford University Press in the UK and in certain other countries

© Sonia E. Sultan 2015

The moral rights of the author have been asserted

First Edition published in 2015

Impression: 1

All rights reserved. No part of this publication may be reproduced, stored in a retrieval system, or transmitted, in any form or by any means, without the prior permission in writing of Oxford University Press, or as expressly permitted by law, by licence or under terms agreed with the appropriate reprographics rights organization. Enquiries concerning reproduction outside the scope of the above should be sent to the Rights Department, Oxford University Press, at the address above

You must not circulate this work in any other form
and you must impose this same condition on any acquirer

Published in the United States of America by Oxford University Press
198 Madison Avenue, New York, NY 10016, United States of America

British Library Cataloguing in Publication Data
Data available

Library of Congress Control Number: 2015943404

ISBN 978–0–19–958707–0 (hbk.)
ISBN 978–0–19–958706–3 (pbk.)

Printed in Great Britain by
Clays Ltd, St Ives plc

Links to third party websites are provided by Oxford in good faith and
for information only. Oxford disclaims any responsibility for the materials
contained in any third party website referenced in this work.

This book is dedicated to my teachers, Fakhri Bazzaz and Dick Lewontin,

and to Kendall and Jasmine, with love

Acknowledgments

I am very grateful to a number of individuals and organizations whose contributions strengthened and supported the book. The following colleagues kindly offered expert comments on parts of the text: Ann C. Burke, Robert Dorit, James Griesemer, Jason Herman, Felicia Low, and William F. Morris. I am especially grateful to Ben Kerr, Peter Gluckman, and Hamish Spencer, each of whom read extended parts of the manuscript and spent hours with me in intensive discussion. The insight, rigor, and generosity of all of these early readers are deeply appreciated. I also thank Kevin Laland, Felicia Low, and Dylan Schwilk for suggesting references in specific areas; Mary Rumpho for feedback on the description of *Elysia chlorotica*; Scott Gilbert; Eva Jablonka; H. Frederik Nijhout; Dominique Pestre; Johanna Schmitt; and Mary Jane West-Eberhard for advice and encouragement during the course of the writing. Scientists, nature bloggers, and others who have generously allowed use of photographs and other images are credited individually in figure captions; I very much appreciate their contributions. Hea-Ream Lee carefully formatted the reference list, and William Fraker assisted in compiling images and copyright information. I also thank Christine Boylan, Elizabeth Farrell, Kaarkuzhali Gunasekaran, and the production staff at Oxford for their gracious and skillful assistance.

A year's fellowship at the extraordinary Wissenschaftskolleg zu Berlin (2012–2013) provided the time and freedom (in every sense) to investigate and develop these ideas. My deepest thanks to the Fellows who inspired and drank with me, and to the dedicated and able staff. I also thank The Eppley Foundation for Research, which generously funded a semester's leave devoted to writing. I am grateful to the scientific staff of The Liggins Institute, University of Auckland, for stimulating and extremely fruitful discussion, and to the institute itself and Wesleyan University's Center for Global Studies for funding a 2014 visit.

Finally, my thanks to the general editors of the Oxford Series in Ecology and Evolution, Robert May and Paul Harvey, for their initial interest in the project, and Charles Godfray for his constructive comments on an early chapter. Editors Lucy Nash and Ian Sherman have provided support throughout, and I am grateful for their patience.

Contents

Introduction

A flood of new research insights has revealed as never before the intimate causal relationship in which organisms and their environments are bound together. Development is modulated—in astonishingly complex ways—by induced epigenetic modifications, parental nutrient and stress levels, chemical and physical conditions at the cellular level, and a host of other environmental factors and feedbacks. Just as the environment participates in shaping the organism, the organism's presence, activities, products, and byproducts modify its environment. By virtue of these environmental modifications, organisms influence their own functional and selective circumstances and, through effects on shared habitats, those of cohabiting neighbors. At this point, these phenomena are solidly established. The challenge is to expand our understanding to encompass these complex causal interactions and feedbacks.

This book is a close re-examination of the organism–environment relationship. It draws on a rich and diverse biological literature to explore how environments influence individual phenotypes and how these realized phenotypes in turn influence their environments. The book goes on to investigate the broader ecological and selective consequences of these reciprocal effects. Examples and case studies from across the biodiversity spectrum are discussed. Throughout the book, salient research questions and potential approaches are highlighted to make clear how this integrative perspective can generate testable hypotheses; an overall research agenda and some practical applications are laid out in an epilogue.

In focusing on the organism–environment relationship, the book builds on, and synthesizes, two recently emerged disciplines. First, recognition of the environment's role in shaping gene expression, and, hence, individual phenotypes, has led to the emergence of *ecological developmental biology* (S. Gilbert 2001; Sultan 2007; S. Gilbert and Epel 2009). In an "eco-devo" approach, developmental mechanisms and outcomes are explicitly studied in the context of environmental conditions. Flexible individual responses to the environment, or *plasticity*, are included within this broad, unified approach to environmentally mediated development.

The second newly emerged discipline focuses on the other side of the organism–environment relationship—that is, the effects of organisms on the environment. Individually and collectively, subtly or dramatically, organisms inevitably alter their abiotic and biotic circumstances. These organismic impacts, and the resulting changes in selection pressures, are known as *niche construction* (Lewontin 1983, 2000; Odling-Smee et al. 2003; Scott-Phillips et al. 2013; or, alternatively, *eco-evolutionary feedbacks*, Pelletier et al. 2009; Travis et al. 2014). Each of these exciting new areas of study focuses on only one of the causal arrows that constitute the organism–environment relationship. To more fully understand this dynamic, reciprocal relationship, the influences of organisms and environments on each other must *both* be taken into account. Accordingly, the book further aims to bring together eco-devo insights with those from niche construction, to determine the joint consequences of these processes for ecological interactions and natural selection. The goal is to frame a more integrated understanding in light of current developmental, ecological, and evolutionary knowledge.

The book begins with a chapter on development. This initial chapter examines the regulatory roles of epigenetic and environmental factors to argue

Organism and Environment. First Edition. Sonia E. Sultan.
© Sonia E. Sultan. 2015. Published 2015 by Oxford University Press.

for a unified eco-devo approach based on the *norm of reaction*. With this eco-devo approach as a starting point, the second chapter builds a conceptual framework for understanding the adaptive encounter between the individual organism and its environment. The third chapter describes the precise cues and transduction mechanisms that underlie plastic eco-devo responses. The subsequent two chapters explore in detail how such phenotypic responses modify the environmental conditions that organisms experience, and how diverse organisms modify their external environments across temporal and spatial scales. The sixth chapter discusses the consequences of individual eco-devo responses (*trait-mediated interactions)* and niche-constructing impacts for the structure and function of ecological communities. A substantial final chapter considers natural selection in the light of the organism–environment relationship, including a discussion of transient and inherited environmental effects on phenotypes (such as induced epigenetic changes), and selective feedbacks that result from the effects of organisms on their environments. These mutual influences make clear that adaptation and, consequently, natural selection result from the functional and fitness consequences of the two-way organism–environment relationship.

The book aims to offer a shift in emphasis and focus that may lead to new and different insights, fresh ways to interpret known phenomena, and research questions that might not otherwise arise—a "way of seeing" (as Richard Dawkins described his influential 1982 book, *The Extended Phenotype*). The reader is invited to think through a diverse array of biological scenarios and to consider in his or her own terms several key issues that are raised as open questions. Throughout, the goal is to discover where this shift in focus might lead.

The opportunity to develop an integrated understanding of the organism–environment relationship comes at a critical time and, indeed, not a moment too soon. The emergent field of ecological developmental biology has directed attention to the mechanisms and outcomes of individual development in the context of real-world environmental conditions. As a result of escalating human activities, natural habitats are being disrupted and transformed at an unprecedented rate. Biological researchers are keenly aware of the interdependence of organisms and their environments. We seek urgently to know how altered climatic, atmospheric, and biotic conditions will affect the development and function of organisms—their phenotypes as expressed in the real world—and how these effects will in turn alter their persistence, ecological communities, and evolutionary trajectories into the future. These questions provide powerful and immediate motivation to work toward an updated conceptual scheme for biological inquiry, in which the complex interactive relation of organisms to their environments occupies the central place.

CHAPTER 1

The environmental context of development

The view that genes control all cellular activities in development was initially promoted by early twentieth-century biologists such as T. H. Morgan and William Bateson, at a time when the mechanics of inheritance were first being worked out (Sarkar 2006). In the post-Watson and Crick version of this view, DNA functioned as a "master molecule" that controlled development from within the nucleus by giving "commands" to differentiating cells. Because this model made it possible to explain development strictly in terms of the information content of genes, it held out the intoxicating promise that molecular tools would enable biologists to fully "solve" organisms (including ourselves) by knowing their DNA sequences and tracing determinate pathways of gene expression. One relatively trivial legacy of this view is the popular habit of describing the style of a sports car, the ethos of a corporation, or the strategy of a football team as "in its DNA"; other cultural and social legacies of genetic determinism are considerably more complex (see Kevles and Hood 1992). In biology, this model underlies the one-to-one mapping of phenotype onto genotype that was implicit in the mid-century New Synthesis framework for evolution as allele frequency change (V. Lynch and Wagner 2008). It also logically pointed to a strictly gene-based approach to the study of individual development.

A view of DNA as the unique source of developmental information arose through foundational genetics studies in which both the environmental and the genomic contexts of specific allelic variants were held constant (Neumann-Held 2006). Under those conditions, certain DNA sequence variants can be consistently associated with aspects of phenotypic variation. Of course, when only one factor is experimentally varied, different outcomes are attributable to that factor, but this does not mean that other factors are irrelevant under natural conditions, where they too vary. Concluding that genes alone determine phenotypes based on these allelic variation studies is circular reasoning, to be sure. Yet, by a process of conceptual shortcuts, what began as a useful research program for basic transmission genetics has resulted in a deeply entrenched, unique status for genes as quintessential biological information. Until recently, this consensus view was accompanied by a neglect, in mainstream research, of contextual factors in gene expression (Griffiths 2006; Oyama 2006). For instance, in 2010 the National Center for Biotechnology Information (NCBI) website of the US National Institutes of Health offered the following "science primer":

> Every organism, including humans, has a genome that contains all of the biological information needed to build and maintain a living example of that organism. The biological information contained in a genome is encoded in its *deoxyribonucleic acid* (*DNA*) and is divided into discrete units called *genes*. Genes code for proteins that attach to the genome at the appropriate positions and switch on a series of reactions called gene expression . . . Today we know that a single gene consists of a unique sequence of DNA that provides the complete instructions to make a functional product, called a protein. Genes instruct each cell type—such as skin, brain, and liver—to make discrete sets of proteins at just the right times, and it is through this specificity that unique organisms arise. (NCBI 2010)

As of early 2015, a reorganized website explained that the DNA molecule "holds the blueprint for

Organism and Environment. First Edition. Sonia E. Sultan.
© Sonia E. Sultan. 2015. Published 2015 by Oxford University Press.

how living organisms are built" (National Library of Medicine and National Human Genome Institute 2015). If each gene's DNA sequence "provides the complete instructions" for the organism's growth and differentiation, genes comprise the only source of developmental information, and the genome can be studied as an assemblage of these "discrete" informational units (National Library of Medicine and National Human Genome Institute 2015).

In reality, DNA sequences do not provide this kind of self-contained instruction but affect development differently depending on conditions at several biological levels. Contemporary biologists face the central challenge of restoring the genomic, epigenetic, and environmental context dependence of gene expression to our understanding of development. Meeting this challenge calls for two key changes to the prevailing model: first, seeing developmental outcomes as emerging from regulatory interactions in real time rather than as scripted in advance by gene sequences; and, second, expanding this model to admit additional sources of information to the process of individual development. Together, these changes allow development to be studied in the richer organismic and real-world context we aim to understand. This chapter explores these two conceptual shifts through a brief overview of recent findings about developmental mechanisms and outcomes that lays the foundation for a norm of reaction approach, which is introduced in the final section.

1.1 The genome as a regulatory system

According to the *New York Times* (Angier 2003), the DNA molecule experienced its "midlife crisis" in 2003, the 50th anniversary of its structure's publication by Watson and Crick and the year that the complete human genome sequence was released. This crisis point has grown even more urgent, as new findings have brought increasingly into question the mainstream view of DNA sequence as a directive blueprint for gene function and hence organismic form. Is phenotypic expression in fact the determinate readout of a coded, self-contained program? Do genes specify all of the essential information for development, or do developmental instructions emerge from a system in which genes play a part? After a decade of extraordinary progress in molecular genetics and developmental biology, we can move forward from this crisis point to an expanded understanding of genes in their genomic, epigenetic, and environmental contexts.

There is a great deal of new and, in some cases, astonishing information to consider as we move toward this new understanding. In just the past decade, complete genomic sequences have become available for over 250 eukaryotes, from the common mouse, chicken, and housefly to the flying lemur and the Chinese pangolin (Kanchisa et al. 2014; The Genome Institute at Washington University 2014). These sequences reveal not, as expected, a straightforward text coding for discrete functional genes with neutral, noncoding sections interspersed, but a surprisingly high proportion of mobile genetic elements (such as transposons), transcription factors, and selectively constrained noncoding sequences, all pointing to complex, interactive, regulatory mechanisms (references in Szathmary et al. 2001; Sarkar 2006; Pagel and Pomiankowski 2008; Garfield and Wray 2010; Lenhard et al. 2012; K. Morris and Mattick 2014). For example, the results of the multiyear ENCODE project, which aimed to identify all functional elements in the human "blueprint of life," show that much of the human genome codes for regulatory elements such as noncoding RNAs (ENCODE Project Consortium 2012). With the advent of sophisticated biochemical tools to directly investigate the functional protein–protein and DNA–protein interactions that regulate gene transcription and translation, as well as in situ hybridization and immunohistochemistry techniques for visualizing the precise location and timing of gene activity and studies of transcriptomes that permit its quantification, a richly complicated picture has taken shape to replace the "wonderfully simple" one-way informational pathways from genes to proteins envisioned in the early decades of molecular biology (Keller 2000; see also S. B. Carroll et al. 2005). This picture reveals a functional complexity and pliancy, which as yet are far from understood, in the path from gene to organism (ENCODE Project Consortium 2012; Qu and Fang 2013). At this point, it has become abundantly clear that phenotypic outcomes are not rigidly predetermined by the organism's DNA sequence; indeed, in eukaryotes, the very premise that the presence of a specific DNA sequence necessarily codes for a specific protein has begun to "unravel" (Sarkar 2006; detailed history provided by K. Morris and Mattick 2014). The genome has been revealed not as a string of set instructions but as a remarkably dynamic system of coactive signals and feedbacks.

In her 1983 Nobel prize acceptance remarks (quoted by Keller 2000, 34), the iconoclastic geneticist Barbara McClintock described the genome as "a highly sensitive organ of the cell." Recent insights confirm this view of the genome in its cellular context as a *developmental system*—a complex of interacting factors that can be both robust and flexible in response to myriad internal and external inputs.[1] This system property explains not only epistasis and pleiotropy but also a number of otherwise unaccountable observations: why inactivating single genes has little or no phenotypic impact for the great majority of loci in eukaryotic genomes, including those of yeast, nematodes, angiosperms, and mammals (Pagel and Pom 2008 and references therein); why the expression of most mutations varies in different genetic backgrounds and environmental conditions (Lewontin 2000; Remold and Lenski 2004; Brem et al. 2005); and why there is no correspondence between the number of genes and phenotypic complexity (Szathmary et al. 2001; Gluckman and Hanson 2005). Furthermore, the multiple interacting signals and chemical switches that characterize gene regulatory pathways in eukaryotes lend their developmental systems a diffuse interdependence termed *weak linkage* to create an inherently flexible type of regulatory organization (Kirschner and Gerhart 1998).

Our point of departure, then, is that phenotypic expression is guided not by DNA sequence per se but by the genome's highly resilient systemic regulatory processes, which shape the extraordinarily precise tissue- and stage-specific expression of genes. In fact, one of the key insights of contemporary molecular genetics is that the majority of genes participate in such regulatory interactions rather than code for proteins (Mattick 2012). This regulatory modulation explains how evolutionarily conserved coding sequences such as specific *Hox* genes may lead to entirely different phenotypic outcomes, depending on the phylogenetically distinct "developmental system" in which they occur (Gottlieb 2004). The genes of the *Ultrabithorax* and *abdominal-A* complexes, for instance, are present in all arthropods (including insects, crustaceans, and myriapods) as well as in their more ancient sister taxon, the *Onycophora*, or velvet worms. Evolved changes in the regulation of these shared *Hox* genes along the body axis result in distinct expression domains, and consequently body plans, in these different types of arthropod: *Ultrabithorax* and *abdominal-A* are expressed in the abdominal segment in the fruit fly *Drosophila*, in the thorax of the crustacean *Artemia*, in all body segments except the head in the centipede, and only in the hindmost tip of the worm *Onycophora* (Grenier et al. 1997; Figure 1.1). Regulatory changes have also evolved in the effects of these genes on downstream target genes; their products repress the expression of *Distall-less* to prevent limb formation in the insect abdomen but not in their expression domains in the other arthropod taxa. The contributions these genes make to developmental specification are thus entirely dependent on the regulatory—that is, genomic—context in which they occur.

Such coordinated effects on gene expression result from the combined influence of multiple regulatory elements and events (e.g., Lenhard et al. 2012; for comprehensive discussions, see S. B. Carroll et al. 2005 and Davidson 2006). Identifying these factors and understanding how their enzyme kinetics and other real-time chemical interactions condition gene expression and hence development have proved complex beyond all expectation. *Cis*-regulatory elements located within or near a structural gene locus serve as regulatory sites to which activator or repressor proteins specifically bind to mediate transcriptional activity at that locus. Simultaneous binding of particular combinations of transcription factors may be needed to activate transcription and hence gene expression (Davidson et al. 2002). Different *cis*-elements come into play at

[1] This recognition by biologists follows two decades of prescient argument by philosophers of biology in favor of a "developmental systems approach" to both heredity and evolution, an approach built on a growing critique of gene-centric conceptual models; for an excellent introduction to this literature, see the edited volume by Oyama et al. 2001.

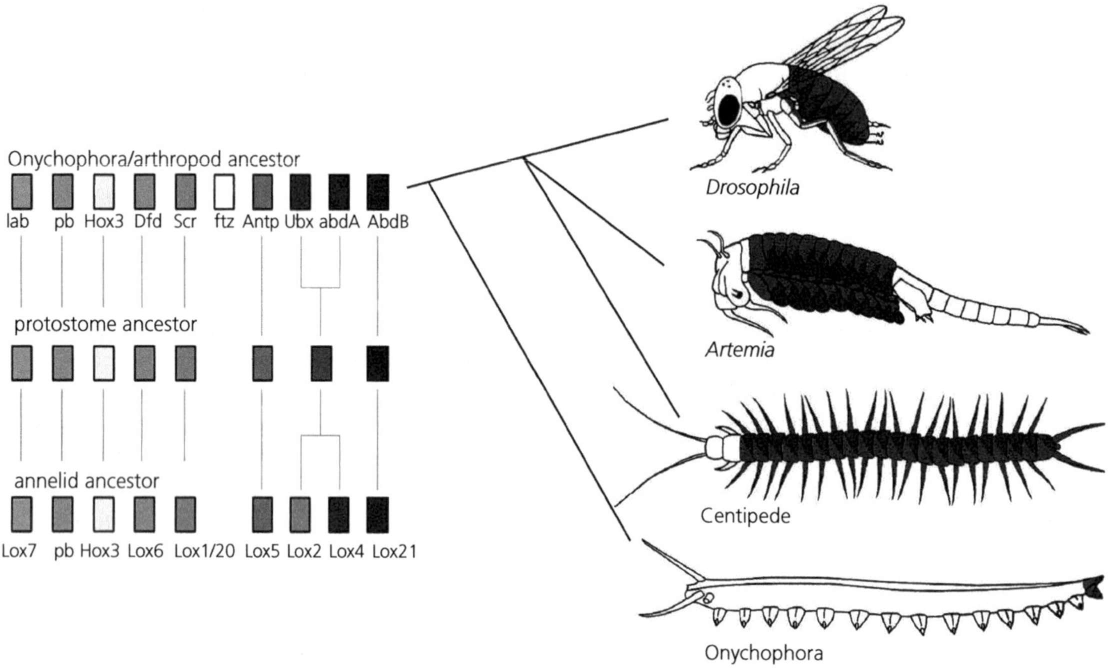

Figure 1.1 The genes of the *Ultrabithorax* (*Ubx*) and *abdominal-A* (*abdA*) complexes are present in all arthropods and in their ancient sister taxon, the velvet worms. Body-plan diversity in different arthropod groups reflects evolved regulatory changes that cause different expression domains of these shared *Hox* genes along the body axis. *Ultrabithorax* (indicated in gray) and *abdominal-A* (indicated in black) are expressed in the abdominal segment in the fruit fly *Drosophila*; in the thorax of the crustacean *Artemia*; in all body segments except the head in the centipede (a myriapod); and only in the hindmost tip of the velvet worm, *Onycophora*. Image courtesy of Jen Grenier and Steve Paddock. From Grenier et al. 1997 (caption modified), reproduced with permission of Elsevier Publishers.

different developmental stages or locations to influence the gene's transcription level, often depending on signals from multiple transcription factors (S. B. Carroll et al. 2005). At the same time, a single transcription factor or other signaling pathway component can play a number of different roles within a given organism by regulating distinct target genes in diverse tissues and developmental stages (S. B. Carroll 2008; V. Lynch and Wagner 2008). As a result of these complex interactions, subtle changes in *cis*-regulatory elements can have important and diverse pleiotropic effects (S. B. Carroll et al. 2005; G. Wagner and Zhang 2011) or, conversely, can lead to strong but more modular (i.e., localized) phenotypic impacts (Wray 2007 and references therein).

The astounding complexity of these *cis*-regulatory networks is illustrated by the exceptionally well-characterized signaling events that activate embryonic expression of the protein-coding gene *endo-16* in the sea urchin *Strongylocentrotus purpuratus*. The Endo-16 protein contributes to the specification of a particular embryonic cell lineage to form the endomesoderm tissue in the early embryo, tissue that will eventually form the larval gut lining and skeletal rods (Davidson et al. 2002; Balhoff and Wray 2005; Oliveri et al. 2008). The endomesoderm forms from the differentiation of cell lineages that derive from a ring of very early (sixth-cleavage) embryonic cells (Figure 1.2a). Based on an impressive body of molecular developmental, gene expression, and experimental embryological data, a network model has been constructed showing the *cis*-regulatory gene interactions that underlie this single developmental step (Figure 1.2b). Note that this system was chosen as the subject of a complete "genomic regulatory network" study precisely because of its relative *simplicity*: rather few genes are expressed in the sea urchin embryo; the embryo produces a morphologically simple larva with few distinct cell types; and there are relatively few regulatory steps between gene expression in the embryo and final cell fate (Davidson et al. 2002). The most important result of this intensive collaborative study may be not the specific biochemical events uncovered but the "deep, layered and hierarchical" regulatory complexities (Davidson 2010, 912) revealed to underlie even this *simple* developmental transition.

Further complexity is introduced to genomic systems by the existence of another major aspect of gene interaction. *Trans*-regulatory proteins and RNAs derived from more distant sites in the genome can also initiate or block a gene's transcription and can influence its mRNA dynamics and stability (see Lemos et al. 2008 for an excellent overview). Unlike the more modular *cis*-regulatory elements, these *trans*-acting enhancers and repressors can occur thousands of base pairs from a gene's promoter site or even on a separate chromosome and are inherited independently of the gene they regulate. But the utility of this *cis/trans* distinction is limited, because *trans*-acting factors often interact with or bind to *cis*-regulatory sequences to jointly regulate gene expression, as has been found in yeast, humans, insects, and plants (Lemos et al. 2008). Even more broadly, a strict distinction between structural and regulatory genes has broken down as the ubiquity of both direct and indirect gene–gene interactions has been revealed (Yukilevich et al. 2008). It is perhaps most accurate to see all genes as "regulatory"—that is, as components of an inherently epistatic developmental system rather than as discrete, fixed bits of information. Interestingly, this is equally true at the level of quantitative trait loci (QTLs), heritable factors typically composed of multiple DNA regions which jointly influence continuously varying (or *quantitative*) traits: the expression of a particular QTL is contingent on the rest of the genome, and QTL–QTL epistasis is well known (Mackay 2013; e.g., Weinig and Schmitt 2004; Bloom et al. 2013).

These regulatory systems provide for stunning evolutionary lability, in part because their many interacting elements provide numerous sites where mutation can create novel phenotypic effects, and hence where natural selection can act (Sultan and Stearns 2005; Garfield and Wray 2010; Moczek et al. 2011). For example, nucleotide substitutions in *cis*-regulatory regions can add or remove binding sites for particular transcription factors or change the intensity of protein binding at those sites, and new regulatory functions can be acquired if binding sites are relocated to new target genes due to recombination (Balhoff and Wray 2005). Evidence suggests that sequence variation affecting gene expression is abundant across genomes (Rifkin et al. 2005) and, further, that under artificial selection substantial differences in gene expression can evolve in very few generations (e.g., Toma et al. 2002). In fact, in contrast to expectations for neutral evolution of noncoding sequences, because of their roles in regulatory interactions such sequences can in fact be subject to intense selection (Rifkin et al. 2005; Hemberg et al. 2012; and references therein), which leads either to the evolution of highly conserved developmental mechanisms or to morphological diversification. Although mutations that alter developmental outcomes can arise in either regulatory or coding regions (Galant and Carroll 2002), *cis*-regulatory regions may constitute evolutionary "hot spots," where diversifying regulatory mutations accumulate while protein-coding regions remain stable (Stern and Orgogozo 2009). Among species of *Drosophila*, for example, there are many cryptic sequence changes that create different *cis*-binding sites for a given set of transcription factors that are essential to body patterning (Ludwig et al. 1998).

Based on these insights, a consensus is emerging that the key to patterns of trait conservation and diversity in multicellular organisms is not change in DNA sequence at functional genes but rather the evolution of regulatory interactions (S. B. Carroll 2008; V. Lynch and Wagner 2008; Garfield and Wray 2010). Evolutionary developmental (*evo-devo*) biologists explicitly study change in regulatory pathways as the basis of morphological innovations, an approach that restores developmental processes to the study of evolution after a long period of exclusion (Amundsen 2001). Evo-devo studies have produced a number of key insights to the origins of phylogenetic diversity (Wagner 2000). For instance, humans share a number of noncoding regulatory elements (such as enhancers that mediate tissue-specific gene expression) with distant vertebrate relations such as zebrafish and pufferfish, teleosts with whom humans last shared a common ancestor

(a)

(b)

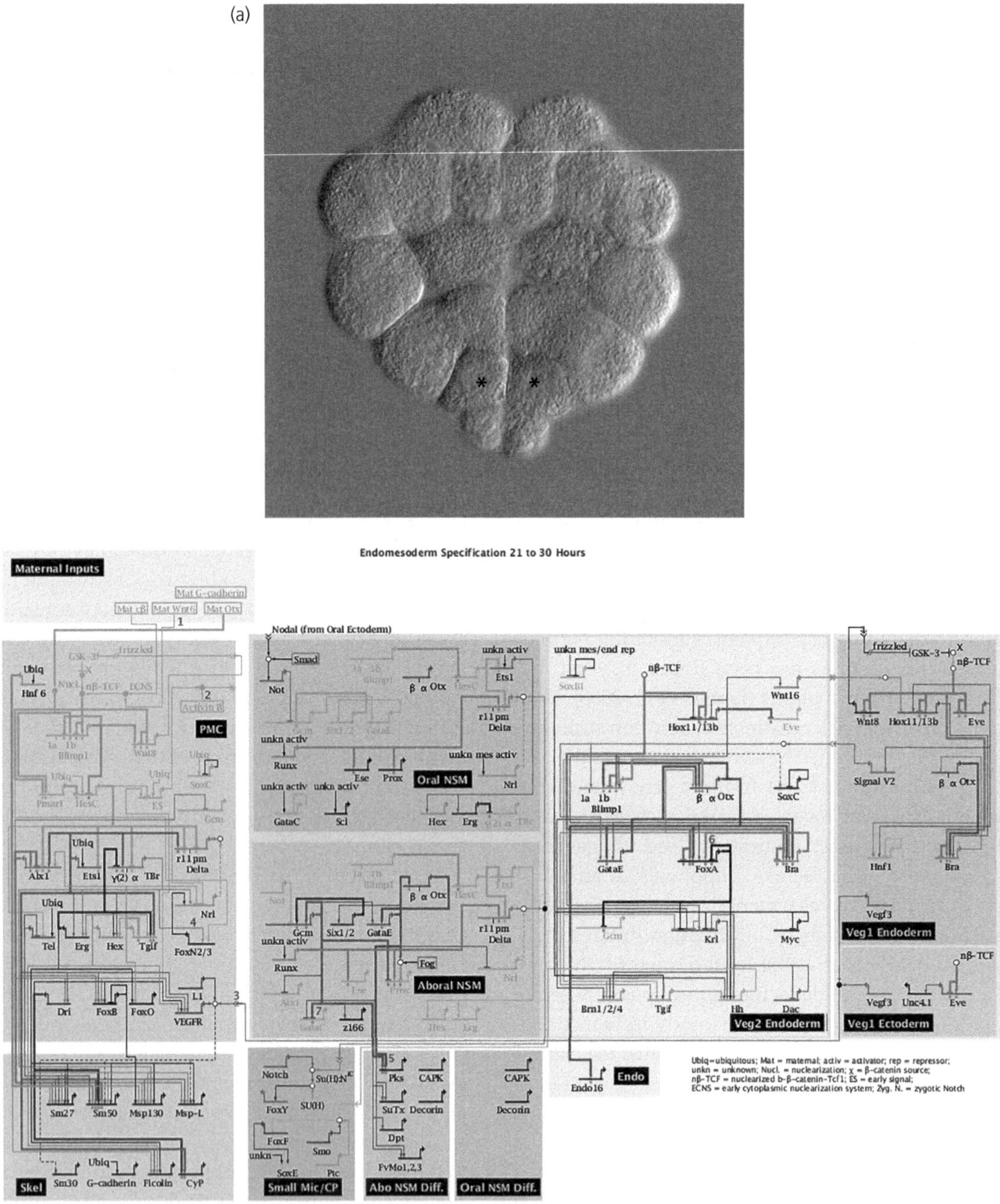

Figure 1.2 A sea urchin embryo (a) is shown at a very early (sixth-cleavage) stage of development. Four of the embryo's micromere cells (two of which are visible here, marked with asterisks) give rise to the endomesoderm cell lineage that subsequently differentiates into the animal's gut lining and skeletal rods. Image courtesy of Andrew Ransick, California Institute of Technology. In (b), a data-based network model shows the complex *cis*-regulatory gene interactions that lead to the differentiation of these endomesoderm cells. Arrows and solid lines indicate gene activation or repression; rectangles show downstream differentiation genes. The signaling pathways that activate *endo-16* expression are shown in the second panel from the right. Details are given by Davidson et al. (2002) and Peter and Davidson (2010, 2011). Image downloaded 2/22/2015 from the Davidson Lab website (http://sugp.caltech.edu/endomes/); reproduced (with modified caption) by permission of Eric Davidson, California Institute of Technology (copyright Hamid Bolouri and Eric Davidson). For the color image, see Plate 1.

almost 450 million years ago (Venkatesh et al. 2006; Figure 1.3a). The extraordinary evolutionary stability of these regulatory regions provides evidence that such regions can indeed be selectively maintained. Surprisingly, though, humans share an even larger number of conserved noncoding elements with our even more remote relatives, the cartilaginous fishes (exemplified by the elephant shark, *Callorhinchus milii*; Figure 1.3b), although we last shared a common ancestor with this early group of jawed vertebrates well over 500 million years ago (Venkatesh et al. 2006). Evidently the radiation of the teleost fishes, the largest and most diverse vertebrate group, was characterized by diversification of these various regulatory elements.

The shift from studying macroevolution as change in DNA sequence to a focus instead on changes in gene *regulation* is both exciting and daunting. Perhaps an even more difficult step is to incorporate gene regulatory dynamics into population–genetic studies of microevolution. The particular challenges of studying selection on regulatory variation arise from the epistatic complexity and pliancy of developmental systems (Garfield and Wray 2010). Genetic variation in *trans*-acting molecules such as transcription factors can be particularly difficult to identify, since these effects often arise jointly from several sites across the genome (Lemos et al. 2008). Even sophisticated genomic mapping approaches are problematic in traits involving this

Figure 1.3 Noncoding, regulatory genetic elements can be highly evolutionarily conserved. (a) The pufferfish *Fugu rubripes*, a teleost fish, shares with humans many noncoding enhancer regions that mediate tissue-specific gene expression. These regions are evidently under strong stabilizing selection. Image courtesy of Byrappa Venkatesh. (b) The genome of the elephant shark, *Callorhinchus milii*, a member of the yet more evolutionarily remote class of cartilaginous fishes, contains an even greater number of regulatory elements in common with the human genome. Image credit and copyright Doug Perrine, SeaPics.com.

kind of complex genetic architecture (Shimizu and Purugganan 2005).

One approach to this "confusing maze of revealed connections" (Koonin and Wolf 2008, 15) is to analyze networks of molecular gene regulation as complex systems of sequential interconnections and feedbacks (Davidson 2010 and references therein). Perhaps paradoxically, these analytical network studies can illuminate certain fundamental properties of genomes as evolved, biological systems. When gene regulatory networks are modeled as sets of "nodes" (i.e., functional genes) and "edges" (i.e., interactions with other genes or shared regulatory factors), a rich diversity of possible outcomes is generated, just as a given genome can result in diverse individual phenotypes. This diversity is produced even when model regulatory networks are greatly simplified by allowing only *on* and *off* nodal states and one-way edges (Lemos et al. 2008). Network approaches to actual gene expression data also provide key insights to the nature and evolution of genomes. A meta-analysis of network topologies, based on published genome sequence data and expression profiles across a broad taxonomic sample, showed significantly more highly connected gene loci compared with random gene networks (S. Bergmann et al. 2004). Using available knockout data, this study also found that the most highly connected genes tended to be implicated in essential aspects of function and were evolutionarily the most conserved (S. Bergmann et al. 2004). This result confirmed a model by A. Wagner (1996) showing that more densely connected networks were less sensitive to disruption by mutation, that is, more evolutionarily stable. Both theoretical and empirical studies thus indicate that dense interconnectedness of regulatory interactions may be a fundamental property of genomes as evolving, robust developmental systems.

A further layer of dynamic complexity must be considered in the initial step from gene sequences to phenotypic outcomes: at the posttranscriptional stage, *alternative splicing* and editing of exons (coding sequences) can produce different mRNA transcripts and hence different proteins from a single gene at different times and locations (Maniatis 1991; Mazin et al. 2013; and references therein). Remarkably, then, even the amino acid sequence in a polypeptide product may not be determined by a gene's DNA sequence (Lewontin 2000). This mechanism can produce considerable functional diversity: for example, alternative splicing of exons composed of tandem arrays in a single *Drosophila* axon guidance receptor gene, *Dscam*, can potentially generate 38,016 different protein isoforms (Crayton et al. 2006), and even more extreme cases are known. Although initially considered a rare genetic quirk, alternative splicing is now recognized as a widespread mechanism for tissue-, sex-, and stage-specific regulation of gene expression in eukaryotes (V. Lynch and Wagner 2008). In humans, for example, estimates of the proportion of genes that produce multiple (and largely tissue-specific) products due to alternative mRNA splicing have risen from approximately 10% a decade ago to between 92% and 94% (E. Wang et al. 2008). In particular, studies of the human brain reveal that, during both postnatal development and aging, widespread splicing changes occur in genes associated with ontogenetically specific events such as synapse formation and with brain function and neurodegenerative conditions (Mazin et al. 2013).

The regulation of these functionally precise "variable readings" of eukaryote genes is not yet fully understood, although several mechanisms have been implicated. These include RNA binding factors that interact with *cis*-acting RNA regions, "riboswitches" that bind to small metabolites to act as regulatory sensors, and noncoding small RNAs that silence or degrade specific mRNA sequences (Lemos et al. 2008; E. Wang et al. 2008; Mazin et al. 2013; also see Section 1.2 on epigenetic regulation by noncoding RNAs). Alternative splicing introduces a wholly unexpected openness to the process of DNA-based information transfer: the final mRNA transcripts produced by editing and splicing are emergent, short-lived entities that "do not reside on the chromosome" and may even be assembled after the transcript has entered the cytoplasm rather than in the nucleus (Keller 2000, 64). In addition, functional diversity can be generated from a given coding DNA sequence by posttranscriptional editing of RNA base-pair sequences (Mattick and Mehler 2008 and references therein). Although it has been found in diverse organisms (and indeed was first identified in a protozoan; Covello and Gray 1989), this type of RNA editing appears to be most common in vertebrates, where it may play an important role in regulating gene expression associated with developmental and functional brain plasticity (Mattick and Mehler 2008). However, most RNA editing appears to occur in noncoding RNAs, which (although still poorly understood) have been increasingly implicated as regulators of gene expression (Ha and Kim 2014). In this light, it seems likely that RNA editing is of primary importance as a modulator of RNA-mediated epigenetic regulation (Nishikura 2010 and references therein; also see Section 1.2).

1.2 Extragenomic developmental information: epigenetics and environment

These insights into gene regulation suggest a very different view of the developmental role of an individual's DNA sequence than Watson and Crick could have envisioned. Instead of a coded text consisting of pure bits of information that are faithfully transmitted across generations, the genome now appears as a stunningly dynamic network of biochemical interactions—not a "blueprint" for a determined phenotypic outcome but rather an agile "information management system" (Pagel and Pomiankowski 2008). And yet this system is not bounded by the nucleus but is itself embedded in a larger and even more complex informational context. To the many interacting genomic factors that influence phenotypic expression must be added three further sources of developmental information: epigenetic processes and the organism's cellular and external environments.

1.2.1 Variability and heredity expanded: epigenetic regulation of gene expression

Epigenetic effects at the molecular level can be defined as biochemical mechanisms that shape patterns of gene expression in the absence of any change in nucleotide sequence; these mechanisms act primarily by altering the accessibility of DNA to the transcriptional machinery (C.-T. Wu and Morris 2001; Bateson and Gluckman 2011; see Duncan et al. 2014 for a brief overview).[2] Although initially seen as a rare embellishment to Mendelian variation, early whole-genome studies in "model" organisms such as *Arabidopsis* and in humans revealed molecular epigenetic modifications to be unexpectedly widespread (e.g., Heintzman et al. 2007; X. Zhang et al. 2007). These modifications are now understood to be a fundamentally important source of variation in gene expression. In the past decade, a remarkably exciting, if bewildering, array of epigenetic mechanisms has been uncovered. To date, such mechanisms are known to include DNA methylation and hydroxymethylation (Holliday and Pugh 1975; Cedar and Bergman 2009; P. Jones 2012), histone modifications such as acetylation and phosphorylation (Bannister and Kouzarides 2011), and diverse activities of small RNA molecules (Ha and Kim 2014; K. Morris and Mattick 2014), but the list is rapidly expanding as new phenomena come to light. Although biologists are just beginning to understand these novel regulatory processes, it is clear that the epigenome is an essential source of developmental information within and across generations, in both unicellular and multicellular taxa (Jaenisch and Bird 2003).

Epigenetic mechanisms were initially investigated as regulators of cell differentiation during ontogeny (Reik 2007). In many organisms, methylation and histone modifications play an important role in the correct timing of tissue-specific gene expression patterns as ontogeny proceeds (Heard 2013; Boland et al. 2014), and multiple epigenetic effects shape the developing brain in vertebrates (Meaney and Ferguson-Smith 2010). Substantial epigenetic developmental information is evidently transferred to offspring via sperm cells in humans and other mammals, and via pollen in plants (Hammoud et al. 2009; Herrera et al. 2014; Ihara et al. 2014). Epigenetic mechanisms also underlie the process of genomic imprinting, a complex mode

[2] Alternatively, the term *epigenetics* is sometimes used in a broader sense to include the entire environmental and cellular context of development (Waddington 1957; see discussion of alternative definitions in Bird 2007).

of inheritance in which certain genes are expressed in a parent-of-origin-specific manner and which is common in mammals, flowering plants, and insects (MacDonald 2012). The discovery of genomic imprinting demonstrated that gene expression patterns could be inherited based on epigenetic silencing of one set of alleles, depending on the gender of the parent that contributed them, without any change in DNA sequence. In mammals, imprinted genes are often central to fetal development (Gluckman et al. 2007; Plasschaert and Bartolemei 2014).

In addition to mediating characteristic ontogenetic trajectories, epigenetic changes that affect phenotypes can also be variably induced in individuals, either by specific environmental factors or spontaneously (references in Jaenisch and Bird 2003; Gluckman et al. 2007; Nelson et al. 2012; Duncan et al. 2014; also see Section 1.2.2 on epigenetic mechanisms and environmental response). Although work in natural systems has been limited to date, epigenetic modifications are evidently a source of substantial phenotypic variation in nature that does not arise directly from allelic variants (e.g., Herrera and Bazaga 2010; Lira-Medeiros et al. 2010; Becker et al. 2011; C. Richards, Schrey, et al. 2012). Because epigenetic effects on gene expression can be maintained across mitotic and, in some cases, meiotic divisions, they can be long term, but to varying degrees (Gill et al. 2012; Turck and Coupland 2013; Herrera et al. 2014). These biochemical modifications are well documented to persist through cell lineages within an individual and (when they occur in germ cells or in organisms with no segregation between germ and somatic cells) through several but not all subsequent generations; at some point they are biochemically "reset" (Holliday 1990; Johannes et al. 2009; Burton et al. 2011; Seong et al. 2011). Although such reversals can be experimentally induced by environmental or dietary changes that undo DNA methylation or histone modifications, it is not yet known whether in natural systems such reversals are typically stochastic or are induced by specific conditions, or how long epigenetic modifications persist outside of the laboratory (Klironomos et al. 2013; Duncan et al. 2014; Heard and Martienssen 2014). Even laboratory data on long-term epigenetic persistence are uncertain, since the reported number of generations may simply reflect the duration of an experiment rather than an observed reversal.

Evidence for inherited epigenetic effects on human health has opened a major new avenue for biomedical investigation. Epigenetically mediated effects of poor maternal nutrition on offspring development and physiology may strongly influence the incidence of adult obesity and cardiovascular disease (Gluckman et al. 2009; Jiménez-Chillarón et al. 2012; and references therein); heritable epigenetic variation has been implicated in other disease states, including cancer (Nelson et al. 2012; Varley et al. 2013). In studies with mice, dietary factors such as folate that directly affect the biochemical process of methylation create, in the subsequent generation, paternal effects that are associated with developmental malformations. These effects are caused by diet-induced changes to sperm-cell epigenomes; such changes can take place either when males are in utero (and the maternal diet is manipulated) or during adult spermatogenesis (Lambrot et al. 2013). Indeed, the prevalence and impact of these nonallelic (and potentially environmentally induced) transgenerational effects on individual health may be a key reason why genome-wide association studies have shown such limited success in identifying DNA sequence variants that in themselves explain the distribution and inheritance of disease phenotypes (Miklos 2005; Manolio et al. 2009; Slatkin 2009; Drong et al. 2012; Mattick 2012; Nelson et al. 2012). More broadly, a great deal remains to be learned about both the induction and resetting of these various effects and their modes of heritable transmission in various systems (Lambrot et al. 2013; Herman et al. 2014). It has just recently been shown that regulatory RNA molecules released from somatic cells (in this case, human tumor cells grafted into mice) mediate the transfer of information via the bloodstream to germ cells and thence to sperm, thus providing a route for molecular epigenetic information to become heritable in a mammal (Cossetti et al. 2014; see Soubry et al. 2014 for a review of paternal transmission of epigenetic marks).

Although in some cases epigenetic effects on gene expression can be heritable for between a few to several dozen generations or more (Jablonka and Raz 2009; Mattick 2012), their persistence varies because they are also reversible. In other words, they simply "do not adhere to the rules of Mendelian inheritance" (Lemos et al. 2008, 91; Becker et al. 2011; Cossetti et al. 2014). Indeed, epigenetic modifications of phenotypic expression are sometimes considered to be "Lamarckian" because they can be transmitted to subsequent generations after being acquired. Not surprisingly, then, it has taken decades for these molecular effects to be accepted as a part of mainstream genetics. Contemporary awareness of molecular epigenetics has expanded the neo-Darwinian view of DNA sequence as the fundamental mode of inherited developmental information (Jablonka and Lamb 2002; Mattick 2012), placing even the initial phase of gene expression squarely in a dynamic cellular, organismic, and environmental context.

At the mechanistic level, epigenetic modifications shape gene expression by altering protein–gene interactions that determine the accessibility of DNA to the biochemical machinery of gene transcription. DNA molecules exist in the cell nucleus wrapped around core histone proteins in nucleosomes, the repeating unit of the chromatin complex. Each histone contains many chemical sites available for posttranslational chemical modifications; in the textual language of genetic information, such sites are termed *marks*. These marks affect the binding activity of regulatory proteins to the DNA and thus specify whether particular regions of DNA are to be activated or silenced in specific cell lineages. Chromatin enriched for acetylated histones is accessible to transcription factors; as a result, genes located in that chromatin can be transcriptionally active, while genes located in "condensed" chromatin (enriched for nonacetylated histones) are inaccessible to transcription factors and are thus inactive (Jablonka and Lamb 2005; Bannister and Kouzarides 2011). Numerous studies have shown that increased methylation at promoter regions also tends to be associated with gene "silencing" or down-regulation, while methylation at intragenic regions can be associated with either repression or activation, depending on the particular sequence in which it occurs (P. Jones 2012; Varley et al. 2013).

As noted above, the effects of chromatin modification mechanisms (i.e., DNA methylation and histone marks) on gene expression can lead to the expression of disease phenotypes. The potential medical importance of these mechanisms is particularly clear with respect to the malfunction in mitotic regulation that underlies cancer. Many tumor cells show aberrant, cell-heritable epigenetic modifications that silence tumor-suppressing genes (Jablonka and Lamb 2002; West and Johnstone 2014). Since human cancers commonly show widespread changes in DNA methylation patterns in early stages, it is hoped that further methylation studies may lead to new techniques for early cancer detection (Kanai and Hirohashi 2007); drugs that alter histone acetylation are already in use as a treatment for certain cancers, with more under clinical trial (West and Johnstone 2014).

Epigenetic mechanisms may also be a heretofore unrecognized source of selectively important phenotypic variation in natural populations. Indeed, the first naturally occurring morphological mutant to be characterized at the molecular level was determined to be an "epimutation" rather than a DNA sequence change (Cubas et al. 1999). In 1749, the Swedish botanist and pioneering systematist Carl Linnaeus described from his own cottage garden outside Uppsala a radially symmetric mutant of "butter-and-eggs," *Linaria vulgaris*, a common European herb with characteristic bilaterally symmetric, snapdragon-like flowers (Figure 1.4). This radical change in floral morphology can be traced to the lack of expression of *Lcyc*, a homolog of the *cycloidea* gene that regulates dorsiventral asymmetry in the related genus *Antirrhinum*. The lack of *Lcyc* expression is due not to a mutation in the gene's DNA sequence but instead to heavy methylation at the *Lcyc* locus, as this methylation transcriptionally silences the gene. This epimutation is heritable but can occasionally revert to the characteristic flower type after several generations, through methylation changes during development that restore *Lcyc* gene expression (Cubas et al. 1999). This species of *Linaria* depends on cross-fertilization by honeybees, so this change in floral form and hence pollinator search image could have substantial fitness consequences—an example of an individually mediated developmental phenomenon of potential evolutionary significance.

Figure 1.4 The common European herb *Linaria vulgaris* produces characteristic bilaterally symmetric flowers (left panel). A naturally occurring "epimutation" of this species with radially symmetrical flowers (right panel) was first described by Linnaeus in 1749. These dramatically altered plants have the same DNA sequence at the floral symmetry locus *Lcyc* but a different pattern of chromatin methylation. Image courtesy of John Innes Centre, Norwich, UK.

It has become clear that heredity is mediated at the molecular level not purely by discrete, stably transmitted DNA sequence variants but also by multiple information-altering mechanisms that lend the process an unlooked-for flexibility. Qualitatively new modes of cross-generational gene regulation are continuing to be found, including several that show gene silencing and other epigenetic roles for noncoding RNA (Bernstein and Allis 2005; Mattick and Mehler 2008; Lenhard et al. 2012; Ha and Kim 2014). For example, *Arabidopsis* mutants that are homozygous for certain point mutations at the *hothead* locus produce a proportion of progeny with wild-type DNA sequences at this locus: evidently the previous nucleotide sequence is restored by an RNA or DNA "archive" (Lolle et al. 2005). The converse has been shown in mice with respect to *Kit*, which encodes a key cell surface receptor protein that binds various growth factors and hormones. Mutations at this locus show pleiotropic (and sometimes drastic) effects on both development and gamete production. Studies of a particular *Kit* mutation that produces a characteristic white-spotted coat revealed that homozygous wild-type mice can produce offspring with this phenotype, even though they lack the mutant allele. This "paramutation" evidently represents information carried over from a previous heterozygote generation via RNA molecules (present in mouse sperm along with the male's haploid genome) that influence embryonic gene expression (Rassoulzadegan et al. 2006; Chandler 2007; and references therein). Non-Mendelian inheritance can also be mediated directly by noncoding RNAs, as was recently found in the unicellular ciliate *Oxytricha trifallax*. In this highly polyploid organism, "RNA-guided recombination" directs the reassembly of coding sequences into stably inherited alternative genome arrangements and regulates chromosome copy number in progeny cells (Nowacki et al. 2008; Yao 2008).

Many genomic sequences that were previously considered "junk" are now known to code for small or "micro" RNAs (and possibly long RNAs as well) that play a regulatory role, for instance by altering enzymatic access to the chromatin by binding to DNA (Koziol and Rinn 2010). These noncoding RNAs can also target and disrupt protein-coding mRNA transcripts to silence gene expression in a tissue- or cell-lineage-specific manner (Ha and Kim 2014; K. Morris and Mattick 2014). Small RNAs found near transcriptional start sites apparently interact with promoters to help regulate transcription, although this process is not yet well understood (Lenhard et al. 2012). Interestingly, noncoding RNAs may carry environmentally induced effects on the phenotype from one generation to the next, including the neurobehavioral effects of social environment. In one recent study, traumatic, unpredictable separation of newborn mice from their mothers altered several aspects of microRNA activity in the pups, including in their hippocampi and other brain structures involved in stress responses. These epigenetic changes were associated with different behavioral responses to aversive conditions such as brightly illuminated maze compartments. When sperm RNA from traumatized males was injected into fertilized wild-type egg cells, these phenotypic effects were reproduced in the F2 generation; this result indicates that RNA can contribute to the transmission of stress-induced traits in mammals (Gapp et al. 2014).

Additional mechanistic complexity arises from the ways epigenetic processes interact with each other to shape gene expression patterns. DNA methylation and histone modifications can jointly regulate transcriptional patterns, and noncoding RNAs can direct DNA methylation to particular sites in the genome to cause highly specific effects on gene activity (Meaney and Ferguson-Smith 2010; Duncan et al. 2014; and references therein). Histone modifications and methylation marks can also alter the three-dimensional conformation of DNA in the nucleus so as to physically influence epistatic interactions among genetic loci (Qu and Fang 2013). These "overlapping and interdependent" epigenetic effects (Bateson and Gluckman 2011, 57) verify that phenotypes emerge from the dynamic interplay of different types of regulatory elements and not simply from the presence or absence of particular DNA sequences. Indeed, the very notion of genes as discrete pieces of developmental information has become open to question (Keller 2000; K. Morris and Mattick 2014).

It is not yet clear how these expanding insights to molecular epigenetics will change our understanding of allelic diversity and its heritability. This understanding must take into account not only the partially heritable effects of interacting epigenetic processes on gene expression but also the reciprocal effects of the genome itself on these epigenetic processes. DNA sequence influences the likelihood and precise location of spontaneous or environmentally induced epigenetic marks: for example, single-nucleotide polymorphisms can alter particular sites so as to block them from being methylated or can change the binding of transcription factors that in turn regulate patterns of methylation (Gutierrez-Arcelus et al. 2013; Teh et al. 2014; and references therein). Such effects of genotype on epigenotype were demonstrated when both types of data were analyzed from a large, ethnically diverse group of newborn babies (Teh et al. 2014). Variation in the infants' genome-wide methylation patterns was strongly associated with in utero factors such as maternal smoking or depression, gestational age, and the baby's birth order, but the precise impact of these environmental factors on methylation varied depending on DNA sequence polymorphisms associated with ethnic group (Teh et al. 2014). The effect of specific epigenetic marks on transcriptional activity can also be sequence dependent (Meaney and Ferguson-Smith 2010). The impact of sequence variation on epigenetic marking and its transmissibility (Herrera et al. 2014) may provide a mechanistic, epigenetic basis for certain aspects of genotype by environment ($G \times E$) interaction (see Section 1.2.2).

As a result of the interplay of genetic and epigenetic variation, the source of heritable variation in individual development is neither genotype alone nor genotype plus independently generated epigenetic modifications but rather a gene regulatory system that arises from the interaction of genotype and epigenotype. As is true for any interactive system, studies of either component alone (i.e., the association of either gene sequence variants or epigenetic variants with phenotypes of interest) will reveal an incomplete picture of the causal dynamics (Meaney and Ferguson-Smith 2010; Teh et al. 2014). Although a great deal remains to be learned about epigenetic mechanisms, it is clear that genetic information is mediated at multiple levels that shape its expression and hence its precise developmental influence. The next step is to situate this intricate gene regulatory system in its environmental context.

1.2.2 The environment as a source of developmental information

Along with genomic and epigenetic regulatory elements, environmental factors critically inform developmental processes. Although these factors can disrupt development (as in the case of extreme temperature stress or toxins), environmental inputs participate in characteristic outcomes as well, often regulating adjustments to particular conditions. Extraordinary tools have recently produced a wealth of data on phenotypic expression pathways, leading to "a virtual revolution... in our knowledge of environmental influences on gene expression" (Gottlieb 2004, 85; also see Chapter 3). A key insight is that normally occurring intracellular, somatic, and external environmental signals influence specific aspects of DNA and mRNA activity (in part via the epigenetic mechanisms discussed above), actively shaping developmental outcomes (discussed by Jablonka and Lamb 1998; S. Gilbert 2012). The environment can now be understood as a source of regulatory information rather than a neutral backdrop.

The environment's substantive informational role has been revealed at several mechanistic and organismic levels. The three-dimensional shape or proper "folding" of gene products (which is essential to protein function) is influenced by conditions within the cell, such as the presence of interacting molecular cofactors; the spatial effects of high concentrations of macromolecules in the cell solution; temperature; and pH (Lewontin 2000; Zhou 2008, 2013). The effects of the intracellular environment on protein folding kinetics is an active area of study, in part because such in vivo information is essential to understanding prion diseases that arise from the altered chemical properties or inactivation of misfolded proteins (Chen et al. 2008).

In some cases, the cell's internal environment can determine both the activity level and the specific developmental role of a given protein, which can change even after the protein is assembled. Such *posttranslational modifications* are exceptionally difficult to study, since these mechanisms can only be investigated in cellular context. One relatively well-studied case is that of the protein β-catenin, which plays two distinct, competing developmental roles depending on its interactions with various molecular binding partners in the nucleus and cytosol (the fluid portion of the cytoplasm). Both roles are critical: β-catenin is a key structural component of cell–cell adhesion during morphogenesis and is also a required activator in the Wnt signaling pathway, which directs tissue-specific cell differentiation patterns (Daugherty and Gottardi 2007). The alternation between these adhesion and signaling functions must be coordinated, since adhesion must be inactivated during times when cells migrate to form newly differentiating tissues (Lilien et al. 2002). Although the precise mechanisms are imperfectly understood, it is clear that β-catenin's functional status is primarily regulated by the phosphorylation of its binding partners, as phosphorylation affects their binding affinity. This biochemical process is "fine-tuned" via signals conditioned by cell cycling and metabolism, cellular growth factors, and possibly extracellular cues as well (Lilien et al. 2002; Verheyen and Gottardi 2010).

Also within the cell, environmental conditions can inform development via an "intracellular conversation" that takes place between nuclear genes and the cell's organelles and which appropriately alters the organism's subsequent development. When a young plant's chloroplasts are stressed by excessive absorption of light energy, they transmit chemical signals that alter the expression of nuclear genes involved in cell division and differentiation. This signaling from plastid to nucleus leads to reduced cotyledon expansion and stem elongation, both of which are developmental responses that minimize the seedling's subsequent light exposure (Lopéz-Juez 2009 and references therein). This information pathway, known as "plastid retrograde signaling," was first observed decades ago but was largely ignored as an aberration from presumed genetic control of development.

The environment surrounding a cell can also influence the rate and extent of gene activity during differentiation (Lewontin 2000). These effects can be somatic (in multicellular organisms) or external (in the case of unicellular taxa). For example, expression of several genes in yeast strains varies depending on the growth medium in which cells are cultured (Lemos et al. 2008), such as media that provide glucose as the carbon source versus those that provide ethanol as the carbon source (E. Smith and Kruglyak 2008).

One difficult-to-classify aspect of an organism's internal environment is its *microbiome:* the community of bacteria, archaea, fungi, and protozoa that cohabit its body cavities, surfaces, and tissues. These symbionts are acquired in part via "vertical transmission" from the maternal parent (e.g., bacteria within egg cells) and in part from the outside environment or other host individuals (McFall-Ngai 2002; Fitzpatrick 2014). While pathogenic microbes may negatively affect host fitness, the full range of possible biological interactions between microbial symbionts and hosts has only recently come to light (Fellous et al. 2011). The microbiome is now understood to play crucial roles in the host organism's normal ontogeny as well as in its nutrient acquisition, metabolism, immune function, and behavior. The dependence of individual development on the presence of particular symbionts—that is, on biochemical information and substances contributed by quite unrelated organisms—appears to be surprisingly widespread in both animals and plants (McFall-Ngai 2002; Wilks 2007; further examples in S. Gilbert and Epel 2009). A particularly well-studied mammalian example (based largely on experiments using microbe-free laboratory mice) is the critical role of bacteria in completing the postnatal development of the gastrointestinal tract. Newborn individuals acquire a complex bacterial gut community by passing through the birth canal or via contact with the mother's skin. These bacteria initiate angiogenesis via signals mediated by specialized epithelial cells lining the small intestine, thus inducing the formation of the finely branched network of blood vessels essential for nutrient absorption by the host (Stappenbeck et al. 2002; Hooper 2004; Wilks 2007; also see Chapter 2, Section 2.3.4).

The finding that development can arise collaboratively through interactions between a plant or animal host and its microbial inhabitants takes us a long way from the notion that an individual's phenotype is a direct readout of its inborn, essential genetic code and has raised interesting questions about the nature of biological individuals, including ourselves (Pradeu 2011; see Chapter 2, Section 2.1). The human body houses several tremendously diverse prokaryote and eukaryote communities that contribute a wealth of biochemical information; it can be said that "we are built not by one genome but by hundreds, if not thousands" of genomes within symbiont cells that far outnumber human ones (S. Gilbert 2011, 66; see Bevins and Salzman 2011). Remarkably, these human microsymbionts play a major signaling role in the determination of immune and metabolic systems and in the function and possibly the development of that quintessentially human organ, the brain (reviewed by Cryan and Dinan 2012).

Finally, external environmental conditions can also affect gene expression in multicellular taxa such as animals and plants as well as in unicellular organisms, in some cases by altering hormonal conditions in the cytoplasm, the nucleus, or both. For example, an individual insect's production of wings is informed not only by spatially and temporally specific expression of *Hox* and *Ubx* gene products but also by various environmental and resulting hormonal signals that are integrated into these gene regulatory pathways to determine whether the adult will be winged or wingless (S. B. Carroll et al. 2005). This kind of *polyphenism* (distinct phenotypes produced by a given genotype) is particularly well studied in ants, which form socially complex colonies based on the differential development of individuals into distinct functional types or "castes." Although the basic signaling network underlying wing production is highly conserved across holometabolous (completely metamorphic) insects such as flies and butterflies, all of the approximately 10,000 extant ant taxa produce at least one wingless worker caste (S. B. Carroll et al. 2005). Evidently, this conserved developmental network is derailed in these individuals, presumably in response to environmental conditions such as seasonal cues and food availability, as these determine the appropriate proportions of the various castes in a colony.

Abouheif and Wray (2002) investigated the switch points regulating individual winglessness in several closely related ant taxa. Surprisingly, the gene expression differences that lead to winged versus wingless phenotypes are species specific, so these signaling pathways are not shared among even closely related ant taxa. When embryonic *Pheidole morrisii* ants experience certain photoperiod and temperature shifts, the level of juvenile hormone surges and they become queens—that is, winged reproductives. The remaining larvae develop into sterile adults; but whether they become soldiers or workers depends on a second environmental signal, a later dietary cue that also influences juvenile hormone levels (Abouheif and Wray 2002). Further caste differentiation can also be environmentally regulated, as in the Florida harvester ant, *Pogonomyrmex badius*. In this species, the hormone levels that effect the dramatic differences between major and minor worker ants are determined, via the insulin signaling pathway, by the quantity and type of food available (C. Smith et al. 2008; see Mutti et al. 2011). In polyphenic social insects, such fundamental aspects of development as life history, body size and plan, and sensory systems are directly regulated by ecological conditions, possibly via environmentally cued epigenetic modification of genes involved in muscle development and neuronal function (Simola et al. 2013; Yan et al. 2014). In plants, too, key external cues regulate gene expression to shape individual life history. In *Arabidopsis*, the *LEAFY* gene encodes a plant-specific transcription factor that binds to several distinct regulatory sequences which collectively determine the transition to flowering in shoot tissues and thus initiate reproduction; this gene is one of several that act to integrate the environmental cues of photoperiod (day length) and vernalization (cold exposure) with these pathways, through an intricate set of transduction mechanisms (Parcy 2005; Amasino and Michaels 2010; Des Marais et al. 2013).

By means of such signals and their precise effects on gene activity, the environment enters into the developmental process as a source of specific regulatory information (see Chapter 3 for detailed case studies). A myriad of environmental factors are implicated as regulatory inputs, depending on the organism (discussed by Gottlieb 2004). In nematodes, the presence or absence of food specifies the level of neuronal *daf-7* mRNA expression to either stimulate or inhibit larval development (Ren et al. 1996). Natural light/dark cycles activate specific genes implicated in circadian rhythms in *Drosophila* (*period* and *timeless* genes; Myers et al. 1996) and in mice (c-fos genes; Smeyne et al. 1992). Light influences numerous functionally important aspects of gene expression in bacteria and yeast; indeed artificial "photoswitches" are of increasing interest as a way to manipulate these microbial systems (Drepper et al. 2011). Gene activity in plants is influenced not only by well-recognized environmental factors such as temperature, humidity, and light but also by vibration, touch, and sound (Braam 2005; Jeong et al. 2008). Incubation temperature determines offspring sex in many reptiles (Sarre et al. 2004), evidently through differential activation of genes encoding steroid-producing enzymes (Crews 2003). Nest temperature can also directly affect the expression of reptile traits with gender-specific functional benefits, such as innate immune function, which is determined by the complement of circulating proteins available to attack pathogens (Freedberg et al. 2008). In the African cichlid fish *Astatotilapia burtoni*, change in a male's social status (a perceived "opportunity to rise in rank") initiates rapid proliferation of neurons in regions of the brain that control complex behaviors (Maruska et al. 2012, 3471). Acoustic, tactile, and visual stimuli all can affect gene expression and RNA synthesis in the brains of birds and mammals (e.g., Ehret and Fisher 1991; Mack and Mack 1992; Mello et al. 1992). Indeed, environmental complexity per se results in increased brain RNA diversity in rats (Gottlieb 2004).

Environmental factors that influence gene expression include behaviors of the organism and interacting individuals; this observation suggests that there can be a complex and temporally changeable play of effects. In one study, infant rats deprived of maternal care on postnatal day 13 showed altered gene expression of corticotropin-releasing hormone receptors in specific brain regions; these changes modified the circuitry of stress responses in these rats as compared to those receiving uninterrupted maternal care (Vazquez et al. 2003). In such cases, inherited differences in stress reactivity across generations can be due to parental behavior rather than to genetic factors (Francis et al. 1999; Champagne 2008). More commonly, parental behavior, like other environmental influences, is likely to interact with genetic variation in its phenotypic impact.

A compelling example of this kind of genotype–environment interaction occurs in rhesus monkeys, *Macaca mulatta*. Like humans, rhesus monkeys are polymorphic for a specific serotonin transporter gene. At this particular locus, the *LS* allele reduces transcriptional efficiency as compared with the alternative *LL* allele. Yet the consequences of this allele for an individual depend on a subtle aspect of its social environment: rearing as a newborn by its mother versus by a peer (Suomi 2004). Peer-reared individuals carrying the *LS* allele do show reduced serotonin metabolism, along with early neurodevelopmental deficits and excess alcohol consumption as adolescents. But if they are mother-reared, individuals with the *LS* allele show none of these metabolic or developmental effects and actually consume *less* alcohol than mother-reared *LL* individuals. In this case, the same genetic factor can have either a psychopathological or an adaptive impact, depending on early maternal attachment conditions (Suomi 2004).

Human behaviors, too, can result in subtle yet biologically powerful environmental effects on gene expression. For example, the psychological stress induced in medical school examinees causes reduced mRNA expression by the interleukin 2 receptor gene, thus altering the immune system response (Glaser et al. 1990). Conversely, both long- and short-term practitioners of relaxation techniques such as meditation, yoga, and qi gong showed significantly altered gene expression profiles: genes associated with cellular metabolism and stress responses (including apoptosis, antigen processing, and mRNA splicing) were consistently either up- or down-regulated (Dusek et al. 2008). These observations provide a cellular mechanism for the long-known health benefits of these ancient behavioral practices.

In many cases, environmental effects on development have been shown to be mediated by epigenetic forms of gene regulation (discussed by Badeaux and Shi 2013; Duncan et al. 2014; K. Morris and Mattick 2014). In the honeybee *Apis mellifera*, for example, feeding larvae with the sophisticated but biochemically mysterious food known as royal jelly reduces DNA methylation in a major metabolic regulatory gene (*ALK*) and in genes that interact with key hormonal pathways in larval tissues, thereby determining the larvae's developmental fate as fertile queens rather than sterile workers (Kucharski et al. 2008; Foret et al. 2012). Diet-induced methylation effects are also important in mammals. In mice with the coat color gene *agouti*, for instance, maternal diet influences gene expression by changing the methylation state of several alleles (Cropley et al. 2006; *agouti* also has pleiotropic effects on metabolism and consequently influences obesity). Environmentally appropriate life-history responses of high-altitude plants to seasonal cues are also mediated by an epigenetic mechanism. Activity of *FLC*, a primary regulator of reproductive onset in response to cold period (vernalization), is controlled by DNA methylation (Jaenisch and Bird 2003). Epigenetic regulation clearly constitutes a significant mechanism for potentially adaptive phenotypic adjustments in response to environmental signals (Jablonka and Lamb 2005; Lemos et al. 2008; Meaney and Ferguson-Smith 2010; Soubry et al. 2014). However, a great deal remains unknown regarding the precise pathways that link external conditions to particular epigenetic modifications and phenotypic effects (Duncan et al. 2014; Gapp et al. 2014; K. Morris and Mattick 2014).

One environmental factor that has been intensively studied in recent years is prenatal nutrition in mammals. Epidemiological, clinical, and experimental work has shown that poor maternal diet is associated not only with low birthweight, as has long been recognized, but also with specific aspects of offspring development and metabolic phenotype that in humans are strongly linked to adult health problems, including obesity, diabetes, and cardiovascular disease (Gluckman et al. 2009; Jiménez-Chillarón et al. 2012; Gluckman et al. 2011; and references therein). In some cases, specific epigenetic pathways for these effects have been identified. Maternal carbohydrate intake early in pregnancy is associated with DNA methylation of the *RXRA* promoter in umbilical cord tissue and with the body composition of offspring at age 9 (Godfrey et al. 2011). In sheep (which are often studied as a model for human reproductive biology), moderate maternal undernutrition around the time of conception leads to a number of methylation and histone modifications linked to decreased expression in the adult hypothalamus of a key gene that regulates energy balance, as well as to increased obesity in adult male offspring (Begum et al. 2013). Similarly, offspring of mother rats that were given a low-protein diet during pregnancy show altered histone acetylation and methylation patterns at receptor sites for metabolically central hepatic genes and developed metabolic and cardiovascular abnormalities (Gluckman et al. 2007 and references therein). Such long-term effects of uterine environment on offspring development, physiology, and health may be of immense biomedical significance.

These new mechanistic insights fundamentally alter our understanding of the environment's role in development. Although it is a biological truism that development takes place in an environmental context, the role accorded to the environment has until recently been a very limited one. The genes have been seen to provide the precise "information" that guides development, while the environment merely provides raw materials to permit the realization of the genetic program or, if extreme, "insults" or interferes with that program's expression (Oyama 2006). Whether or not it is explicitly voiced, this view is the basis for studying development in a single "control" environment, which is meant to be developmentally neutral. This notion of a neutral environment, as well as the research strategy that notion dictates, loses its utility once it is recognized that environmental conditions also provide critical, precise developmental information that is implemented via specific effects on gene expression. Accordingly, environmental and genetic factors play similar roles as determinants of development (Nijhout 1999): both environmental and genetic factors *inform*—that is, lend form to—phenotypic outcomes. In other words, there need be no qualitative distinction between the type of information provided to the developmental system by genetic elements and that provided by environmental factors (West-Eberhard 2003; Leimar 2007).

This point is exemplified by phenomena like the *bithorax* phenotype in *Drosophila*, as this phenotype can be caused either by certain mutations or by exposure to ether (Gibson and Hogness 1996 and references therein), and by color-morph mimics of the Great Barrier Reef fish *Pseudochromis fuscus*, as these mimics are produced by phenotypic plasticity in the absence of any gene sequence differences (Messmer et al. 2005). In some cases, environmental factors can compensate for the developmental/functional impact of unfavorable genetic mutations. In one such case, a gene that codes for a major neurotransmitter was knocked out in a strain of laboratory mice. When these knockout mice were provided with an enriched neural-stimulation environment that included toys and tunnels, they showed no cognitive impairment because their enriched learning environment caused increased synaptic density in a region of the mouse hippocampus (Rampon et al. 2000). Curiously, such environmentally induced outcomes are known as *phenocopies*—mimics of the different "real" phenotypes produced by specific genes.

The primacy thus accorded to genetic factors reflects a deep-seated distinction between the organism's inner, "essential" traits and the "superficial" details of individual variation, details which reflect environmental response (Lewontin 2000). Neo-Darwinian theory reinforced this distinction by emphasizing that only heritable traits (presumed to be determined by genes) were relevant to evolution, consigning environmental effects on development to the category of "noise." For much of the twentieth century, this distinction remained a conceptual stumbling block to the study of phenotypic plasticity (Sultan 1992), as environmental effects on phenotypic outcomes were considered both "uninteresting and unimportant" (Pianka 1994, 126) compared with genetic factors. Plant population biologists, confronted with the dramatic and obvious effects of environment on individual development and fitness, were explicitly exhorted to ignore these effects as meaningless variation (e.g., Stebbins 1980). Generally, traits known to be expressed differently in different conditions were considered at best "a group of interesting exceptions" to the rule of rigid genetic control (Bonner 1988, 154).

This essentialist view shaped research approaches to both development and evolution in key ways. In both disciplines, the choice of "model systems" meant to typify entire classes of organisms was made in part because of their (quite atypically!) canalized, relatively context-independent developmental outcomes (J. Zhang and Lechowicz 1994; J. Bolker 1995; S. Gilbert 2001). Specified strains of these research organisms were studied in uniform laboratory conditions to reveal their "true," purely gene-based traits, unaffected by either environmental or genomic variables. The tiny nematode *Caenorhabditis elegans* was selected as a "tame" model system for developmental studies by Nobel laureate Sydney Brenner in part because it thrives on agar plates and so can be studied entirely outside of a natural environmental context.

Once the substantive role of environmental information is recognized, it becomes clear that the "control" environment is simply one environment among many possible environments, and the phenotype expressed in it, one of several possible phenotypes. The key question then becomes, in which environment or range of conditions do we wish to understand developmental processes? Because environmental conditions provide specific inputs to developmental pathways, the choice of conditions in which to study developmental mechanisms is a critical one. Indeed, even in model organisms, functional gene effects may vary dramatically depending on environmental conditions (Weinig and Schmitt 2004; Mackay 2013). Gene interactions in bacteria, viruses, and yeast can vary from essential to lethal depending on environmental factors such as nutrient medium (Hurst and Pál 2008 and references therein). QTLs that contribute to growth rate in yeast showed different and, in some cases, antagonistic effects and interactions depending on the particular carbon source provided (Bhatia et al. 2014). In *Drosophila melanogaster*, alleles at QTLs that increase female life span at high larval densities *reduce* life span at low population density (Leips and Mackay 2000); opposing effects were also found for 10 of 17 other markers, depending on temperature and nutrient levels (Vieira et al. 2000). In many cases, specific genetic loci participate in trait expression only under certain environmental conditions (termed *conditional neutrality;* see Chapter 7, Section 7.2.2). For example, the *erecta* mutation in *Arabidopsis thaliana* strongly affects timing of reproductive onset, but only in certain natural sites and times of season (Weinig and Schmitt 2004). It is not yet known how common such conditionally neutral (or *cryptic*) genes are in natural populations (Ledón-Rettig et al. 2014). However, it is intriguing that, in a meta-analysis of 700 studies reporting trait means of plants grown in different experimental environments, nearly 60% of QTLs were not expressed in some conditions (Des Marais et al. 2013; for additional references on both conditional neutrality and environmentally dependent allelic effects on fitness, see Anderson et al. 2013, 2014).

The environmental dependence of gene expression raises an even more fundamental methodological question: if we wish to know about developmental mechanisms and outcomes in the real world, how accurate is information obtained under laboratory conditions? Traditionally, "control" laboratory environments for various model organisms have been based on convenience rather than designed to recreate natural levels of key ecological factors. These often highly unnatural lab conditions seemed to make sense when the environment was seen as simply permitting genetically preprogrammed development to unfold; they are deeply problematic once environmental inputs are recognized to inform the process (S. Gilbert 2001; Anderson et al. 2014).

It is now widely recognized that phenotypes and their underlying genetic architectures "may differ substantially between natural and artificial growth conditions" (Shimizu and Purugganun 2005, 583). For this reason, lab studies often fail to identify as functionally important those genes that are in fact essential under naturally occurring levels of stress or resource availability (Hurst and Pál 2008). For example, Weinig et al. (2003) found that different loci contributed to variation in flowering time in recombinant *Arabidopsis* lines in field versus controlled settings: certain loci were involved under field conditions but not when photoperiod was controlled in the lab, and the key QTL for flowering time identified in controlled mutant and transgenic studies played no significant role whatsoever in natural conditions. Interestingly, the favorable temperature and moisture levels in which *Arabidopsis* genetic variants are generally screened fall entirely outside the climatic conditions that occur within the species' substantial geographic range in nature (Hoffmann 2002). The environmental context dependence of gene regulatory networks also makes it impossible to generalize about network properties based on topology alone: although gene "hubs" with many connections are more likely to be functionally indispensable, this cannot be assumed without tests in realistic conditions (Hurst and Pal 2008).

How can we rethink development so as to implement this new understanding? Biologists understand evolutionary outcomes as uniquely emerging from the episodic interactions of multiple factors over multigenerational time rather than from the inevitable unfolding of some internally programmed design. In just the same way, developmental outcomes should be seen as emerging in ontogenetic time from the transient interactions of genetic, genomic, epigenetic, and environmental factors. Through these interactions, ontogeny becomes a Markovian process—a sequence of events where each determines the starting point for the next—in this case, a chain of complex feedbacks between the organism's previous state and its current environmental experience (Lewontin 1985; Sultan 2003a; Gluckman and Hanson 2005). The consistent presence of certain genetic elements, genomic architectures, epigenetic marks, and environmental factors can jointly bring out like phenotypic outcomes across generations; like ecological succession, such a process can be "self-organizing" in the absence of a fixed controlling blueprint (Griffiths and Gray 1994). Phenotypic diversity within and among individuals can result from variation in any of these sources of regulatory information—genetic, epigenetic, or environmental—all of which are equally "important and interesting" for understanding both development and evolution.

1.3 Plasticity, ecological development, and the norm of reaction

An expanded understanding of development, as shaped by regulatory interactions among external as well as internal factors, suggests an interdisciplinary research approach known as *ecological developmental biology* (S. Gilbert 2001). The rationale is quite straightforward: because environmental conditions inform the developmental process in specific ways, we can best understand this process, and predict its outcomes, in ecological context. Accordingly, ecological development, or "eco-devo," seeks to explicitly include the organism's particular environment in studying both the signaling pathways and the ecological and fitness consequences of phenotypic expression (Sultan 2003, 2010). In so doing, eco-devo research departs from the conventional practice of studying development in a single controlled environment meant to reveal "pure" gene expression. (This approach remains appropriate for certain research goals, such as initially characterizing interactions between specific gene products.) Instead, eco-devo studies aim to characterize the mechanisms and outcomes of phenotypic expression across the range of abiotic and biotic conditions that occur in natural populations. This research approach entails two important changes, one conceptual and one practical. The conceptual change is a shift to thinking about the genotype as a *repertoire of environmentally contingent possibilities* rather than a single determined outcome. The practical innovation is to bring into experimental design the environmental variability that has been intentionally excluded from studies of both development and genetic variation under a strictly gene-based model of phenotypic determination. Together, these changes integrate development with ecology and evolutionary biology in a newly coherent way.

The integrative eco-devo approach has emerged from the confluence of two recent currents in biological thought; these currents originated in distinct disciplines but share a focus on the environmental context of development. First, molecular developmental biologist Scott Gilbert and colleagues coined the term *eco-devo* to emphasize the importance of investigating developmental processes and outcomes in the context of "real-world" environments, in addition to the generally unnatural conditions of lab studies (S. Gilbert 2001; S. Gilbert and Bolker 2003; S. Gilbert and Epel 2009). This goal has become genuinely urgent as we seek to understand the developmental and selective consequences of the novel, disrupted, and contaminated abiotic and biotic conditions caused by human activities (Sultan 2007; Ledón-Rettig et al. 2014).

Second, following on Anthony Bradshaw's pioneering 1965 review of *phenotypic plasticity* and on influential work on life-history flexibility published in the 1970s and 1980s (e.g., Jain 1979; Stearns 1982, 1989; Scheiner and Goodnight 1984; Schlichting 1986; Sultan 1987), evolutionary ecologists have focused increasingly on individual phenotypic response to environmental heterogeneity as a source of both adaptive and maladaptive variation in functional and fitness traits. This research program has produced a substantial base of knowledge about environmentally contingent phenotypic expression and its potential evolutionary impact (see Scheiner 1993; Schlichting and Pigliucci 1998; Pigliucci 2001; West-Eberhard 2003; DeWitt and Scheiner 2004; Sultan and Stearns 2005; Bateson and Gluckman 2011; and references therein). Individual developmental plasticity has now been documented in such diverse taxa as algae and land plants, lichens, marine and freshwater invertebrates, insects, fish, amphibians, reptiles, and mammals (references in Sultan 2000; S. Gilbert 2001). Indeed, such plasticity is now clearly seen to be "the rule, rather than the exception" (S. Gilbert and Bolker 2003, 6). Eco-devo provides a single integrated framework for investigating phenotypic expression, whether plastic or inflexible, in relation to environmental variables of interest.

In eco-devo, then, both developmental and evolutionary biologists aim to examine the environmental context dependence of phenotypic expression. Despite its recently coined name (S. Gilbert 2001), eco-devo actually represents a return to the more holistic approach to individual development embraced by early twentieth-century researchers in embryology and genetics, such as E. E. Just in the United States and Wilhelm Johannsen in Denmark (Sultan 1992, 2003; Byrnes and Eckberg 2006). In evolutionary biology, too, before the eventual dominance of Fisher's "genetical theory of natural selection," Sewall Wright (1931), J. B. S. Haldane (1946), and I. I. Schmalhausen (1949) included individual environmental response as a primary cause of phenotypic variation and consequently selection. What is distinctly new about eco-devo, and what distinguishes it from the two decades of plasticity research that led up to its emergence, is its multidisciplinary focus. Eco-devo investigations extend from the molecular and cellular mechanisms of environmental signal transduction to phenotypic outcomes at the ecological level (S. Gilbert 2001; Sultan 2007). This inclusive research approach will undoubtedly require—and inspire—new scientific collaborations as it bridges the long-standing disjunction between development and evolution (discussed by Amundsen 2001).

1.3.1 The norm of reaction

When individual development is studied in a range of ecologically relevant conditions rather than a single control condition—that is, in an eco-devo framework—a picture emerges of surprisingly flexible and diverse outcomes. In the vast majority of cases where organisms are studied in more than one environment, each genotype expresses several alternative phenotypes rather than a single fixed outcome. Even in the extreme case of a lethal mutation, the precise course and duration of developmental events leading to mortality often vary from one set of conditions to another. This individual repertoire of phenotypic expression is termed the *norm of reaction*, defined as *the set of phenotypes produced by a given genotype in a specified range of developmental environments* (Woltereck 1909; Johannsen 1911; for an excellent historical overview see Sarkar 2004). The century-old norm of reaction concept (with its somewhat cumbersome name, translated from the German *Reaktionsnorm*) provides the essential data for eco-devo investigations.

In practice, norms of reaction are determined by raising cloned or inbred replicates of individual genotypes in a set of controlled treatments and then measuring phenotypic traits of interest (examples among many include Gupta and Lewontin 1982; Sultan and Bazzaz 1993a–c; Biddle and Eales 1999; Lindroth et al. 2001; Kingsolver et al. 2004; Fuller et al. 2005; E. Smith and Kruglyak 2008; Pennekamp et al. 2014). Traits can be measured at any phenotypic level, from short-term protein synthesis or metabolic rates to specific aspects of anatomy, morphology, life history, physiology, or fitness. Norms of reaction are shown graphically by plotting, for each genotype, the phenotypic state it expresses in each experimental treatment for a given trait (Figure 1.5a). Depending on the genotype, the environments tested, and the trait, these norms can be linear or nonlinear; they may be relatively flat across environments (canalized; Figure 1.5b), differ continuously from one environment to another (plastic; Figure 1.5c), or constitute discrete, often dramatically different alternatives elicited in response to environmental thresholds (polyphenism; Figure 1.5d; Haldane 1946; Scheiner 1993; Schlichting and Pigliucci 1998; Nijhout 1999; Moczek et al. 2011).

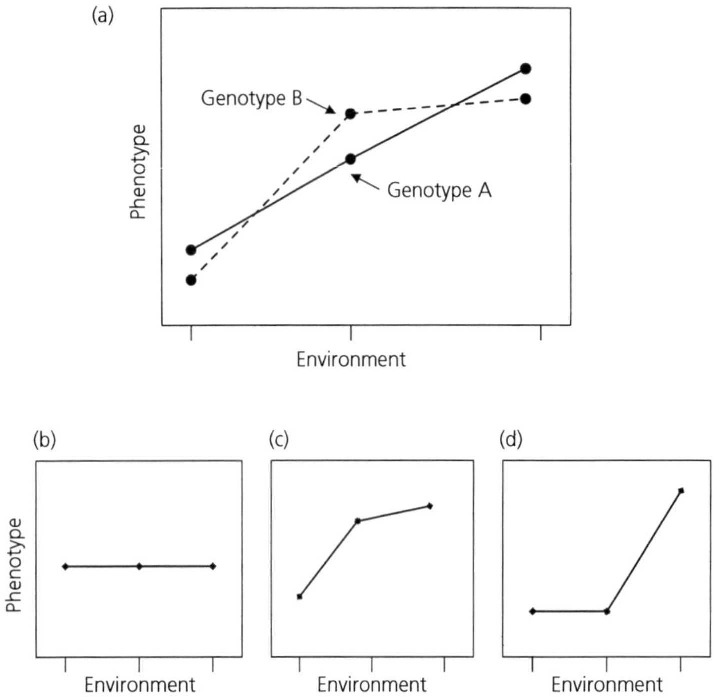

Figure 1.5 A genotype's repertoire of phenotypic expression can be determined by raising cloned or inbred replicates in each of several test environments and then measuring phenotypic traits of interest expressed in each environment. (a) The norm of reaction can be linear (e.g., genotype A) or nonlinear (e.g., genotype B). (b–d) Norms of reaction can be flat across environments (canalized; b), vary continuously from one environment to another (plastic; c), or consist of discrete alternatives elicited at different environmental thresholds (polyphenic; d).

Note that these characteristic patterns of environmental response are genomic properties; as such they are inherited and evolve (Schmalhausen 1949; Via 1987; Stearns 1989; Gomulkiewicz and Kirkpatrick 1992; Scheiner 1993; Schlichting and Pigliucci 1998; Scheiner 2002; e.g., Windig 1994; Nussey et al. 2005; Sultan et al. 2013; see Scheiner 2013 and references therein for theoretical models). As products of evolution, genotypic norms of reaction are shaped by the forces of natural selection and genetic drift and reflect both phylogenetic history and genetic constraints. In consequence, norms of reaction vary among genotypes, populations, and related species (e.g., T. Griffith and Sultan 2005; Husby et al. 2010; J. Hutchings 2011; Matesanz et al. 2012; average genotype-based norms of reaction can be characterized at the population and species levels for higher-order comparisons; Sultan and Stearns 2005 and references therein). An individual's norm of reaction may also be conditioned by cytoplasmically inherited factors and molecular epigenetic effects on gene activity that affect its phenotypic responses to the environment (see Chapter 7, Section 7.2.1). In practice, the genome and epigenome will likely be treated as a single (i.e., experimentally confounded) entity, although it is possible to separately identify genetic (sequence-based) and epigenetic effects on development and their interactions (Meaney and Ferguson-Smith 2010; Teh et al. 2014). The possible evolutionary impact of transient epigenetic effects on realized norms of reaction is not yet well understood (see Chapter 7, Section 7.1.1–7.1.2).

1.3.2 Norm of reaction diversity and genotype–environment interaction

Norm of reaction data for groups of genotypes also reveal a very different picture of genetic diversity than is seen in a single common environment. Again depending on the range of test environments and the traits measured, norms of reaction for a given set of genotypes may show one of several diversity patterns (see Haldane 1946). Genotypes may express identical response norms across the environmental range, for instance, in the case of either strong developmental constraint or previously strong selection in those environments (Figure 1.6a). Norms of reaction may differ in parallel

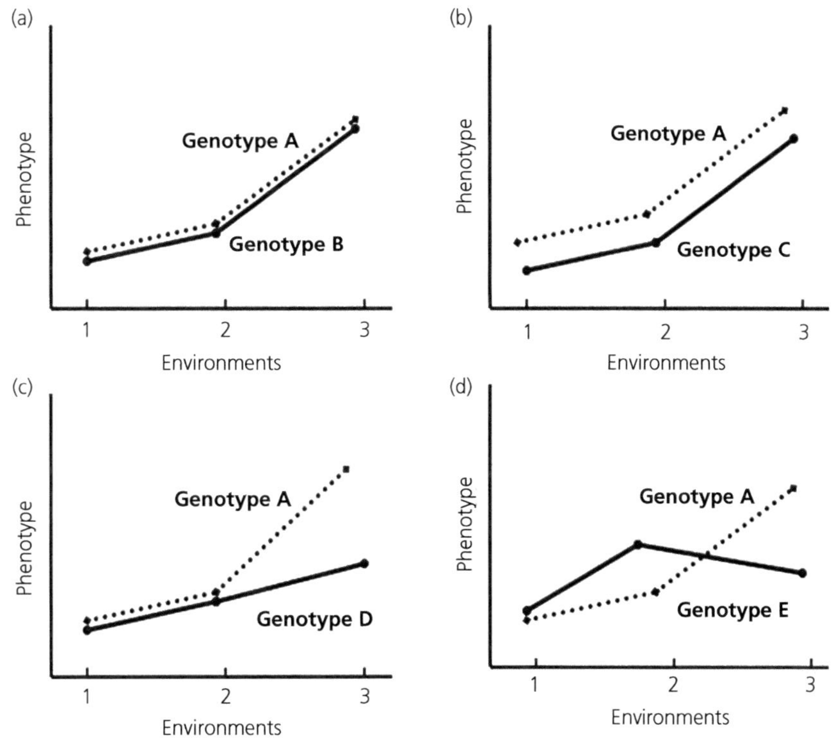

Figure 1.6 Possible patterns of norm of reaction diversity are shown for hypothetical genotypes A–D. In each frame, the phenotypic values expressed by each genotype in three alternative environments are plotted for a trait of interest. Norms of reaction in response to a given set of environments may be identical (a), be parallel (b), or show genotype by environment interaction ($G \times E$), differing in the magnitude (c) or direction (d) of response. Modified from Sultan 2011.

across environments (Figure 1.6b), although this pattern is evidently rare in natural systems. Most commonly, naturally evolved genotypes show *G × E interaction variation*, or nonparallel response patterns across a shared range of environments (Figure 1.6c–d; Haldane 1946; Falconer and Mackay 1996; and references in Section 1.3.1). When genotypes are replicated within each developmental environment, norm of reaction data can be analyzed statistically using standard quantitative genetics models to test for average ("main") effects of *genotype* and of *environment* on phenotypic values, as well as for the presence of *G × E interaction* variance.

In biological terms, nonparallel norms indicate that the *magnitude* of phenotypic adjustment from one environment to another, the *direction* of that adjustment, or both differ among genotypes (Figure 1.6c and 1.6d, respectively). In addition, *G × E* variation is particularly important from an evolutionary point of view (see Chapter 7, Section 7.2.2, for full discussion). When genotypes in a population express different but nonparallel norms of reaction for functional and fitness traits, the amount of genetic variation available to selection (i.e., that is phenotypically expressed) as well as the rank order of genotypic fitnesses can *differ from one environmental state to another* (Figures 1.5a and 1.6c–d). Whenever *G × E* variation is present, then—which is apparently the great majority of cases in natural systems—selective outcomes will depend on the distribution of environmental states encountered (Lewontin 1957; Via and Lande 1985; Mitchell-Olds and Rutledge 1986; Sultan 1987; Gillespie and Turelli 1989; Scheiner 1993; Nager et al. 2000; Lewontin 2001; Sultan 2007). In other words, in the presence of *G × E* variation, neither the amount of available genetic variation nor the ranking of genotypic fitness values can be considered intrinsic properties of a population.

Genotype × environment interaction is also a central concept at the level of individual development. Norm of reaction data show how the specific interactions of genomic and environmental factors jointly shape phenotypic outcomes: the expression of a given genotype depends on the developmental environment, and the phenotypic effect of a given environment depends on the genotype in question (see Lewontin 1985, 2001). As a general example, consider the following: For an individual sunflower plant, the size, shape, functional anatomy, and physiological behavior of its leaves will be different if it grows in a sunny, open spot versus a shaded one or if it receives ample water during growth versus being drought-stressed—the expression of this plant's genotype with respect to leaf traits is environmentally contingent. But a genetically distinct sunflower individual, or a plant of another species, is likely to make different adjustments in these traits in response to sun versus shade or dry versus moist conditions. It seems counterintuitive, but the precise effects of environment on phenotypic expression are properties of the genome.

A dramatic example of genotype–environment interaction is provided by the leopard gecko, *Eublepharis macularius*, which like many reptiles (and certain fish) has temperature-based environmental sex determination. Yet the effects of temperature on offspring sex vary among genotypes. In this species, increasing nest temperature from 30.0°C to 32.5°C can either raise or lower the proportion of male offspring produced, depending on maternal genotype (Janes and Wayne 2006). Such norm of reaction data provide a critical insight regarding developmental causation: every phenotype expressed by an individual organism can be understood as the result of a unique *interaction* of genetic and environmental factors. In other words, the effects of genotype and environment are not simply additive; the effect on the phenotype of either genotype or environment depends on the state of the other (see Gupta and Lewontin 1983). Note that these interactive effects can only be revealed by an experimental design in which both elements are varied (Fisher and Mackenzie 1923).

In areas of medical science such as diabetes and cancer research, such genotype–environment interactions are being actively studied as a paradigm for disease causation. Here, dramatic examples of the interplay of "nature" and "nurture" are changing the way medical scientists understand the role of genes. With a few very rare exceptions, for instance, type II diabetes is caused by the interaction of genetic polymorphisms that affect the cell's sensitivity to insulin, and environmental factors such as lack of exercise and a high-fat diet. The genotype does not in itself cause the disease; it "merely changes the sensitivity to the environmental interaction" (Gluckman and Hanson 2005, 18). Likewise, these environmental lifestyle factors are likely to lead to disease only in the more sensitive genotypes. The conclusion to be drawn may seem paradoxical: type II diabetes can best be prevented in individuals at *genetic* risk by altering their *environment*. Similarly, in the case of well-studied cancers, genetic mutations that consistently elevate risk (i.e., across a range of human environments) are believed to account for only 1%–5% of cases at most, while specific gene–environment interactions have been identified as the predominant causes of several major cancers (Mucci et al. 2001; e.g., Neslund-Dudas et al. 2014; Schoep et al. 2014).

A concrete data set will illustrate these key norm of reaction insights about causation. The data are from a molecular epidemiological study of lung cancer, a disease for which specific genetic as well as environmental factors have long been identified. Known genetic factors associated with higher incidence of lung cancer include null alleles at several polymorphic loci that code for glutathione S-transferases (GSTs), metabolic enzymes that detoxify carcinogens such as those in tobacco smoke. The production of GST detoxification enzymes is also activated by isothiocyanates (ITCs), chemicals introduced to the body by eating certain foods. ITCs are particularly abundant in broccoli and other vegetables in the mustard or crucifer family (Brassicaceae). Accordingly, in this case both the genetic and environmental factors operate on the same biochemical mechanism (Mucci et al. 2001), although that need not be the case for the interaction of such factors to affect phenotypic expression. The norms of reaction in Figure 1.7 show how two well-studied

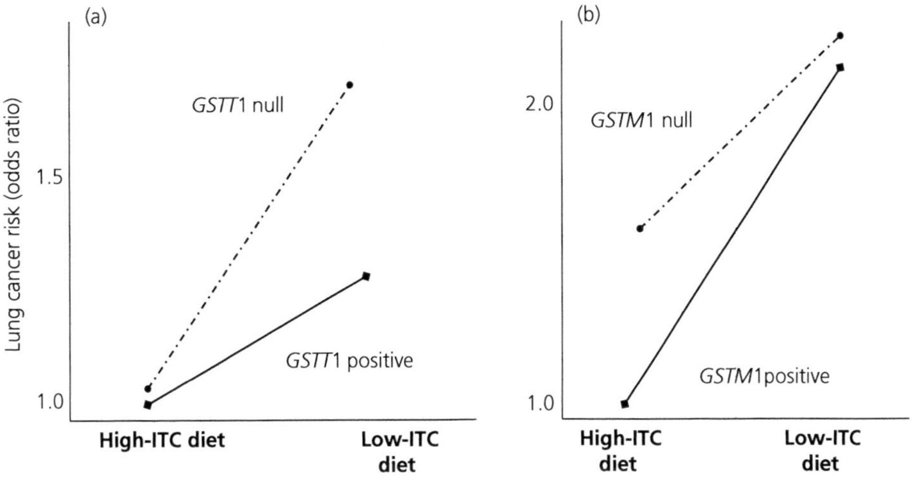

Figure 1.7 The probability of former smokers developing lung cancer is influenced jointly by the genotype at the *GSTT1* metabolic detoxification locus and by dietary intake of cruciferous vegetables high in GST-activating isothiocyanates (ITC). (a) The increased risk due to the homozygous null genotype is nonsignificant in the high-ITC dietary environment (1.0 for *GSTT1* positive vs. 1.06 for *GSTT1* null) but substantial in the low-ITC environment (1.29 for *GSTT1* positive vs. 1.79 for *GSTT1* null). (b) In the case of a second GST locus, *GSTM1*, the effect of the null genotype is insignificant in a low-ITC dietary environment, as both the null and the positive genotypes are associated with a high cancer probability in the context of such a diet but the presence of the null genotype increases the lung cancer odds ratio for former smokers from 1.0 (for the *GSTM1*-positive genotype) to 1.55 in the context of a high-ITC diet. For *GSTM1*, the improved diet does not fully offset the metabolic impact of the null genotype. However, notice that the increased risk due to a low-ITC diet is greater for both *GSTM1* genotypes than is the increased risk due to the null allele in either dietary environment. Data redrawn from Spitz et al. 2000.

genetic loci interact with dietary environment to determine lung cancer risk in former smokers (details in Spitz et al. 2000).[3] With a high-broccoli diet, individuals who are homozygous for the null *GSTT1* mutant allele, as well as those who carry one or two functional alleles at this locus, have similar, very low probabilities of disease (virtually identical risk in these former smokers as in the population at large; see Figure 1.7a). In this instance, as is often the case, a particular environment is relatively favorable for all genotypes. However, the relative probability of developing lung cancer increases markedly for former smokers with both null and functional genotypes in the ITC-poor (low-broccoli) dietary environment. In conjunction with this general response pattern, notice that these two types of genotype show $G \times E$ variation (nonparallel norms): the magnitude of the environmental effect is much greater for individuals carrying two null alleles, as the risk for such individuals increases especially sharply with a low-ITC diet (Figure 1.7a). In such cases, norm of reaction data make clear that the impact of a genetic variant depends on environmental circumstances, and vice versa.

Genetic polymorphism at a second locus affecting GST production, *GSTM1*, also interacts with this aspect of diet to influence the probability that a former smoker will develop lung cancer. In this case as well, cancer risk is elevated steeply in the ITC-poor environment for individuals with both null and functional genotypes (Figure 1.7b). By contrast, though, the null *GSTM1* genotype increases risk to individuals in the favorable ITC-rich environment (Figure 1.7b). Interestingly, the increased risk due to a low-ITC diet is greater for both null and functional *GSTM1* genotypes than the increased risk due to the null genotype in either dietary environment (diet-related risk increase is 70%–110%, compared

[3] In this case a second, behaviorally mediated environmental factor, the previous smoking history of the study's subjects, adds another dimension to the interaction. The *GST* × ITC–level norms of reaction are different for those who continue to smoke and for those who have never smoked, compared to the former smokers shown (details in Spitz et al. 2000).

to allele-related increases of 10%–50%). In fact, norm of reaction data often reveal that the average phenotypic differences between environments for all genotypes are greater than genotypic differences expressed within any environment (references in Sultan 2003). Note that this can be true even when there is a statistically significant average effect of genotype, as in this case (Spitz et al. 2000). Such patterns are clearly inconsistent with the causal primacy generally attributed to genetic factors. Norm of reaction data provide a biological scale on which the relative impact of genotypic and environmental variation among individuals can be assessed. The lack of this kind of scale is one reason why development studies that compare phenotypes produced by different genotypes in only one environment have contributed to an overemphasis on genetic sources of variation. Consider the conclusions a researcher would draw if she compared the cancer risk of a homozygous null *GSTT1* genotype to that of functional *GSTT1* genotypes within just the low-broccoli environment (Figure 1.7a): that study would point to a statistically significant genetic effect, without revealing that an even more pronounced reduction of risk occurs when the same null genotype experiences a change in diet.

In the case of disease phenotypes, this organismic calibration can provide insights that inform therapeutic approaches and public policy. In the lung cancer case discussed above, although it is certainly accurate to say that the disease "has a genetic component," the data would seem to support a preventative approach based on an inexpensive and effective environmental change—eating more broccoli—rather than the more difficult and costly task of identifying and directly treating individuals with null *GST* genotypes (see Vitzthum 2003). Because realized phenotypes reflect the influence of specific conditions on molecular processes of specific genomes, the view that "nature" (genes) and "nurture" (environment) are alternative causes of individual variation in any species is well recognized as scientifically obsolete, although this false causal dichotomy remains powerful in social and behavioral science (Hyman 2004). Recognizing the biological interaction of genomes with environments argues powerfully for an integrated research approach to studying phenotypes, in place of one that varies either factor alone while holding the other constant.

1.3.3 Implementing a norm of reaction approach

An eco-devo, norm of reaction approach to development clearly requires some basic changes to experimental design, because each genotype of interest must be studied in several developmental environments. Moreover, the choice of precisely *which* environments and genotypes to study becomes critical, since phenotypic outcomes are specific to particular $G \times E$ interactions. As a result of this specificity, it is impossible to simply extrapolate to natural systems research findings for genotypes and environments that were chosen based on arbitrary research conventions or convenience (S. Gilbert 2001). Because eco-devo studies aim to produce data that are meaningful for organisms in the real world, they require that environmental treatments be ecologically relevant, that is, based on single or multiple factors that vary in nature and, ideally, on naturally occurring levels of those factors (Miner et al. 2005).[4] Unavoidably, even experimental treatments that replicate naturalistic combinations of several environmental variables will be drastic simplifications of real habitats; but an informed choice of one or more essential variables can ensure that, at the least, findings are relevant to such habitats.

In some cases, salient abiotic factors to be considered are quite obvious, such as light, mineral, or moisture availability for plants, and temperature, aquatic pH, or food type and abundance for animals. More subtle aspects of the abiotic environment may be equally critical, such as the spectral quality of available light, or the pattern of temporal variation in temperature. Such variables can be challenging to recognize, as well as to manipulate experimentally. Interacting environmental variables can also be tricky to implement, but multifactorial studies (e.g., Sultan 2001; Langerhans et al. 2007) can provide further insights to development in nature, where organisms experience

[4] The terms "in nature" and "naturally occurring" refer to field (as opposed to controlled) conditions, rather than pristine habitats that are unaffected by human influence, since such habitats arguably no longer exist.

highly complex environments. Of course, real environments also include a number of biotic factors that influence development, such as competitors, facilitators, parasites, predators, pathogens, and prey. Variation in the identity and density of biotic interactors can be important (although demanding) to test in norm of reaction studies (e.g., Relyea 2004). And abiotic and biotic factors can interact to elicit unexpected phenotypic effects. In amphibians, low-concentration herbicide residues were found to interact with the presence of predators to disrupt development in drastic ways not predicted by norms of reaction to either of these stresses alone (Relyea and Hoverman 2006).

Clearly, choosing appropriate factors and factor levels for eco-devo research demands a good understanding of the organism's real-world environmental experience: not only which external factors measurably vary in the organism's habitats but which aspects of this variation matter to the organism. A recent study of plasticity in the Panamanian leaf-cutting ant *Acromyrmex echinatior* manipulated a key aspect of the social environment: the presence of large worker individuals. Patrilineal genotypes varied in their propensity to develop into large workers in response to the removal of this caste from the colony (W. Hughes and Boomsma 2007). A similar finding with respect to the presence of foragers in honeybee colonies (N. Chapman et al. 2007) suggests that diverse norms of reaction in response to colony composition may be common in social insects (further references in W. Hughes and Boomsma 2007).

The dimensionality of relevant environments for norm of reaction studies is further increased by the fact that conditions experienced early in life can alter an individual's developmental, physiological, or behavioral responses to later cues and conditions (Gluckman et al. 2007; Bateson et al. 2014). In other words, the norm of reaction is itself environmentally contingent across the life cycle: dynamic rather than static. In the nematode *Caenorhabditis elegans*, individuals exposed to a favorable chemical cue early in development show a stronger migratory response when they encounter the same olfactory stimulus as adults, an effect termed *olfactory imprinting* (Remy 2010). In mammals (including humans), the infant's experience of parental care (along with early nutritional conditions) is well known to influence its responses to stress as an adult; such effects of early parental care on adult norms of reaction for brain function are evidently mediated by specific epigenetic changes to a critical glucocorticoid receptor promoter in the hippocampus (Lucassen et al. 2013; T. Zhang et al. 2013). Even earlier influences on mammalian norms of reaction occur in utero, when maternal nutritional status affects fetal development in ways that strongly shape both juvenile and adult physiological and behavioral responses to diet. Offspring of undernourished mother rats develop very differently than those of genetically similar but well-fed mothers when both sets of juveniles are given a high-fat diet, exhibiting hyperphagia (overeating) along with elevated insulin and leptin levels, higher blood pressure, and obesity (Vickers et al. 2000). In humans as well, the fetal environment can alter how individuals respond to both high-calorie and low-calorie postnatal environments in ways that are critically important to health outcomes throughout life (Gluckman and Hanson 2005). The latter case was recently demonstrated in a study of children who experience severe malnutrition. Those children who were born smaller due to poor prenatal nutrition responded to famine later in childhood by developing marasmus, a malnutrition syndrome which is characterized by "thrifty" patterns of protein, amino acid, and lipid metabolism that promote survival (Forrester et al. 2012). In contrast, children in famine conditions whose mothers had been well nourished during pregnancy were more likely to develop kwashiorkor, a form of malnutrition characterized by a relative inability to mobilize stored energy, and consequently had a lower likelihood of survival (Forrester et al. 2012).

Norms of reaction can also be shaped by environmental conditions during previous generations. Maternal and sometimes paternal environment can specifically inform phenotypic expression in progeny of both animals and plants, often in ways that interact adaptively with the individual's own developmental environment (references in Mousseau and Fox 1998; Gottlieb 2004; Maestripieri and Mateo 2009; Herman and Sultan 2011; Salinas and Munch 2012; see Chapter 7, Sections 7.1–7.3). Although egg- and ovule-based effects such as nutrient and hormone provisioning reflect only the maternal environment, transgenerational environmental effects that are mediated by persistent epigenetic modifications can grant a developmental role to grandparental or even more remote past environments (e.g., Remy 2010; Herman et al. 2012). Persistence across multiple generations can also be maintained by parental behaviors that cause offspring to recreate the same behaviors in every generation (e.g., Champagne 2008).

Such transgenerational effects may cause offspring to express an adaptive trait that was initially induced in a previous generation. The induced enhanced behavioral response of *Caenorhabditis elegans* individuals to chemical cues described above has been found to persist for more than 40 generations in the laboratory (Remy 2010). In the yellow monkeyflower plant, *Mimulus guttatus*, simulated herbivory causes epigenetic changes that alter the expression of *MYB MIXTA-like 8*, a gene involved in the production of leaf hairs, a structural defense against chewing. This altered gene expression pattern is inherited (via both maternal and paternal contributions) such that the progeny of damaged plants produce denser leaf hairs even when their own tissues have not been damaged (Scoville et al. 2011). Similarly, when a mother daphnia (a tiny aquatic invertebrate) receives a chemical cue indicating the presence of a predator, her offspring develop a defensive helmet even when they themselves do not encounter the predator cue (Agrawal 2001). Notice that, in such cases, the distinction between "plastic" and "constitutive" traits becomes slippery, since the original induced response is constitutively expressed in one or more subsequent generations. More pragmatically, since transgenerational environmental effects heritably alter individual norms of reaction, experimental studies may confound genotype and past environment as sources of norm of reaction diversity.

Along with informed choice of environments, attention should be paid to the choice of genotypes in norm of reaction studies that aim to understand the nature and diversity of developmental systems. It is well recognized in medical studies of disease and other human phenotypes that this element of a study's design is central to the utility and generality of the results, and the same principle holds for other organisms. Genotypic samples should ideally be drawn from natural populations rather than from laboratory-bred strains, which have undergone both sustained inbreeding and (deliberate or unintended) artificial selection and for which the patterns of response to major environmental variables may be very different from those of natural populations (e.g., Nakagawa et al. 2012).

With ecologically informed design of both environmental treatments and genotypic samples, norm of reaction studies can test how particular factors and factor thresholds affect individual development and fitness in naturally evolved systems, allowing greater insight to the past, current, and prospective success of those organisms in various field contexts. Eco-devo data obtained with these design considerations in mind will be especially useful for understanding and perhaps addressing two daunting issues: how human-altered environments elicit developmental or life-history effects that threaten the survival of particular taxa, and the explosive spread of certain nonnative organisms in introduced ranges that comprise novel biotic and abiotic conditions.

1.3.4 A unified view of development

This chapter's overview of individual development leads to an important new jumping-off place. When development was understood as straightforward gene expression, cases of phenotypic plasticity (i.e., flexible norms of reaction) were seen as quirky exceptions to the general model, presumed to arise from distinctively complex regulatory mechanisms or special "plasticity genes" that allowed for environmental inputs but also entailed unique costs (carefully examined by DeWitt et al. 1998). This was perhaps a natural response to what appeared to be a new and surprising phenomenon. The 15 years of work discussed in this chapter lead to a startlingly different conclusion: that phenotypic expression *in general* is shaped by an extraordinarily complex and multilayered set of gene regulatory interactions that include environmental information at various levels and time points. As attention has shifted from identifying specific genes to elucidating these regulatory processes, for the first time it has become possible to unify our understanding of development as a single environmentally contingent process.

In a unified eco-devo approach, the norm of reaction can take the place of separate developmental categories, such as plastic versus canalized traits, which were initially believed to reflect distinct underlying genetic architectures. After Waddington (1960) proposed that canalizing mechanisms evolved to buffer genotypes against genetic and environmental perturbations, canalization came to be viewed as a distinctive type of developmental system. Certainly all genomes are "canalized" for certain traits, with respect to many molecular and environmental events during ontogeny that occur robustly despite variable inputs (Bateson and Gluckman 2011 and references therein). Similarly, all genomes are "plastic" with respect to certain traits and environmental variables. Indeed, plasticity in underlying traits can permit other aspects of phenotypic expression to remain constant across environments; the two patterns of variation are "intertwined" (Bateson and Gluckman 2011).

A unified framework for development makes it possible to examine contrasting patterns of expression such as canalized, plastic, and polyphenic simply as alternative norms of reaction: relatively constant or flexible outcomes across a given range of environments that result from *common* genetic architectures and modes of regulation. Nijhout (2003) has proposed a plausible evolutionary account for these diverse types of response pattern. He argues that the direct and generally maladaptive effect that variation in such environmental factors as temperature and pH has on developmental processes causes regulatory pathways to evolve in two ways: either selection will refine this environmental sensitivity so as to generate alternative adaptive phenotypes, thus producing plastic or polyphenic expression patterns, or it will favor mechanisms of physiological and developmental homeostasis that buffer the effects of environmental variation so as to produce more uniform phenotypic outcomes, or canalized expression. In this view, canalization and plasticity can be seen as "two sides of the same evolving developmental coin"—rather than ascribe to each a specialized *kind* of gene regulation, both can be seen *as* gene regulation (Sultan and Stearns 2005, 314; see also Stearns 2003). In this light, Nijhout (2003) notes that developmental responses that are relatively invariant are not somehow "more genetic" than developmental outcomes that vary from one environment to another. With this conceptual step—the common regulatory basis of all developmental outcomes, whether constant or variable—the "nature versus nurture" duality disappears for good.

With respect to describing norms of reaction as phenotypic response patterns, a further caveat is necessary. Whether a given trait appears to be plastic or canalized depends on both the environmental range and the set of genotypes in which it has been studied (Sultan and Stearns 2005 and references therein). In general, a lack of complete norm of reaction data creates the risk that such categories of response may simply reify experimental limits. A case of polyphenism, or distinct alternative phenotypes, may reflect either a major developmental threshold or experimental (or natural) environmental discontinuities that reveal a small group of discrete phenotypes and fail to accurately capture an underlying continuous norm of reaction (Karan et al. 2000; Stearns and Hoekstra 2000; Nijhout 2003). Indeed, most norm of reaction studies test relatively few environmental states, and few have characterized entire reaction norms (Karan et al. 2000; but see Windig 1994 for an exceptionally comprehensive case). By the same token, a trait may be quite constantly expressed across a certain range of environments, expressing a canalized pattern, but then show dramatic plastic adjustment when the range includes other, perhaps more extreme or complex, environments.

In a unified developmental approach, plasticity and canalization do not demand distinct causal mechanisms. The ecological and selective consequences of environmentally labile versus buffered expression patterns remain of fundamental importance and interest; the key questions in evolutionary ecology continue to center on the relation of expressed phenotypes to the varying conditions they encounter. Yet this insight suggests some changes in research emphasis and direction. For example, the possible costs of plasticity can be studied within the broader framework of gene regulation rather than assumed to exist as the unavoidable result of a unique, and uniquely cumbersome, genetic architecture (Sultan and Stearns 2005). The absence of such a dedicated developmental–genetic system may explain why evidence for specific costs of plasticity has remained elusive despite considerable research efforts (Sultan 1992; Scheiner and Holt 2012; see also Auld et al. 2010). Instead of seeking presumed costs of plasticity, researchers might ask a more general question: how have particular response patterns been shaped by key evolutionary factors, such as the existence of reliable environmental signals, phylogenetic constraints on or innovations in underlying transduction and regulatory systems, and fitness consequences of alternative expression patterns in natural populations? The norm of reaction approach also suggests a different way to frame questions about ecological and evolutionary diversity at the level of genotypes, populations, and taxa. Most importantly, this eco-devo approach makes possible integrated studies of individual development and evolution as dynamic processes that unfold in the context of real environments.

CHAPTER 2

The organism–environment relationship: Ecological niches, adaptation, eco-devo, and niche construction

Conceptualizing the relationship between organisms and their environments is pivotal for both ecological and evolutionary investigation. In both disciplines, this relationship is generally seen as an interaction between separate entities, in the sense that an individual whose traits are *internally* (i.e., genetically) determined confronts an *externally* defined and measurable environment. Because this well-defined confrontation results in the relative success or failure of the individual in its population, it effects a sorting process that matches environments with suitable genotypes. It is through this population-level sorting process (aka *natural selection*) that the external environment is understood to shape the genomes of future individuals. In ecological terms, a taxon's inherent traits determine the *niche* that taxon occupies in environmental space, and consequently its geographic distribution.

On closer examination, however, the environment extends into the organism, and the organism into its environment, in ways that obscure the boundary between them and lead to biologically intimate, causally multidirectional interactions. This insight reveals deeply imbedded tensions in ideas about ecological niche and evolutionary adaptation and points to an altered understanding based on two ways that organisms and their environments interpenetrate: *ecological development* or *eco-devo* (environmental influences on the features of organisms; see Chapter 1, Sections 1.2.2–1.3; also see Chapters 3 and 4), and *niche construction* (the shaping by organisms of their environments; Chapters 4–6).[1] Through these phenomena (and their underlying mechanisms of phenotypic determination, environmental response, and ecological causation), individual organisms play active roles in their own success or failure, creating developmental and ecological feedback in the process of evolution (Lewontin 2000; Odling-Smee et al. 2003; Post and Palkovics 2009). Alongside the excitement of this recognition comes a major new research challenge: how is it possible to measure and investigate environments, phenotypes, and the causes and patterns of their variation if the boundary separating organisms from their environments is porous rather than solid, and if causality runs in both directions?

[1] As explained previously (Chapter 1, Section 1.3), *ecological development* refers to the influence of environmental factors on the expression of phenotypic traits, including morphology, physiology, behavior, life history, etc. *Phenotypic plasticity* (including *adaptive plasticity*) is a major aspect of ecological development. *Plasticity* refers specifically to trait expression patterns that differ from one environmental state to another. *Ecological development* is a more inclusive, and functionally neutral, term that denotes the environmental context dependency of phenotypic expression, whether the outcome is stable or plastic with respect to particular traits. Note that plasticity in one trait (such as dense winter fur in a mammal, or plant shoot elongation in response to flooding) can allow stable expression of a functionally related trait, such as body temperature or leaf photosynthetic rate.

Organism and Environment. First Edition. Sonia E. Sultan.
© Sonia E. Sultan. 2015. Published 2015 by Oxford University Press.

2.1 The case of the green sea slug

The curious case of a small marine mollusk exemplifies the interpenetrating relationship of organism and environment. The tiny Atlantic sea slug *Elysia chlorotica* feeds on the filamentous yellow-green alga *Vaucheria litorea*, sucking out the contents of each long, multinucleate algal cell but retaining only the chloroplasts, which are incorporated into cells that line the animal's digestive tract. There the chloroplasts carry out perfectly successful photosynthesis, providing the sea slug with metabolic energy and carbon products to build its protective mucous sheath (Rumpho et al. 2000, 2008; Venn et al. 2008; Rumpho 2011; and references therein). This "kleptoplastic" (plastid-grabbing) symbiosis is unusual because the chloroplasts are incorporated into the sea slug's body as naked organelles that remain functional rather than as unicellular symbionts, so it is the animal's *own* cells that carry out photosynthesis. It is not yet clear how chloroplasts can continue to function within these mollusk cells, since there is no evidence for horizontal gene transfer of algal components into *Elysia* DNA; one possibility is that key algal genes are somehow incorporated extrachromosomally into the animal's cells to transiently regulate photosynthetic function (Bhattacharya et al. 2013; de Vries et al. 2013).

Despite their essential functional role, chloroplasts are not transmitted to *Elysia* offspring via eggs. Instead, the planktonic mollusk larvae must individually encounter filaments of *Vaucheria* algae and feed on them for several days as juveniles to acquire photosynthetic function and, indeed, to complete metamorphosis and development (Rumpho et al. 2011; Pelletreau et al. 2012). The ingested chloroplasts turn the juvenile animal a brilliant green color (the source of its scientific and common names). From this point on, the animal can sustain itself entirely by photosynthetic carbon fixation, for all purposes living as a plant for the duration of its 10-month life span (but opportunistically feeding again if it should find any *Vaucheria*). Weirder still, the adult animal is actually shaped like a leaf with a small head (Figure 2.1); because of its flat, blade-like parapodia, it is hypothesized to regulate its light interception by altering its positional angle, as leaves do on a plant's stem (Rumpho et al. 2011).

Figure 2.1 The kleptoplastic (plastid-grabbing) green sea slug *Elysia chlorotica* is a brilliant green color, owing to chloroplasts it acquires as a juvenile by feeding on the alga *Vaucheria litorea*. These ingested chloroplasts are incorporated as functional organelles into the sea slug's own tissues, allowing it to support itself through active photosynthesis for the duration of its 10-month life span. Mature adults such as the one shown here are typically 2–3 cm in length. Patrick J. Krug, photo credit and copyright. For the color image, see Plate 2.

So, in this case (and those of many other kleptoplastic sacoglossan species), the organism's essential functionality—its mode of adaptation to the environment—is itself environmentally acquired. Is the alga part of the mollusk's environment, or is it part of the animal itself? Both, actually. The animal incorporates its (biotic) environment into its own development in the most profound way possible. The resulting phenotype defines the environment it inhabits to be that of a plant, characterized primarily by light availability rather than heterotrophic resources. The environment constructs the green sea slug, and the green sea slug constructs its environment.

Although the details of this scenario are unusual, this *mutual construction* is a universal feature of the organism–environment relationship (Lewontin 1985). Elements of the external environment, from whole microorganisms (S. Gilbert and Epel 2009; Kerney et al. 2011; Nicholson et al. 2012) to nutrient or signaling molecules (see the references in

Chapter 3), are intimately incorporated in any biological individual as both developmental information and physiological resources. Conversely, the resulting organism determines the nature and scale of the abiotic and biotic variables that constitute its relevant environment and influences the actual states of those variables. As the philosopher Russell Gray has noted, the "co-construction" of the organism and its environment raises the question as to whether internal and external factors exist independently of each other in a biologically meaningful way (Gray 1992) and brings us back to the notion of a porous boundary between the two, with causation running reciprocally between organism and environment. This "co-construction" also suggests a changed basis for research approaches, which have hitherto aimed to separately characterize the organism and its environment in order to define ecological niches and selective pressures. Is it useful to focus on the organism–environment *relationship* rather than on the two as separate, independent entities—and in so doing, integrate evolution, ecology, and development? Below, this relationship is examined, using the long-standing concepts of *ecological niche* and *adaptation* as points of departure.

2.2 Resolving the ecological niche

The term *niche* is an architectural metaphor that suggests a preexisting place of some kind into which an organism properly fits (Lewontin 2000). This central ecological concept has long retained its pluralistic nature. The niche can be understood as a set of habitat parameters that delimit a species' distribution (following Joseph Grinnell), as the species' ecological role in its community (following Charles Elton), or as G. E. Hutchinson's "*n*-dimensional hyperspace" (Hutchinson 1957), an abstract volume circumscribed by environmental axes for all of the conditions and resources that influence the species' performance (see Colwell 1992 and Griesemer 1992 for historical discussion and references). Ecologists' persistent "love–hate relationship" with the niche concept (Hairston 1995) reflects the fact that these multiple definitions reflect a fundamental and still-unresolved ambiguity. Do niches have an independent existence outside of organisms, as subsets of environmental conditions, ways of functioning in natural systems, or both? If so, an infinite number of preexisting "vacant niches" can be discovered for potential occupation (see critique and references in Herbold and Moyle 1986). Alternatively, does the niche exist solely as a property of a specific organism, the ecological realization of its particular physiology, form, life history, and interactions with other organisms?

Rather than generate all possible combinations of functions and features, to posit ecological vacancies for carnivorous trees, photosynthetic cats, and so on—a task demanding considerable imagination, considering that actual niches include such combinations as colonial, frog-hunting, flying nocturnal mammals that roost in hollow logs (e.g., the fringe-lipped bat, *Trachops cirrhosus*; M. Cramer et al. 2001)—many ecologists agree that niches exist only by virtue of real organisms. Yet even so, the tension between organismic and external aspects of the niche results in a maddening paradox, succinctly stated by Robert Holt (2009): "It is true that niches cannot be defined independent of organisms; it is also true that it makes no sense to talk about niches without reference to external environments." In other words, niches do not exist independently of the organism, but they can only be studied as if they did.

In practice, this paradox leads to an unresolved version of the concept, in which a species' general physiology is seen to determine *which* external factors are relevant to its function and distribution (e.g., temperature, pH, or topography), while the niche itself is measured in absolute units of those factors in sites the species occupies (e.g., Kearney and Porter 2004; Knouft et al. 2006). With this approach, niche space characterized along externally quantifiable axes can be used to map a species' geographic range or the limits of its potential spread (e.g., Hargreaves et al. 2014). Niche-based *species distribution models* predict range limits by describing the abiotic conditions in which a species is currently found and then projecting those "environmental predictors" (viewed as ecophysiologically limiting factors) onto a landscape to create maps of suitable habitat (Guisan and Thuiller 2005 and references therein). This site-based approach provides a practical way to assess potential geographic distribution, although it does not clarify the problematic relation between a species' ecological niche and its actual range (which is also shaped by history, dispersal, and temporal fluctuations in both climatic and biotic conditions; Gaston 2009; Hargreaves et al. 2014). More fundamentally, it does not resolve the organismic versus external nature of the niche itself.

Field ecologists recognize that a species' niche space as defined in external units will vary dynamically among populations and even among individuals across the species' range, depending on shifting combinations of varying abiotic conditions, interacting organisms, and genetic differences for environmental responses (e.g., Ashton et al. 2010; Emery and Ackerly 2014; see discussion of the "realized environment" in Ackerly 2003). This variability in constraints, resources, and phenotypes suggests that the niche can best be understood as a *joint* property of the organism and its environment—an understanding that resolves the paradox by locating the niche at the interface between the two. Studying a species' niche would then entail characterizing a range of organism–environment relationships that reflect the variability at this individual scale, instead of aiming to define a single, fixed Hutchinsonian container. Of course, this does not make the niche any easier to measure; indeed, it raises new challenges both conceptual and operational.

Recent efforts to formulate a biologically coherent niche concept have been informed by two important insights that are consistent with this change in perspective. The first insight is that niche models must take into account the variable impact individual organisms make on their environments by means of resource depletion, habitat modification, and other activities (Leibold 1995). Sophisticated empirical approaches to quantifying species' niches add to external environmental parameters a focus on these organismic impacts, including demographic factors such as positive density dependence, competitive and population dynamics, and fine-scale environmental heterogeneity (references in Holt 2009). A second insight is that, as a result of such variable organism–environment encounters (that is, due to the "realized niche"), the niche itself evolves (Ackerman and Doebeli 2004), a process that can create feedback for other aspects of adaptive evolution. For instance, the evolution of greater ecological breadth along certain axes will lead to new abiotic and biotic encounters, further shaping the evolution of a species' environmental responses and resulting geographic distribution (e.g., Emery et al. 2012; T. Griffith and Sultan 2012). Despite this recognition, however, little is known at present about the potential for selective change in the environmental parameters of species distributions (Hargreaves et al. 2014).

Although these insights constitute key steps forward, they do not encompass the individual level of organism–environment interaction where the niche is actually realized. This requires adding a third insight: that how a given environment is experienced (i.e., as favorable or unfavorable) depends on the individual's phenotype. This point is amusingly illustrated by a study published online by an anonymous blogger (Seopher 2006). The aim was to determine the blogger's "best working environment," measured as the number of computer keystrokes and mouse clicks per minute. He tested background sound as the major environmental variable, comparing his response to 30 minutes of a silent "control" treatment versus 30 minutes of exposure to then-popular television shows and styles of music. For this self-declared fan of heavy metal rock, substantially higher rates of keyboard activity occurred in response to either the provocative cartoon *South Park* or intense "metalcore" music than with gentler television comedies such as *Seinfeld* (the expected favorite) or "snooze-inducing" classical music. Of course, for a lover of classical music, a Mozart concerto might be the most productive environment, while a more easily distracted person might work best in silence; another individual's productivity might be more heavily influenced by another aspect of the environment altogether, such as artificial versus natural light, the presence or absence of coworkers, or the availability of a caffeine supply; no doubt each person's productivity could be maximized by a particular combination of elements. This rather frivolous case serves as a useful parable: it is the individual's *experience of* its environment, an experience which is determined jointly by its phenotype and the nexus of interacting external conditions, that determines its functional success and hence, in nature, its ecological distribution.

These insights lead to further, quite broad, questions. Since niches can evolve, taxa and populations may be more or less constrained in this aspect of evolutionary lability, depending upon genetic variation and architecture, developmental constraints and plasticity, gene flow and dispersal, and effects of organisms on their (selective) biotic and abiotic environments (Holt 2009; Emery and Ackerly 2014). The evolution of the niche, then, is actually the evolution of the developmental, physiological, and behavioral responses that comprise the organism's relationship to its environment. In other words, our discussion of the niche has become a discussion about adaptation—the evolutionary frame for this relationship.

2.3 Adaptation

Although Darwin did not include *adaptation* as one of the "principal scientific terms" defined in the glossary for the *Origin of Species* (Darwin 1859), he wrote a great deal about "the exquisite adaptations of one part of the organization to another part, and to the conditions of life, and of one distinct organic being to another" (Darwin 1859, 114). It was in the *Origin* that the term was first used in a biological sense, to describe, for example, how the woodpecker's characteristic zygodactyl feet, strong chisel-tipped beak, and long, barbed tongue are "so admirably adapted to catch insects under the bark of trees" (Darwin, 1859, 114; see Figure 2.2). To explain how species came to possess such "beautiful adaptations," Darwin inferred the process of natural selection, an invisible hand choosing, over time, those variants whose fit to their "conditions of life" was most perfect. Eighty years later, in his foundational New Synthesis work *Genetics and the Origin of Species*, population geneticist Theodosius Dobzhansky (1937, 170) defined adaptation in a similarly broad way as "a harmony between the organism and its environment."

However, the term *adaptation* implies something more than mutual harmony. The word derives from the Latin *ad* (a prefix meaning 'to') and *aptare* (the verb 'to fit'), so the precise meaning is to be suited *to* something, that is, to "a given situation or set of circumstances" (Allen 1990). In other words, "the concept of adaptation implies a preexisting world

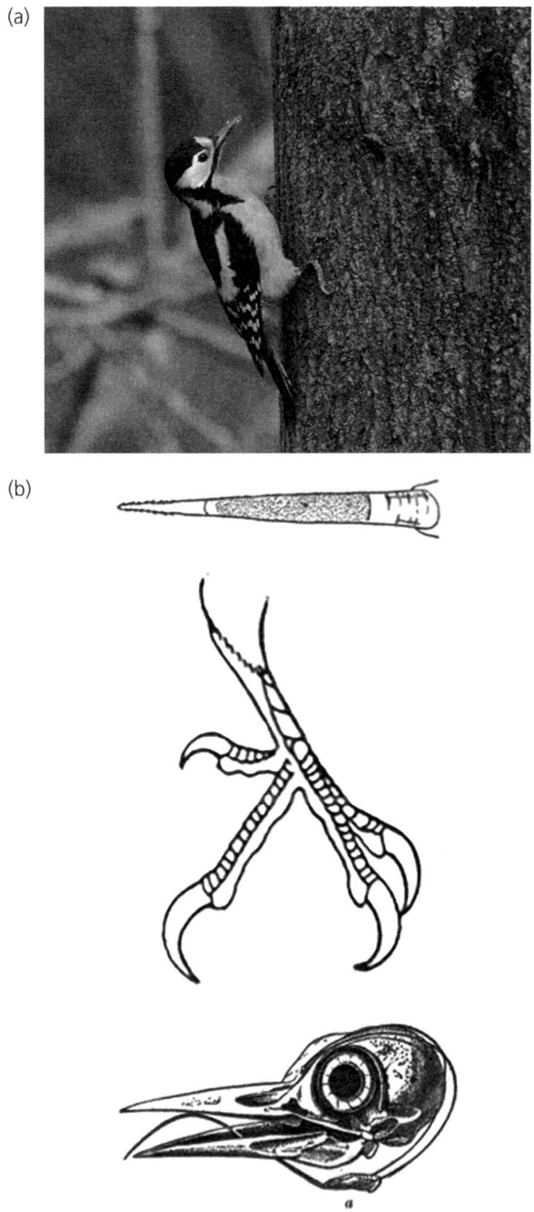

Figure 2.2 (a) The great spotted woodpecker, *Dendrocopos major*, climbs trees to extract its insect prey from crevices in the bark. Photograph by Darrel Birkett. (b) In *The Origin of Species*, Darwin developed the concept of "adaptations" by noting the (presumed) functional fit of an organism's traits to its "conditions of life," using the woodpecker as an example. Illustrations from a 1901 book highlight as "the woodpecker's tools" its skull with a straight, pointed beak and long hyoid apparatus, its barbed tongue, and its zygodactylous feet. Reproduced from Fannie H. Eckstorm, 1901, *The Woodpeckers*, Cambridge MA, Riverside Press; illustrations by Louis Agassiz Fuertes (Courtesy of Project Gutenberg).

that poses a problem to which an adaptation is the solution" (Lewontin 1978). Defining adaptation as the evolutionary process of becoming fitted *to* preexisting challenges is consistent with a view of organisms as passive products of their genes, which confront external selective forces (Lewontin 1985). The organisms themselves are seen to define this relationship only insofar as their inherited properties dictate which aspects of the environment impose these selective pressures (Lewontin 1978). In this neo-Darwinian view, then, adaptation is seen as a causally *asymmetric*, one-way process through which external forces shape organisms to match "pre-established environmental templates" (Odling-Smee 2007, 277; see also Laland and Sterelny 2006; e.g., Harvey and Pagel 1991; G. Williams 1992; Anderson et al. 2014). Adaptive diversity is also conceived primarily at the level of species' constitutive features rather than individual phenotypes, a view that reflects traditional emphases in evolutionary biology on (a) comparative anatomy as the key approach to diversity (dating back to Georges Cuvier at the start of the nineteenth century) and (b) studies of birds and mammals, taxa in which ecological developmental responses occur largely via behavior and life history (see e.g., Ernst Mayr's authoritative (1970) synthesis of neo-Darwinism, *Populations, Species, and Evolution*, an abridgment of his earlier *Animal Species and Evolution*).

Although it is certainly true that the harmonious and often marvelous fit of organisms to their environments remains for biologists a powerfully motivating observation, defining adaptation as the evolution of solutions to purely external problems turns out to be "a slippery concept" (Reeve and Sherman 1993; see also Mayr 1983). Indeed, despite enormous research effort, there are surprisingly few complete cases that document all three components of such adaptation in natural systems: (a) gene-based heritable traits; (b) ecological information pointing to preexistent, measurable selective agents; and (c) historical evidence that selection has occurred in response to those particular environmental factors (Endler 1986; Brandon 1990; Jain 1990; Orr and Coyne 1992; see also Gaut 2012). The adaptive (functional or fitness) consequences of alleles in ecological context are often presumed but only rarely rigorously demonstrated (Barrett and Hoekstra 2011; Fournier-Level et al. 2011). There are two causes of this lack of solidity, both of which arise from the porous interface between organisms and their environments.

2.3.1 Ecological development and the adapted phenotype

As discussed in Chapter 1 (Sections 1.2.2 and 1.3), the environment enters directly into developmental as well as physiological and behavioral pathways. As a result, the degree to which an individual organism is adapted to its circumstances is profoundly shaped (either positively or negatively) by environmental influences on functional trait expression, correlations among traits, and both absolute and relative fitness (Mitchell-Olds and Shaw 1987; Grafen 1988; Kingsolver and Schemske 1991; Rausher 1992; Travis 1994; Scheiner et al. 2002; Stinchcombe et al. 2002; Kruuk et al. 2008; Kingsolver et al. 2012; Anderson et al. 2014; and references therein). Hence, a meaningful definition of adaptation—of the organism's harmonious fit to its "conditions of life"—must include the ways that those conditions themselves enter into the process of phenotypic expression (Lewontin 2000; Sultan 2003a).

The realized phenotype is contingent upon such abiotic and biotic aspects of the external environment as temperature, pH, resource levels, and the presence, identity, and density of prey, pathogens, competitors, and mutualists (details in Chapter 3; for further case studies, see S. Gilbert and Epel 2009). An organism's phenotype depends as well on its internal environment—its physical and biochemical states at the cell, tissue, and whole-body levels, regulatory agents such as hormones and cell–cell interactions, and dynamic coadaptive adjustment among developmental, physiological, and behavioral traits. Note that the latter, internal parameters are also shaped by the external environment, via direct effects as well as by virtue of the organism's ecological developmental responses to it. As a result of both cytoplasmic and epigenetic inheritance, environmental effects across one or more preceding generations can also contribute to an individual's phenotypic state (reviewed by

Badyaev and Uller 2009; Jablonka and Raz 2009; Herman and Sultan 2011). Consequently, at each point in an individual's ontogeny, its functional adaptedness to its environment (as well as its resulting fitness) arises from a phenotypic encounter with a complex nexus of environmental factors, an encounter conditioned by a chain of previous phenotypic states, each of which was the product of a prior interaction with the environment (see full discussion and references in Chapter 1, Section 1.3).

The organism with its beautiful adaptations cannot then so readily be confined within its skin; the environment participates in a dynamic process of phenotypic expression by means of repeated feedback loops during individual development and, indeed, throughout the life cycle (Sultan and Stearns 2005; Oyama 2006). As with the niche concept, the causal directionality mediating this organism–environment interaction is difficult to pin down: does the environment affect the organism's phenotype, or does the organism respond phenotypically to its environment? Perhaps both interpretations are equally true; what is essential is that the adaptedness of organisms is determined in part through plasticity and other environmentally contingent aspects of phenotypic expression, rather than dictated from the nucleus. Because their realized features are the joint outcome of genetically bounded, evolved phenotypic expression patterns and external conditions that elicit particular outcomes, organisms as adapted products of evolution are not entirely separable from their environments.

2.3.2 The principle of niche construction

By the same token, biologically meaningful environments are not separable from organisms; organisms do not simply confront preexisting environmental scenarios (G. Simpson 1953; Lewontin 2000). Rather, by virtue of their expressed phenotypes, organisms themselves shape or "construct" (1) the *external* conditions of their existence, and (2) their *experience* of those external conditions. As a result of these two modes of *niche construction*, individual organisms play an active role in their own adaptation by shaping the very selective conditions that contribute to present fitness and subsequent evolution.[2]

In their influential 2003 book, Odling-Smee et al. examine how animals and other organisms can shape the conditions of their existence, either by means of metabolic processes, activities, and behaviors that physically alter environmental variables (which they term "perturbation") or by movement in space or time that results in encountering different sets of such variables ("relocation," discussed in historical context by Bateson and Gluckman 2011). Similarly, Donohue (2003, 2005) includes both environmental modifications (such as resource depletion) and "habitat choice" (by means of dispersal and germination timing) as means of plant niche construction. The concept of niche construction can be further extended to include phenotypic adjustments that permit the organism to *experience* a given set of conditions as more favorable, without either changing those conditions or moving to different ones.[3] These various individual-level effects provide an additional means of adaptive organism–environment matching: in addition to natural selection altering the organism to fit the environment, by means of niche construction the organism can change its environment (whether external or experienced) so as to fit itself (Odling-Smee et al. 2003). Indeed, some authors argue that the term should be applied only to beneficial effects of organisms on their environments (Dawkins 2004; Brodie 2005). Because both positive and negative niche-constructing effects result from the morphology, life history, physiology, and behavior that organisms express in their particular circumstances,

[2] The term *niche construction* is used in the original sense of Lewontin (1985, 2000), in conformity with the recent literature. Some authors, particularly in the United States, refer to these same phenomena as *eco-evo feedbacks* (see Chapter 7, Section 7.4).

[3] Changes made by organisms to external environments are here termed *habitat construction*, a category that corresponds broadly to the "perturbation niche construction" of Odling-Smee et al. (2003). In the ecological literature, such changes to external environments are sometimes termed *ecosystem engineering* (e.g., C. Jones et al. 1997; Hastings et al. 2007). *Experiential niche construction* is a new subcategory that includes "relocation" or "habitat choice" (sensu Donohue 2003; Odling-Smee et al. 2003; Donohue 2005) as well as other phenotypic adjustments that change the individual's experience of the environment without altering that environment.

these effects are themselves to some extent environmentally conditioned: ecological development and niche construction are interwoven.

Individually mediated niche-constructing effects of the organism on its (external or experienced) environment can alter the selection pressures acting on the organism and, consequently, its subsequent evolution. It is because of this selective feedback that niche construction can change evolutionary outcomes (Odling-Smee et al. 2003; Laland and Sterelny 2006; see Chapter 7, Section 7.4, for full discussion). A few familiar examples will briefly illustrate this general point. In many animals, bower or nest building constitutes a dramatic form of environmental modification. For instance, over 11,000 species of ants and termites build nests that transform soil mineral availability, organic content, and water-holding properties so as to facilitate the insects' regulation of temperature, humidity, and gas exchange (de Bruyn and Conacher 1990; additional references in Laland and Sterelny 2006). Clearly, these "constructed" conditions shape subsequent selection pressures on functional and social traits in these animals. Similarly, in the many thousands of birds, fish, and other vertebrate species that build nests, burrows, or bowers, this type of habitat construction has led to the evolution of complex associated behaviors such as nest guarding and courtship rituals (Odling-Smee et al. 2003 and references therein). Niche construction via phenotypic changes that allow a given environment to be experienced as less severe, such as the seasonal production of long, heat-trapping fur by Arctic foxes, *Alopex lagopus*, likewise mediate selection pressures on the animal's other traits. Because the combination of highly insulating winter fur and a curled-up resting position in snow allows the animal to conserve body heat without reducing metabolic rates during prolonged periods of intense cold, this species has not evolved to hibernate during the polar winter like other Arctic mammals but instead maintains its normal metabolic levels year round (Prestrud 1991).

Niche-constructing effects by one type of organism on the external environment may alter selection on other organisms that occupy the altered habitat (Odling-Smee et al. 2003). It may be less intuitively obvious that experiential niche-constructing effects such as the Arctic fox's long winter fur will also affect selection on co-occurring organisms by influencing the niche-constructing organism's interactions with its environment. For instance, because the fox's seasonal coat allows it to remain alert rather than torpid during winter, it is an elusive prey item for potential predators and a ready hunter whenever conditions allow. The effects of one organism's niche-constructing activities on its community can in turn feed back to influence biotic and abiotic selection pressures on the niche-constructing organism itself, in a type of diffuse coevolution (see Chapter 7, Section 7.4.4).

2.3.3 Niche construction via modifications of the external environment

Organisms inevitably modify their external environments, via various physical manipulations as well as simply by virtue of their presence, resource uptake, and waste deposition. These modifications can affect abiotic and biotic conditions in ways both subtle and impressive. The following few examples point to the broad range of potential niche-constructing effects on the external environment, effects which are here termed *habitat construction* (see Chapter 5 for detailed case studies). Although these impacts can be measured instrumentally as external to the organisms, note that they both result from and contribute to essential features of the organisms themselves, their eco-devo responses, and their adaptive functionality.

For microorganisms, the niche is often defined in terms of performance on a single environmental axis, such as the pH or carbohydrate source in a given growth medium. However, bacteria also construct their own niches (and those of successively occurring taxa) by means of phenotypic effects on these external, biochemically measurable aspects of the environment (Marco 2008). Indeed, like the photosynthetic sea slug discussed above, bacterial phenotypes can create entirely new niches through reciprocal interactions with their environments. This has been well documented in *Pseudomonas fluorescens* (Rainey and Travisano 1998). As a result of their respiratory activities, experimental populations of this aerobic soil bacterium generate vertical chemical gradients in liquid culture medium, transforming a homogeneous environment "into one containing . . . high and low O_2 regions" that constitute "new microhabitats" (Koza et al. 2011, 665, 668). This habitat construction alters both ecological and evolutionary dynamics in this system. New mutants can exploit these diverse ecological opportunities in ways that lead to adaptive radiation as well as successional changes. Cells of the *wrinkly spreader* morph, for instance, produce attachment factors such as polysaccharides that glue them to each other and to the vial walls. This allows them to extend dense mats across the liquid surface in order to exploit the nutrient-rich, high-oxygen interface between air and broth (Koza et al. 2011). This bacterial mat eventually becomes so heavy that it sinks to a low-oxygen level within the culture, at which point, alternative morphs (which may already have arisen within the mat; Rainey and Rainey 2003) are favored. Similarly, the complex communities or "biofilms" that constitute dental plaque are initiated by colonizing bacteria that provide attachment surfaces and alter oxygen and pH conditions in ways that create ecological opportunities for colonization by other types of bacteria (J. Li et al. 2004; Marco 2008).

Plants and animals can also change the levels and variability of critical external resources. For desert species, substrate moisture is such a resource. Several species of nocturnal beetles have been observed to build trenches on bare sand dunes of the coastal Namib Desert just before and during nighttime fogs. Because these trenches are oriented to be perpendicular to prevailing winds, they collect moisture from the air, doubling the moisture content within the ridges the beetles inhabit. This locally enhanced moisture environment results in increased beetle water content (Seely and Hamilton 1976).

Desert plants can likewise construct more mesic microclimates. A recently studied example is the rosette herb *Rheum palaestinum*, which is endemic to the central Negev desert. In marked contrast to the shallow, branched root systems and small, water-conserving leaves of most desert plants, individuals of this species produce a single deep root and enormous leaves nearly 40 cm in diameter (Lev-Yadun et al. 2009). These leaves have a thick, hydrophobic surface cuticle and an unusual, highly ridged texture, with veins that form pronounced channels oriented toward the leaf base (Figure 2.3). As a result of this morphology, rainwater hitting the leaf surface collects and flows to the base of the plant, irrigating soil in the root zone from two to ten times more deeply than rain that hits the surrounding, bare soil. Infiltration of water into the soil is also facilitated by the shrinking of the primary root, which creates a physical space between root and soil. *Rheum palaestinum* root and shoot morphology thus create a "mini-oasis" in which the plant effectively transforms its climate from one in which the average annual rainfall is approximately 75 mm/m^2 to one with the equivalent of over 400 mm/m^2, a level which is more typical of a Mediterranean climate than a desert (Lev-Yadun et al. 2009).

To some extent, all rosette plants, and indeed all plants, intercept and concentrate rainwater in this

Figure 2.3 The desert rhubarb, *Rheum palaestinum*, modifies its environment to mitigate a key stress. Due to the unusual shape, thick cuticle, and channeled surface of the leaves, over the course of a year *Rheum palaestinum* plants (which are endemic to the central Negev desert) each collect and deposit over four liters of rainwater around their own roots. As a result, these "self-irrigating" plants inhabit soil moisture conditions equal to those found in climates that receive nearly six times as much rain as the Negev (details from Lev-Yadun et al. 2009). Photo courtesy of Gidi Ne'eman, University of Haifa-Oranim. For the color image, see Plate 3.

way, exemplifying the penetration of organisms into their external environments. This problematic boundary raises an important operational question about environmental measurements. In this case, for example, is it more meaningful to characterize a plant's habitat by sampling moisture conditions at microsites lacking plant individuals or by sampling soil near plants to reflect the plant's participation in constructing its external environment? A third alternative might be to deliberately sample soil moisture in *both* ways (as was done in the *Rheum* study) and then compare the results to quantify the niche-constructing impact of plants in a given site. Using this approach, sampling strategies could be designed specifically to examine the environmental impact of a particular (native or introduced) species or of various cohorts of interest, such as annual herbs, deciduous trees, or nonnative grasses. Field studies of this type could provide important predictive insights as anthropogenic effects cause particular taxa to be introduced to, or lost from, biological communities (see Chapter 5 for examples and references).

A more unexpected aspect of habitat construction by plants is their impact on wildfires, a key aspect of forest ecology that, like local climate, might initially be considered as purely extrinsic. In a novel manipulative experiment, Schwilk (2003) showed that artificially pruning pine trees of their dead lower branches (which are ordinarily retained) reduced the temperature of subsequent fires, leading to decreased postfire seedling emergence of a co-occurring tree species. Surprisingly, then, even wildfires are not a purely external feature of the environment of forest plants but are influenced in frequency, intensity, and dynamics by plant traits such as branch morphology, retention versus self-pruning, leaf size, and the production of volatile secondary chemicals (Schwilk 2003; Schwilk and Caprio 2011). These flammability traits exemplify how an organism's features influence its external environment so as to create a selective feedback. The view that niche-constructing effects on flammability have indeed caused this kind of evolutionary feedback is consistent with the finding that, in the genus *Pinus*, dead branch retention is phylogenetically associated with fire-adaptive traits such as the fire-dependent release of seeds from cones (Schwilk and Ackerly 2001). This research program further exemplifies how a niche-construction perspective can suggest new and different biological questions and investigative approaches.

When the environment is understood to include biotic as well as abiotic elements, habitat construction can take the form of biologically intimate manipulation of other organisms. Pathogenic fungi, for example, can create novel, favorable conditions within the bodies of flowering plants by co-opting elements of the plants' own signaling apparatus. Fungal hyphae of powdery mildew, *Erysiphales* spp., are able to penetrate the epidermal cells of flowering plants (angiosperms) by exploiting the highly conserved *Feronia* signaling pathway unique to this major plant clade (S. Kessler et al. 2010). The proteins produced by this pathway have evolved to allow a growing pollen tube to penetrate the embryo sac (which contains the ovule), where it will release its two sperm so as to carry out the "double fertilization" characteristic of angiosperm reproduction. A germinating fungal spore is able to penetrate plant epidermal cells by eliciting expression of these same proteins, biochemically altering its external, biotic environment so as to allow access to host tissues (S. Kessler et al. 2010).

Certain fungi can alter their external environments in even more ecologically complex ways. Members of the Brassicaceae (Mustard) family of herbaceous plants in the genus *Boechera* are commonly infected by a rust fungus, *Puccinia monoica*, that develops within the plant body. Specifically, this fungus invades meristematic tissues so as to radically alter subsequent plant development: *Puccinia* causes its host to produce elongated stems that terminate in clusters of bright yellow leaves instead of the shorter flowering shoots that develop in uninfected plants (Roy 1993; Cano et al. 2013; see Figure 2.4). The fungally induced "pseudoflowers" do not resemble the plant's actual flowers—that is, its own reproductive structures—which are primarily self-fertilizing. Rather, their shape, color, and scent closely mimic the insect-pollinated flowers of completely unrelated plants that occur in the same alpine meadow communities, such as buttercups (Raguso and Roy 1998). The "pseudoflowers" even

Figure 2.4 A parasitic fungus radically alters the body of its plant host in a biotic form of niche construction. (a) An uninfected individual of the alpine meadow plant *Boechera stricta* (Brassicaceae), with the species' characteristic pale, cross-shaped flowers. (b) A *Boechera stricta* plant that is infected with the parasitic rust fungus *Puccinia monoica*. The fungus causes the plant to produce elongated, bright yellow shoots instead of the plant's own reproductive structures. The surfaces of these insect-attracting "pseudoflowers" are covered with fungal spermatogonia. Photos courtesy of Tom Lesback. For the color images, see Plate 4.

produce a sugary exudate that functions like nectar to reward insect visitors to their surfaces, which consist of *fungal* reproductive structures. Normally plant-pollinating insects that are attracted to these false flowers carry fungal spermatia (gametes) between different *Puccinia* mating types within or among different host plant individuals, successfully effecting fungal fertilization.

Puccinia effects multiple changes to its external, biotic environment. By co-opting host plant development, it dramatically reshapes above-ground *Boechera* morphology (Figure 2.4), with unknown effects on other organisms that either utilize or compete with these plants. The pseudoflowers produced by host shoots alter the behavior of pollinating insects in the community, and consequently the reproductive dynamics of co-occurring flowering plants as well as of host individuals (Roy 1993, 1994). Such radiating community impacts are a general feature of habitat construction, since changes made by any species to abiotic and biotic conditions will directly or indirectly affect cohabiting organisms (see Chapter 6 for further examples). Notice that, by presenting their gametes in false flowers, *Puccinia* fungi change the way they experience the alpine environment, because by virtue of this phenotype they interact with local pollinating insects. The organism's phenotype determines its experience of a given set of external conditions, a point which suggests a second mode of niche construction.

2.3.4 Niche construction via modifications of the environmental experience

As noted above, organisms inevitably modify their abiotic and biotic environments in ways that profoundly influence their own adaptive circumstances and those of co-occurring species. An equally universal aspect of niche construction is the way that an individual's realized phenotype—including its morphology, physiology, and behavior in a given environment—shapes and transforms how the individual experiences that environment, apart from any measurable effects on external parameters. Largely through phenotypic plasticity and other ecological developmental adjustments, individual organisms not only respond to the environments they encounter, they change the encounter itself, in many cases to one more favorable for their own function (see Chapter 4 for detailed case studies). It is in this aspect of the organism–environment relationship that a meaningful boundary between the two all but disappears.

For example, one well-known aspect of developmental plasticity in plants is the production by the same genetic individual of large, thin "shade leaves" and narrow, thick "sun leaves" under different light conditions (Figure 2.5; Sultan and Bazzaz 1993b; Briggs and Walters 1997; Sultan 2010 and references therein). This developmental response does not alter the intensity or spectral quality of light incident on the plant, so it cannot be said to change the external environment the way the plant's gas exchange activity measurably reduces levels of carbon dioxide and enhances the oxygen concentration in the surrounding air, or the way that nutrient uptake by roots depletes the surrounding soil of mobile mineral ions. However, the production of shade leaves does alter the light environment that the plant *experiences* to one with a greater quantity of available photosynthetically active light, because the larger surface area intercepts more photons. In general, plastic changes expressed in response to resource limits, including higher biomass allocation to tissues that collect a limiting resource (Sultan 2003b), appropriate timing, placement, and proliferation of foraging organs such as roots (Bell and Sultan 1999; Hodge 2004), and morphological or size adjustments to feeding structures that maximize intake (e.g., in sea urchin larvae; Podolsky and McAlister 2005), effectively increase the availability of the limiting resource and hence transform the environment that the organism experiences to a more favorable one. Conversely, the narrow leaves produced

(a) (b)

Figure 2.5 Ecological developmental responses such as leaf plasticity can change the environment that an organism experiences. (a) Under high-light glasshouse conditions (100% of summer sun), *Polygonum persicaria* plants produce narrow, thick "sun" leaves that minimize transpirational water losses from the leaf surface. (b) When grown in reduced light (24% of summer sun), genetically identical plants instead produce broad, thin "shade" leaves. Because the expanded surface area of these "shade" leaves catches more photons, this developmental response increases the amount of available light. Images are shown at the same scale (leaves are approximately 8 cm long). Photo courtesy of Dan B. Sloan and S. E. Sultan. For the color images, see Plate 5.

in full sun (Figure 2.5a) reduce the surface area the plant exposes to the comparatively dry atmosphere, and consequently its transpirational water loss (Fitter and Hay 2002). In terms of niche construction, this phenotypic adjustment transforms the plant's environment to one that is less desiccating.

Plastic responses can also change the individual's experience of environmental variability, again without altering the external environment as such. Behavioral plasticity, such as basking in the sun versus staying in shaded locations, allows an individual reptile to experience mesic temperature conditions rather than extremes of cold or heat. In sessile organisms, developmental and physiological adjustments can smooth out temporal and spatial variation, resulting in an environment that is experienced as more homogeneous. For instance, the proliferation of roots into nutrient-rich soil patches, or after temporal nutrient pulses, provides a plant with consistent access to plentiful soil resources despite the spatial and temporal heterogeneity of such resources in terrestrial soils (Bazzaz 1996; Fitter and Hay 2002; see Chapter 4, Section 4.3.2). Clonal plants that consist of interconnected growth modules, such as species of common goldenrod, *Solidago* spp., can smooth out spatial resource variability by means of physiological integration. Specific examples include the translocation of photosynthetic assimilates from ramets (shoot modules) exposed to sunlight to those shaded by neighbors (Hartnett and Bazzaz 1983; Steufer et al. 1994), and the movement of water to modules located in hypersaline microsites from physiologically connected modules in less stressful locations (Pennings and Callaway 2000).

Like an organism's measurable niche-constructing impacts on external conditions, eco-devo responses that alter the conditions an organism experiences raise operational questions about environmental measurement. Consider the example discussed above of enlarged "shade" leaves that increase a plant's photon capture. Light availability is generally measured as the flux density of photons in the air surrounding a plant, but one might better characterize the light environment the plant actually experiences with a mediated estimate, such as the number of photons intercepted due to both flux density and leaf area. Such an approach could lead to new ways of investigating adaptive diversity. In this case, measuring photon interception rather than ambient flux makes it possible to quantify the different degrees of light limitation a given habitat imposes on taxa that have different repertoires of leaf plasticity. This measurement approach could also produce a changed picture of light variability since, along with temporal or microsite variation in photon flux density due to patchy shade, the timing and extent of leaf plastic responses will also influence photon availability to plants. Depending on norm of reaction diversity, leaf plasticity could either enlarge or even out the spatial and temporal variation in light availability experienced by individuals in a given population; indeed it could shape light availability differently in populations or taxa that have different plasticity patterns. Adding such organism-mediated measurements of environmental conditions to conventional ecological site sampling could lead to new insights, including insights into the severity of novel environmental challenges as they are actually experienced by particular organisms.

Defensive phenotypes (whether induced or constitutive) that reduce the frequency or impact of attacks (or both) are one way that many types of organisms can experience their environment as less dangerous, although such phenotypes may not reduce predator or pathogen abundance. Numerous bacterial species can effect an analogous change in their experience of a threatening environment by expressing a putatively plastic response in which a fraction of individual cells in a population switch to a slow-growing "persistence" phenotype that is insensitive to antibiotics (Balaban et al. 2004; Kussell et al. 2005). This physiological switch is spontaneous and reversible, but it is not yet known whether it occurs at higher rates in response to antibiotic exposure or other specific cues (Kussell et al. 2005).

Organisms can also change their environmental experience simply by relocating to more favorable conditions (Odling-Smee et al. 2003). Such "habitat choice" (Donohue 2005) is mediated by phenotypic responses to unfavorable conditions that can be developmental, behavioral, physiological, or a combination thereof. For example, in the well-studied tropical butterfly *Bicyclus anynana*, larvae that have been poorly nourished emerge from their pupae with more strongly developed thoracic muscles, so that as adults they are stronger fliers and presumably able to reach better habitat (Saastamoinen et al. 2010). Some taxa "relocate" temporally to favorable conditions within a given location, for instance, through induced or seasonal physiological quiescence (i.e., egg or seed dormancy, leaf drop, estivation, or hibernation) during periods of intense drought or cold.

In contrast to these beneficial aspects of eco-devo response, some responses can negatively affect how an organism experiences a given environment. Certain plants may respond to extremely severe light limitation by *reducing* the size of leaves, further limiting their access to photosynthetic energy (Sultan and Bazzaz 1993a). Deleterious phenotypic responses such as reduced growth under resource limits are sometimes viewed as "passive" or "inevitable," as distinct from functionally beneficial responses that presumably evolved as adaptations (discussed by Sultan 1995; Dorn et al. 2000; Callahan and Pigliucci 2002; Donohue 2003; Sultan 2003a). The essential point is that both favorable and unfavorable changes to the environment that an individual experiences shape the individual's growth rate, size, resource use, and realized fitness, as well as its status as a competitor, predator, or mutualist. Consequently, even though they do not directly alter external conditions, phenotypically mediated changes to an individual's environmental experience also affect that individual's habitat-constructing impacts on the abiotic and biotic environment and hence on the selective pressures in the system.

Conditions encountered by maternal animal and plant individuals can affect the external environments of their offspring by changing the timing or location of offspring emergence (e.g., Donohue 2005; Braendle et al. 2006; Galloway and Etterson 2007). Because parental environmental conditions also influence offspring traits (often through effects on resource provisioning, epigenetic state, or hormones; see Chapter 4, Section 4.4), the parental environment can determine *both* the external environment and the environmental experience of the offspring. For instance, in response to changes in day length, pregnant meadow voles, *Microtus pennsylvanicus*, alter a hormonal signal to their developing offspring so that pups born in autumn have thicker coats (Lee et al. 1987; discussed by Gluckman and Hanson 2005). Autumn-born vole pups encounter a different external environment than offspring born in summer, but their heavier fur modulates their experience of this seasonal difference. Similarly, in many animals the parent's prenatal and postnatal behavior mediates development (Gottlieb 2004) so as to alter the offspring's experience of its environment, as well as the environment per se. In such cases, it may not be possible to distinguish niche construction effects on external versus experienced environment.

In fact, upon closer examination, this distinction is often difficult to enforce. Even the many purely external structures built by animals, such as tunnels, nests, webs, and mounds, profoundly influence the environmental experience of their builders. Indeed, these structures can be viewed as "parts of the animals" themselves, since they act as physiological organs to mediate "the flows of matter, energy and information through the organism and between the organism and its environment" (J. Turner 2000, 2). Similarly, many niche-constructing activities of individual plants simultaneously affect both their external and their experienced environments. For example, roots of certain taxa synthesize and exude surfactant chemicals, such as lecithin, which both alter soil chemistry and allow the plant to draw up more water per unit of energy used by root cells and hence experience the soil as less dry (Read et al. 2003). With respect to biotic effects, in many systems the cells sloughed off the root cap as roots penetrate into the soil create a chemical decoy for nematodes that would otherwise attack the root itself, altering both the soil community and the plant's experience of it (Rodger et al. 2003). More broadly, as noted above, eco-devo adjustments that modify the individual's experienced environment result in different patterns of activity, energy use, resource uptake, and waste production, ultimately altering a host of external habitat factors.

The distinction between habitat-constructing and experiential aspects of niche construction is problematic precisely because of the difficulty of identifying a meaningful boundary between the organism and its environment. A growing recognition that development is not internally determined and that an individual's phenotype is influenced by the environment throughout life (see Chapter 1) has led a number of biologists to re-examine the spatial and temporal boundaries that formerly defined individual development (Pradeu et al. 2011). Recent discoveries about (largely obligate) symbioses in an enormous range of systems highlight the difficulty of defining a clear boundary between the organism as a developmental and functional unit and the environment from which it obtains "foreign" but essential elements—surprisingly, the green sea slug turns out to be quite typical in this respect. In nearly all animal systems, bacteria ingested or otherwise incorporated into the body participate in critical aspects of "normal" development and physiology, including digestive and immune function (Turnbaugh et al. 2007; S. Gilbert and Epel 2009; Nicholson et al. 2012; Clements et al. 2014; and references therein). Similarly, developmental and physiological roles are played in the vast majority of plants by mycorrhizal associations between roots and symbiotic fungi, as well as by endophytic bacteria (Hardoim et al. 2008). The approximately 5% of angiosperm species that do not show these microbial associations include parasitic and carnivorous plants, which receive developmental and physiological inputs from other intimate biotic partners (Brundrett 2009).

In humans and other mammals, certain Gram-negative gut bacteria (such as *Bacteroides thetaiotamicron*) induce the expression of genes such as angiogenin-4, which codes for a protein known to induce blood vessel development. Introduction of either *Bacteroides thetaiotamicron* or a normal gut flora to the intestine of a previously microbe-free mouse results in completion of normal capillary development in the host's gut (references in S. Gilbert and Epel 2009; Pradeu 2011). In this case, information essential for "normal" development is provided not by the nuclei of the host's cells but rather by an evolutionarily distant symbiont. Interestingly, this gene has a very different effect on the symbiont's own environment within its mammalian host: the gut protein produced due to angiogenin-4 up-regulation by *Bacteroides thetaiotamicron* acts as a strong, specific bactericide against potentially competing Gram-positive bacteria that might colonize the gut (and incidentally are pathogenic to mammals; S. Gilbert and Epel 2009). Like other mammals, humans are "mixed selves" (Pradeu 2011); like all organisms, we exist as a nexus of more or less intimate environmental interactions rather than as closed systems determined solely by internal factors. This view points to a rather different way of formulating conceptual and empirical questions about adaptation as well as development, including the key question, what is it that evolves?

2.4 Focusing on the organism–environment relationship

Individual phenotypes inevitably affect both the external environment and the organism's experience of that environment, just as environments inevitably shape phenotypic expression. Half a century ago, the eminent New Synthesis paleontologist George Gaylord Simpson proposed that, in studying evolution, "it is more useful to focus neither on environment nor on organisms but on the complex interrelationship in which they are not really separable" (G. Simpson 1953, 199). If, as Simpson proposed, the process of adaptation is *the evolution of this relationship*, rather than a genetic change, internal to organisms, that results from a confrontation with external selective forces, then it is important to study how organisms and their environments interpenetrate and shape each other.

By recognizing this cyclic causation, and the selective feedback it creates, this approach goes beyond simply "extending" the phenotype further out from the genes, as suggested by Dawkins (1982). Indeed, the reciprocal co-construction of organism and environment is a very different conception from the one-way causal arrow that Dawkins confidently roots in the "small genetic fragments" inside the nucleus. In his view, the organism's body, behaviors, and environmental impacts all emanate from these genes or "replicators," which are themselves the beneficiaries of adaptive evolution (Dawkins 1982, 4). This strategy-based notion of genes as the true evolutionary agents logically leads Dawkins to question why organisms (seen as simply "vehicles" for the replicating genes) should exist as individuals, or how they do so as integrated functional entities (Dawkins 1982, Chapter 14). In contrast to an extended phenotype understood to arise from the gene's "long reach," the formative influence of organism and environment on each other grants the organism, with its flexible developmental repertoire, an active role in its own evolutionary trajectory.

How precisely does adopting Simpson's advice to focus on this "complex interrelationship" change the way organisms are studied? An eco-devo/niche-construction focus explicitly promotes the integration of development, ecology, and evolution by directing attention (a) to organismic causes and scales of environmental variability, (b) to the developmental systems and expression pathways that mediate response to that variability, (c) to the reciprocal (i.e., feedback) effects that characterize the organism–environment interaction, and (d) to the study of specific adaptive traits in the precise context of that interaction. For natural systems, such research avenues can add new dimensions of understanding to insights based on existing evolutionary and ecological approaches.

However, there are two caveats to be mentioned. A pragmatic drawback to this altered focus is the extremely difficult mathematics that will likely be required to incorporate the complexities of environmental effects on phenotypes (including fitness), as well as the feedback cycles introduced by niche construction, into theoretical models (but see Chapter 7, Section 7.4.3). A second, less concrete worry is that framing the organism–environment relationship as an object of empirical study could replace neo-Darwinian rigor with what Ernst Mayr once termed a "stultifying . . . holism" (Mayr 1983, 329)—a fear that, by giving up the straight arrow from genotype to fitness, biologists will lose the ability to effectively study adaptations. However, that worry may be overstated. The numerous published examples throughout this text make clear that, with careful design, scientific rigor can be brought productively to bear at the organism–environment interface. That being said, "much work remains to be done" before it is known exactly how an organism–environment focus might inform new ways of studying adaptation, selection, and community dynamics (Stearns 2009, 507), and what specific benefits or pitfalls might arise through the development of such approaches. What is certain is that, in order to generate methods that can move the field forward, it is necessary to think through current ideas and findings about the ways organisms interact with their environments.

For example, researchers investigating adaptive diversity among congeners or co-occurring species generally assess trait variation (based on each species' mean phenotypic value) in relation to measurably distinct environmental states or resources, which are understood as diversifying selective pressures (Schluter 2000a and references therein). However, the resolution of such studies is often limited by a lack of precise information regarding ecological opportunities and overlap (Schluter 2000b), which are both powerfully mediated by individual plasticity (K. Pfennig and Pfennig 2009; D. Pfennig and Pfennig 2012). An alternative approach would be to compare eco-devo response patterns among taxa of interest (based in this case on the mean norm of reaction) for traits that shape the organisms' experiences of environmental conditions that affect their fitness and ecological distribution (e.g., Bell and Sultan 1999; Semchenko et al. 2007; Sultan et al. 2009). A case in point is the work of Voesenek and colleagues (2004), who compared adaptive diversity in 22 plant species that occupy the Netherlands' hydrologically variable Rhine river floodplain. The researchers focused on a key response to flooding: the ability of shoots to elongate above the water level so as to maintain oxygen availability and photosynthetic gas exchange. Species' differences in elongation response to submergence corresponded to different field distributions with respect to depth and duration of soil flooding, illuminating a dynamic, eco-devo aspect of adaptive functional diversity. Studies of selection on patterns of plastic expression could offer new insights into microevolutionary processes as well as into macroevolutionary patterns of adaptive diversity (D. Pfennig et al. 2010 and references), although there are very few such selective studies to date (Kingsolver et al. 2012; e.g., see Donohue et al. 2000). This approach could tie in to mechanistic studies to investigate variation and evolution of the cue perception, transduction, and response pathways that give rise to adaptive patterns of eco-devo expression.

Another potentially fruitful research avenue is suggested by the niche-constructing effects of organisms on their environments. Comparisons of replicate microsites with and without individuals of particular taxa make it possible to quantify the effects of those organisms on otherwise similar external environments (Chapter 5) and communities (Chapter 6). An awareness of these individual niche-constructing effects might suggest different strategies for collecting data on environmental tolerances and field distributions (see Section 2.3.3). Instead of sampling conditions in a given habitat randomly or along spatial transects, field conditions could be precisely characterized according to the patch type or microsites that organisms of interest actually inhabit (e.g., Shavit and Griesemer 2009; Emery et al. 2012) and their probability of dispersal to like conditions in subsequent generations (Jacquemyn et al. 2007). Studies of this kind could be scaled up to characterize distribution at landscape levels or to assess habitat diversity from an organismic perspective. Field distributions might be further examined using organism-based metrics for key environmental factors (see Section 2.3.4), an approach that could connect more directly to related studies of adaptation and variation within and among co-occurring taxa.

Studying putatively adaptive traits in this kind of precise ecological context may help avoid distorted estimates of individual trait effects on fitness, a problem that poses a continuing challenge for evolutionary biologists (Mayr 1983; Lewontin 2000; Kingsolver and Pfennig 2007). More broadly, a core weakness that has long persisted in evolutionary biology is a lack of information regarding "the ecological basis of selection and adaptive evolution" (Kingsolver et al. 2012, 1114; further discussion and references in Wade and Kalisz 1990; MacColl 2011). In other words, despite a highly elaborated theory, as well as thousands of studies on selection in natural populations, we still know surprisingly little about how adaptation actually works at the organism–environment interface. Focusing closely on individual phenotypes in their actual, mediated environments may provide a way to address this fundamental gap in our knowledge.

Is it possible to frame evolutionary questions as Simpson's comments suggest and to study the organism–environment relationship, rather than the organism's genome, as the object of evolution? Such a shift could solve one perplexing problem that arises in measuring selection in natural populations: the fact that microenvironmental variation influences the individual's expression of any putatively adaptive traits, as well as influencing that individual's fitness (Rausher 1992; Stinchcombe et al. 2002). Environmentally based covariance between trait value and fitness distorts the causal signal of interest between a particular functional trait (with its presumed genetic basis) and fitness that would predict selective change (see Chapter 7, Section 7.1). Instead, trait values expressed in poor conditions will be negatively correlated with fitness, even if they are adaptively offsetting those conditions to some extent, and trait values expressed in favorable conditions will be associated with higher fitness, even if they do not contribute causally to it. Clearly, these environmental effects can bias estimates of selection for particular trait states and likely contribute to the consistently poor match between predicted trajectories for selective change and those observed (Kruuk et al. 2003). However, framing this phenomenon as "bias" reflects an implicit view of adaptive traits as constitutive elements that are best defined by controlling for environmental noise. From this point of view, environmental influences on the phenotype and its fitness *obscure* adaptation; in an eco-devo view, they *constitute* adaptation.

CHAPTER 3

Mechanisms of plasticity: Eco-devo pathways as environmental cue and response systems

3.1 Plastic trait expression

A central feature of the organism–environment relationship is that phenotypic expression is conditioned by environmental inputs; realized phenotypes are context-dependent, eco-devo outcomes (see Chapter 1, Sections 1.2.2 and 1.3; also see Chapter 2, Section 2.3.1). That being said, organisms have evolved trait expression pathways that differently incorporate these inputs and are differently sensitive to them. For a given organism in a particular set of environments, certain traits may show flat, constant (or *canalized*) norms of reaction, while the expression of other traits may change from one environment to another (see Chapter 1, Section 1.3.4). The latter, *plastic* reaction norms result when the organism senses a change in some environmental state that informs its phenotypic expression. (The term *polyphenism* is sometimes used to denote cases of animal plasticity in which two or more discrete alternative morphs are produced [e.g., Moczek and Nijhout 2003; S. Simpson et al. 2011; for a comprehensive exegesis of plasticity terms, see Forsman 2014].) In contrast to random developmental variation, such plastic responses are characteristic, repeatable, and specific to the eliciting conditions (see Schmalhausen 1949; Scheiner 2014; and references therein). Following on this insight, the pathways that underlie plastic expression patterns can be generally understood as *cue and response systems*: developmental, physiological, or behavioral adjustments (whether adaptive or maladaptive) that occur when an organism *perceives* some aspect of its environment as a specific piece of information and then *responds* to that cue by expressing particular phenotypic effects. These cue and response systems comprise the mechanistic basis for individual norms of reaction; the focus on such pathways constitutes an important difference between ecological developmental biology (S. Gilbert 2001) and the preexisting research area of phenotypic plasticity (Ackerly and Sultan 2006; Sultan 2007).

Cue and response systems are both fascinating and challenging to study because they are not singular, linear pathways. Rather, these systems comprise signal transduction networks that are embedded in larger regulatory networks. Phenotypic outcomes reflect the integration of these webs of environmental cues, their molecular transduction and regulatory pathways, and the feedback that occurs throughout the life cycle. This leads to a key point about the evolution of phenotypic diversity. In a simple scenario, consistently expressed trait differences among individuals or taxa may reflect major allelic variants that contribute directly to distinct, flat norms of reaction (at least across nonextreme environmental states). A familiar example is the typical "peppered" versus dark "melanic" adult pigmentation of the well-studied moth *Biston betularia*, as this trait is relatively stably expressed in different individuals and the expression is evidently based on alleles at a single locus (Grant 2004). In the case of plastic responses, however, norm of reaction diversity among genotypes, populations, or taxa (i.e., genotype-by-environment interaction,

Organism and Environment. First Edition. Sonia E. Sultan.
© Sonia E. Sultan. 2015. Published 2015 by Oxford University Press.

population-by-environment interaction, or species-by-environment interaction) can reflect an evolved difference *at any point in the system*, from sensory components or sampling behaviors for environmental cue perception, to steps along signal transduction pathways, to the precise expression of phenotypic effects (Sultan and Stearns 2005; Moczek et al. 2011; e.g., Lively et al. 2000; S. Griffith and Sultan 2005).

3.1.1 Adaptive and inevitable aspects of plastic expression

Plastic phenotypic responses may be functionally adaptive to the perceived environment, in which case it is likely that natural selection played a role in their evolution (Nijhout 2003; Scheiner 2006; additional references in Chapter 1, Section 1.3). Alternatively, plastic responses to particular environments may simply reflect the immediate impact of factors that limit developmental and physiological processes, such as low resources, temperature extremes, salinity, or anoxia (Sultan 1995; Scheiner 2006; see discussion of "active" versus "passive" plasticity in Forsman 2014). This distinction is not always clear-cut, as a given phenotype can simultaneously comprise both inevitable and adaptive aspects of response to particular growth limits or stresses.

To take a familiar example, annual plants growing in extremely dark shade perceive an environment in which very few photons are available to power photosynthetic carbon fixation. The plants respond by producing leaves that are both smaller and thinner than those produced by plants grown in high light or, indeed, in moderate shade (e.g., Sultan and Bazzaz 1993a; cf. Chapter 2, Section 2.3.4). The smaller absolute size of leaves produced in deep shade reflects the inevitable, drastic reduction of plant biomass in this severely light-limited environment. Yet, the plant's response also has an adaptive dimension, since thinner leaves maximize the photon-catching surface area that is produced per gram of leaf tissue. Similarly, when reared at high density, Indian meal moths, *Plodia interpunctella*, develop a smaller thorax and have reduced longevity, presumably due directly to resource limits. However, these smaller males also increase relative allocation to reproductive function, producing larger testes and ejaculates that partially offset the negative fitness effects of high density (references in Atkinson and Thorndyke 2001).

As in these examples, growth-limited and adaptive aspects of response are often developmentally joined; the more resource-poor the environment is, the more pronounced the adaptive plastic response will be. Such responses cause a negative statistical association of adaptive plasticity with fitness and can thus complicate interpretation (Mitchell-Olds and Shaw 1987; Rausher 1992; Stinchcombe et al. 2002; also see Chapter 2, Section 2.4). The developmental integration of these dual aspects of environmental response can itself evolve so that resource limits indirectly regulate adaptive alternatives. In the scarab beetle, *Onthophagus taurus*, body-size differences due to food availability are developmentally linked to alternative morphological/behavioral phenotypes: male larvae that encounter limited food supplies develop into small, hornless adults that gain access to mates through "sneaker" behavior; those larvae which instead receive enough food to exceed a species-specific body-size threshold develop an impressive pair of long, curved horns that they use to fight other large, horned males for mates (Moczek and Emlen 1999; Figure 3.1).

Such cases may reflect differently constrained, interacting aspects of phenotypic expression: norms of reaction for certain traits (such as growth rate or body size) may be relatively more directly constrained by suboptimal conditions, while norms for other traits can evolve as partially compensatory adaptive plastic responses to those conditions. Yet, even direct limits to growth will be selectively modulated to be less drastic if genetic variation exists for growth norms of reaction (Nijhout 2003). Moreover, inevitable effects of resource limits on growth can in themselves be beneficial: the small fetus that develops in an undernourished mammalian mother can better survive in that resource-limited uterine environment than a larger one (Gluckman and Hanson 2005). As a general rule, it may be most useful to consider norms of reaction in any particular case as the evolutionary result of natural selection on available patterns of phenotypic expression, given the unavoidable impacts of suboptimal environments on those expression patterns

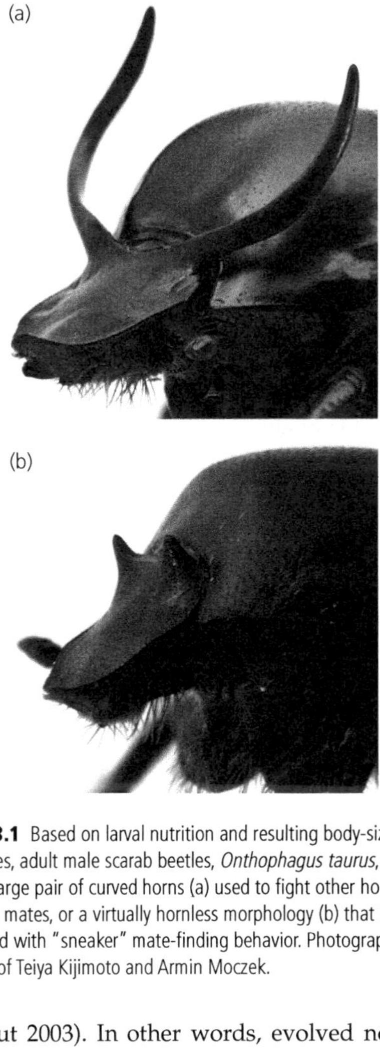

Figure 3.1 Based on larval nutrition and resulting body-size differences, adult male scarab beetles, *Onthophagus taurus*, develop either a large pair of curved horns (a) used to fight other horned males for mates, or a virtually hornless morphology (b) that is associated with "sneaker" mate-finding behavior. Photographs courtesy of Teiya Kijimoto and Armin Moczek.

(Nijhout 2003). In other words, evolved norms of reaction are to some extent *both* adaptive and inevitable. This interpretive approach is consistent with the view eloquently argued by Gould and Lewontin (1979) that all adaptations are to some extent constrained by the developmental systems in which they occur.

3.1.2 Plastic response in animals versus plants

A qualitative distinction is sometimes drawn between "intentional" behavioral responses, which are mediated by a central nervous system, and the kind of growth and movement changes expressed by plants and other non-animal taxa in response to specific environments (e.g., Odling-Smee et al. 2003). This distinction has been challenged by several plant biologists, who have argued that plant eco-devo responses to environmental information constitute *behavior* and, indeed, *intelligence* (respectively, Ballaré 2009 and Trewavas 2014; see Brenner et al. 2006 on "neurobiological" signaling). In a broader framework, all types of environmentally mediated phenotypic expression can be viewed as cue and response systems. Such systems are found in all organisms, although their properties will of course reflect specific organismic features.

Plants are modular organisms that receive diffuse, heterogeneous environmental signals at the level of individual organs such as leaves or roots. Over its lifetime, a plant will continually respond to environmental conditions at this level, as separate meristems differentiate and expand, as well as at the whole-organism level, by integrating these diffuse signals through shared transduction pathways (Walbot 1996; Gilroy and Trewavas 2001; Trewavas 2005; de Kroon et al. 2009; Hodge 2009; Trewavas 2014). In contrast, many animals have irreversibly differentiated cells that take shape in relatively rigid developmental sequences, often (e.g., in bilaterian animals) in the context of a fixed body plan (Walbot 1996; Atkinson and Thorndyke 2001). In such "higher" animals, cue and response systems may be highly integrated at the whole-organism level as a result of centralized nervous and endocrine control; since adult tissues often do not retain plasticity of size and form, environmental responses may shift during the life cycle from developmental to primarily physiological and behavioral (Sultan and Stearns 2005).

However, these very broad differences between animal and plant cue and response systems should not be overstated (Huey et al. 2002). Even highly derived vertebrates can remain developmentally (as well as physiologically and behaviorally) plastic in response to environmental signals throughout the life cycle. For example, rapid, reversible changes in organ size and gut morphology (as well as metabolic rates) are well-known responses to seasonally varying conditions in reptiles, mammals, and birds (references in Piersma and Lindstrom 1997). Adaptive developmental responses to biotic challenges can occur quickly in vertebrates such as the ruddy turnstone, *Arenaria interpres*, a European shorebird whose natural populations undergo surprise attacks by hawks and other raptors. When turnstones in an experimental aviary were briefly exposed to a raptor model at repeated, unpredictable times, within several days they developed a larger pectoral muscle, enabling more effective escape flight (Van den Hout et al. 2006). This rapid plastic response was threat specific: the pectoral muscle mass change did not occur when birds were instead exposed to a model gull, a benign member of their shore community (Van den Hout et al. 2006). More generally, many animals maintain undifferentiated groups of pluripotent cells, such as imaginal disks in insects, and stem cells in vertebrates; these cells, like the meristematic cells in a plant body, provide an extended window for developmental flexibility (Atkinson and Thorndyke 2001).

By the same token, hormonally mediated systemic signaling occurs not only in animals but also in plants, in ways that integrate the responses of separate organs or developmental modules. For example, when only the mature leaves of an *Arabidopsis* plant are experimentally exposed to shade and elevated carbon dioxide, signals are received by the plant's younger leaves that cause them to develop the same stomatal density as young leaves that are directly exposed to those conditions (Coupe et al. 2006 and references therein). This systemic signaling is evidently mediated by interactions between hormones and sugars; microRNAs are also likely to play a role (Coupe et al. 2006 and references therein) More precise similarities between animal and plant cue and response systems can also be found. At the cellular level, plants and animals can express very similar responses to environmental stresses such as salinity or temperature extremes (Roelofs et al. 2008). To a surprising extent, animals and plants may even use the very same signaling molecules along environmental transduction and response pathways (Schultz 2002; Schultz and Appel 2004; Brenner et al. 2006; and references therein; see Section 3.3).

3.2 Environmental cues

Natural environments consist of innumerable abiotic and biotic elements that vary (and partly covary) in space and time. In terrestrial habitats, for instance, such elements include the period, quantity, and spectral quality of light; the chemical composition, relative humidity, and temperature of the air; the availability of various mineral ions and of moisture and oxygen in soil; and the presence and activities of specific hosts, prey, pathogens, predators, symbionts, pollinators, dispersers, competitors, and facilitators. For a terrestrial organism, any or all of these elements can serve as cues for phenotypic response, depending on the organism's sensory apparatus and environmental sampling abilities and behaviors. In this way, usable cues are specifically "constructed" by each organism from the myriad variables in its habitat.

Fish have highly developed senses of taste and olfaction that allow them to utilize specialized and often incredibly attenuated water-borne chemical cues (Brönmark and Pettersson 1994). This chemosensory ability varies among fish species, depending on differences in such traits as the distribution of external taste buds on the body (Gomahr et al. 1992). When a diurnal forager (bobwhite quail, *Colinus virginianus*) and a nocturnal forager (rat, *Rattus norvegicus*) are exposed to both color and taste cues that are followed by an illness-inducing drug, the quail learns to avoid the color cue but not the taste cue, while the rat learns to avoid the taste cue but not the color cue (Hammerstein and Boyd 2012). By virtue of specialized structures on their legs, certain crickets and cicadas perceive the presence of reproductively active conspecific individuals up to several decimeters away, by sensing minute vibrations transmitted via soil or plants in the habitat. These vibrations constitute an environmental signal that other organisms in this community do not perceive (Heldmaier and Werner 2003). By the same token, animals are "blind" to the particular differences in light spectral quality that plants sense by virtue of photoactive pigments (see Section 3.4.1).

It can require some detective work to identify cues that reflect a particular organism's ability to "read" its environment. A case that has long intrigued observers is that of North American temperate periodic cicadas, which emerge synchronously from underground tunnels after a 17-year period of development as nymphs. Since when they are underground they do not experience photoperiod or temperature changes indicating annual seasonal cycles, it has remained a mystery how individuals "know" when exactly 17 years have passed (a phenomenon that may have led to the generic epithet for these species, *Magicicada*). Karban et al. (2000) tested whether cicadas receive a direct or a host plant mediated annual signal. They collected 15-year old nymphs and reared them on roots of genetically identical peach trees (in a common garden) that were conditioned to have either a 12-month seasonal cycle or an accelerated, 6-month cycle. Nymphs that fed on phenologically accelerated hosts emerged after only one further year (i.e., in Year 16 rather than Year 17). This result showed that cicada nymphs neither directly sense degree-day accumulation nor measure the actual time passed but instead use information from their host plant (possibly a seasonal change in the concentration of xylem sap) to count seasonal cycles (Karban et al. 2000).

Another insect case of long-standing interest is the precise cue(s) whereby increased population density induces ordinarily solitary, slow-moving desert locusts, *Schistocerca gregaria*, to develop the high-contrast color patterns and dramatically altered behaviors that characterize the gregarious (swarming) phase, in which individuals become highly attracted to each other and very active (Figure 3.2). Through a complicated series of lab experiments, Tanaka and Nishide (2012) found that, even in the absence of olfactory, tactile, and auditory stimuli associated with crowding, these changes could be induced during the juvenile (nymph) stage by the sight of other live individuals, either in a small plastic cup or on a video screen. Visual stimuli alone also caused nymphs kept in isolation to alter their body proportions to the greater cranial ratio typical of the gregarious form. The purely visual cue was effective even if the individuals seen nearby or on film were crickets or even tadpoles, but still photographs did not elicit the response; movement was evidently an essential part of the cue to indicate numerous living cohabitants. Further studies indicate that behavioral and locomotory "gregarization" of individuals can occur within a matter of hours and that tactile cues can also play a role (Rogers et al. 2014 and references).

As these examples show, environmental cues may be remarkably subtle and can contain extremely precise information. Certain fish and marine invertebrate species can distinguish through waterborne chemical cues not just whether predators are present in their habitat but exactly what those predators have been eating: when individuals of such species encounter predators that have been feeding on members of the same species, the individuals initiate fright or avoidance behaviors; however, if the predators have been eating a different type of prey, the same individuals discern the difference in predator-released metabolic byproducts and express no such response (e.g., Mathis and Smith

Figure 3.2 Individuals of the desert locust, *Schistocerca gregaria*, develop into one of two alternative forms: the slow-moving, lightly pigmented *solitarious* morph, at left, or the darkly pigmented, highly active *gregarious* morph, at right. The gregarious individuals form cohesive, migratory flying swarms. This dramatic morphological and behavioral change is induced by visual and tactile cues that indicate high conspecific density. Photo of laboratory-reared individuals courtesy of Gregory A. Sword. For the color image, see Plate 6.

1993; further references in Brönmark and Pettersson 1994). Individuals of crucian carp, *Carassius carassius*, sense this kind of very precise chemical cue to determine when predatory northern pike, *Esox lucius*, have switched from a macroinvertebrate diet to a piscivorous diet. Here, the cue informs the carp not just that the predator species is present but also that it has entered a particular, threatening, ontogenetic stage (Brönmark and Pettersson 1994). Another subtle cue is perceived by individual plants of certain species that are able to distinguish their own root systems from those of neighboring plants. The mechanism for this self-recognition is not yet known, although it may be related to chemical gradients of root exudates in soil (Gruntman and Novoplansky 2004). The self/non-self distinction leads to developmental adjustments in the plant that can minimize within-individual root competition (Hodge 2009 and references therein). Root exudates also appear to mediate the ability of certain plants to distinguish the roots of siblings from those of unrelated conspecific individuals and reduce competitive lateral-root proliferation in response (Biedrzycki et al. 2010 and references therein).

Environmental cues may be sensed either directly or in the form of internal feedback such as effects on tissue chemistry, growth rate, or activity level. For example, certain amphibians perceive the imminent drying of a pond as a reduction in their own swimming movements (details in Section 3.4.2). Both indirect cues for a given response and cues that rely on very subtle aspects of environmental perception can be difficult to identify without a thorough knowledge of the organism's biology and sensory systems.

In addition, environmental cues can be anticipatory (temporally indirect) rather than immediate. Reliability of such cues as predictors of coming environmental challenges is an important factor in the evolution of plastic response systems (Tufto 2000; Sultan and Spencer 2002; Berrigan and Scheiner 2004; Reed et al. 2010; and references therein). In mammals, developing offspring in utero receive as environmental cues their mother's nutritional status, body composition, and stress levels. In humans, this uterine signal leads to what has been termed a "predictive adaptive response" in the offspring that prepares it for a low-nutrient postnatal environment: children of poorly nourished mothers not only have lower birth weights but also develop a calorie-hoarding phenotype with respect to metabolic function, kidney size, food satiety threshold, and preference for high-fat foods (discussion and references in Bateson et al. 2014). Seasonal cues can also be anticipatory. Many temperate taxa anticipate the onset of cold winter temperatures by tracking seasonal changes in photoperiod rather than directly using ambient temperature (a notoriously unreliable cue) to signal the onset of winter. In small mammals, for example, when nights begin to last longer than 12 hours, production of the hormone *melatonin* (which is produced in the pineal gland during periods of darkness) increases. This internal endocrine signal leads to the timely onset of a suite of behavioral and physiological changes that result in preparation for, and tolerance of, winter conditions such as cold temperatures and scarce food supplies (Heldmaier and Werner 2003). Because the correlation of a given indirect cue with particular ecological conditions can vary among locations or types of habitat, populations may evolve to utilize different cues or sensitivities to initiate the same response (Miner et al. 2005; e.g., Galen et al. 2004; Huber et al. 2004). An important area for future work consists of field studies to estimate correlations between anticipatory environmental cues and the selective conditions to which they presumably have evolved as signals for adaptive response (Leimar 2007).

Environmental cues often consist of multiple interacting elements that are partly redundant; the use of such elements ensures that environments are accurately perceived and that responses are both robust and fine-tuned (Ballaré 2009). Overlapping, combinatorial cues may be particularly typical of functionally essential responses. For instance, the opening and closing of the stomates on a plant's leaves determine its carbon fixation and water loss and hence strongly influence survival and fitness. Stomatal behavior is regulated by a complex suite of extra- and intracellular environmental signals, including light (especially blue light), carbon dioxide concentration, water vapor pressure, fluxes of ions such as potassium and calcium, movement of sugars, and hormones including abscisic acid (reviewed by Hetherington and Woodward 2003; Underwood et al. 2007; Kim and Maik 2010). Although these regulatory pathways and the associated genes are still not fully known, it is clear that multiple simultaneous and partly redundant signals are coordinated in this process, that many feedbacks are involved, and that many of the chemical signaling components are also implicated in other cellular processes (Hetherington and Woodward 2003; Kim and Maik 2010). This complexity indicates that instead of "stand-alone" pathways, this signaling system constitutes a "scale-free network" consisting of many interconnected "nodes" or signal points (Hetherington and Woodward 2003; Kim and Maik 2010). Such networks are characteristically able to coordinate multiple cues and are robust to loss of certain component signals (Strogatz 2001). Another finding that is consistent with a highly reticulate model of stomatal regulation is that plants whose responses are conditioned by exposure to a particular environmental state such as elevated carbon dioxide express different sensitivity to *other* signals, including drought or abscisic acid (references in Casson and Hetherington 2010).

In some cases, what appear to be redundant cues actually come into play at different points in the life cycle. For example, although the quantity of incident light and its spectral quality provide a plant with largely redundant information about the presence of neighbors that cast shade, the red to far-red spectral cue received horizontally as reflected light from neighbors is received earlier in development, triggering a suite of shade-avoidance growth responses, while the quantum reduction cue that leads to shade tolerance responses becomes increasingly intense later on, as the plant's tissues become directly overtopped by vegetation (Ballaré 2009; see Section 3.4.1 for details). Disentangling specific environmental cues from their interacting roles in organismic response systems remains a considerable challenge, particularly since many sensory systems are not yet completely understood. Little is yet known about how plants sense several critical environmental variables such as temperature, atmospheric carbon dioxide concentration, and water availability (Anderson et al. 2014).

3.3 An overview of phenotypic response mechanisms

Once perceived by the organism, environmental information is transduced into internal signals by means of chemical and physical cell interactants. Chemical signaling components include pigments and other receptor molecules; hormones; steroids; metabolites such as sugars, oligosaccharides, fatty acids, aldehydes, alcohols, and proteins; lectins, peptides, amino acids, and nucleic acids; mineral ions; and reactive oxygen species (Schultz 2002; Trewavas 2014). Electrical, osmotic, and mechanical signals can also play a role in signal transduction networks (Blackstone and Bridge 2005; Trewavas 2005). Note that these various signaling components are all found in environmental transduction systems of both animals and plants (Schultz 2002; Brenner et al. 2006). Signals are mediated by cell-, tissue-, organ, and developmental stage- specific sensitivity thresholds and feedbacks and can be influenced as well by external conditions (reviewed by Nijhout 1999; Gilroy and Trewavas 2001; Dufty et al. 2002; Lema and Kitano 2013; Trewavas 2014). Environmental signals can also be mediated through feedback effects on the organism's own physiology and behavior. For instance, in many cases the presence of a predator causes reduced activity in prey individuals and consequently alters their oxygen consumption, food consumption, or both. These effects on the prey's body can feed back to alter growth trajectories so as to bring about defense-related changes in morphology (Andersson et al. 2006; Bourdeau 2010; Bourdeau and Johansson 2012).

Environmental transduction pathways can lead to immediate physiological responses, or to the up- or down-regulation of gene products such as transcription factors and microRNAs, which in turn inform expression patterns of other genes and gene networks in cell-, tissue-, and stage-specific ways (S. B. Carroll et al. 2005; H. Knight and Knight 2001; Murata and Suzuki 2006; de Kroon et al. 2009; G. Cramer et al. 2011). Details of these regulatory networks are provided by "omics" analyses, which include *transcriptomics* (to analyze environmentally induced expression profiles of coding and noncoding RNAs [Lema and Kitano 2013]), *metabolomics* (to reveal environmentally modulated changes in accumulation of sugars, amino acids, etc.), and *proteomics* (to assess protein complexes involved in environmental signal transduction as well as downstream protein profiles [Urano et al. 2010; G. Cramer et al. 2011]). *Epigenomics* techniques can be used to show the locations across the genome of environmentally induced methylation or histone modifications (Duncan et al. 2014).

Hormones are a particularly central component of environmental transduction and response pathways in both animals and plants. The animal endocrine system mediates variation for a broad range of ecologically important developmental traits, regulates the timing of metamorphosis and other life-history transitions, and powerfully influences behavior and offspring phenotype (reviewed by Dufty et al. 2002; Lema and Kitano 2013; e.g., Beckman and Dickhoff 1998; Hofmann 2006; Sommer and Ogawa 2011; Fernald and Maruska 2012; Maher et al. 2013; Dennis et al. 2014). One of the first ecological developmental case studies of the modern era provides an impressive example of animal hormonal regulation. In 1875, the eminent German evolutionary zoologist August Weismann demonstrated that, depending on rearing temperature, individual *Araschnia levana* caterpillars could develop into either a black-spotted, orange adult or a white-banded black morph—butterflies so different in appearance that Linnaeus had classified them as distinct species (discussed in S. Gilbert 2005). Later work revealed that this seasonal polyphenism was mediated by differential production of the insect hormone *ecdysone*, which regulates pigmentation, as cued by spring versus summer temperature and daylight signals (Koch and Buckmann 1987). Further studies have confirmed ecdysone to be one of the two or three critical hormones that regulate polyphenic plasticity in insects (Nijhout 1999).

Both external and internal environmental cues can activate the animal endocrine system to send signals that influence gene transcription, metabolic activities in specific target cells, and biochemical interactions with other growth substances. Within this broad framework, hormone action is astonishingly diverse and fine-tuned. The environment shapes the amount and composition of circulating hormones in several ways—through seasonal rhythms, immediate environmental inputs, and information feedbacks such as proportion of body fat. A particular hormone may play a role in several aspects of phenotypic response, at different locations or stages (Vercken et al. 2007 and references therein). Different types of cells can have different receptors, which have different chemical affinities and are associated with specific responses (Dufty et al. 2002). Hormonal sensitivity is itself an environmentally conditioned response, since the number and type of receptors can be influenced by tissue hormone levels and by the external environment, including the social environment. In mammals such as mice and rats, for instance, the amount and timing of handling by the mother modifies the density and type of hormone receptors that are found in specific parts of the brain in offspring, modulating subsequent responses to stress (Liu et al. 1997; Peña et al. 2013).

Because of the powerful roles played by hormones in environmental response pathways, chemical contaminants introduced to natural habitats that engage these pathways can cause significant problems in animal populations. One such case involves environmental sex determination in reptiles. In egg-laying reptiles such as turtles, lizards, and alligators, the sex of new hatchlings is determined by the external temperature eggs experience during a critical time period (Atkinson and Thorndyke 2001). Incubation temperature acts as a switch by shaping the activity of genes that code for endocrine components, including receptors that determine the impact of estradiol and other hormones (Crews et al. 1994; Lance 2009). This system is disrupted when eggs are exposed to chemical contaminants such as exogenous estradiols and polychlorinated biphenyls (commonly known as PCBs). These chemicals interact with endocrine pathways to cause females to be produced at male-inducing temperatures (references in Vos et al. 2000).

In plants as well as animals, hormones (*phytohormones* or *plant growth substances*) participate in an exceptionally pliable, complex, and diverse set of signaling pathways. Phytohormonal pathways are mediated by biochemical interactants such as calcium ions and sugars, and by tissue- and stage-dependent tissue sensitivities, interactions with environmental factors, and internal feedbacks. As in animals, these pathways help regulate central—and often multiple—aspects of development, life history, and physiology (reviewed by Mopper and Agrawal 2004; Trewavas 2014). For example, the well-studied phytohormone abscisic acid plays a key role in regulating numerous aspects of environmental response, from seed dormancy and germination behavior, to root and shoot development, to cellular responses, to cold and drought stress (G. Cramer et al. 2011). This functional breadth reflects two distinct avenues of abscisic acid regulatory activity. First, direct biochemical effects on ion and water transport allow control of short-term cellular processes such as stomatal opening and closing; second, slower transcriptional effects shape growth trajectories, tissue allocation, and morphology (Farnsworth 2004; G. Cramer et al. 2011).

Signaling pathways in distantly related taxa such as bacteria, plants, and animals often utilize common molecules or types of molecule. Because of these chemically similar or shared pathways, some ecological interactions involve co-opted transduction and response systems (Schultz and Appel 2004). For example, galling sawflies, family Tenthridinidae, order Hymenoptera, synthesize *cytokinins*, a major plant hormone, as the result of a shared insect–plant metabolic pathway. During oviposition, the insect injects cytokinins into host plant tissues along with the egg, causing the plant to form a hollow, tumor-like growth that houses the developing sawfly larvae (Figure 3.3; see Schultz 2002 and references). Sawflies may have acquired these plant signaling molecules from bacteria, which synthesize several key phytohormones that function in their own symbiotic, and often growth-enhancing, interactions with host plants (Long et al. 2008).

Certain modes of environmental input can also directly, indirectly, or both directly and indirectly shape gene activity relevant to phenotypic response (see Chapter 1, Section 1.2.2). For example, the conserved "heat-shock" protein HSP 90 is up-regulated in eukaryotes and bacteria under heat, salinity, drought, freezing, and oxidative stress; this "molecular chaperone" protein interacts with regulatory proteins such as kinases and transcription factors to help protect cellular processes of growth and development under these stressful conditions (Queitsch et al. 2002; W. Wang et al. 2004). Biochemical cues received from microbial symbionts can directly alter gene expression in a host individual; for instance, bacteria in the mammalian gut up-regulate host nuclear genes that are essential to intestinal development and function (Chapter 2,

Figure 3.3 Gall-making sawflies co-opt host plant shoot development by injecting a plant hormone into the tissue of a young branch during oviposition. Here, oviposition by the sawfly *Euura amerinae* has induced a Eurasian bay willow plant, *Salix pentandra*, to produce a large, hollow gall to house the insect's larvae. Photo courtesy of Tommi Nyman, University of Eastern Finland.

Section 2.3.4; also see S. Gilbert and Epel 2009). In the case of diurnal growth rhythms, light-sensitive hormones such as *phytochromes* contribute to gene regulation in ways that integrate endogenous circadian signals with external light conditions (Nozue et al. 2007).

In many animals, changes in social aspects of the environment can lead to surprisingly large-scale effects on the expression of genes that inform behavioral responses (i.e., those expressed in the brain). Males of Atlantic salmon, *Salmo salar*, that develop relatively quickly mature early at small body sizes to employ a "sneaker" life-history strategy that allows them to remain in freshwater to mate, rather than first migrating to the sea before returning to freshwater to mate, as typical large, anadromous males do (references in Aubin-Horth et al. 2005). A comparison of brain transcription profiles of "sneaker" males and age-matched immature (future anadromous) males within the same population and site revealed that these alternative mating tactics led to the differential expression of 15% of the 3,000 neural plasticity and other genes surveyed that are expressed in the brain (Aubin-Horth et al. 2005). The male African cichlid fish *Astatotilapia burtoni* expresses an even more dramatic plastic response to its social conditions: based on the individual's relative success in territorial encounters with other males, it can rapidly switch between a brightly colored, territory-defending "dominant" phenotype and a cryptically colored, nonaggressive, and nonreproductive "subordinate" phenotype (Figure 3.4; Hofmann et al. 1999; Burmeister et al. 2005). These alternative behavioral/life-history phenotypes also have very different patterns of brain cell proliferation (Maruska et al. 2012). The neuroendocrine pathways that regulate this reversible response to social cues strongly alter gene expression, up- or down-regulating approximately 5% of neural genes studied (Renn et al. 2008).

(a)

(b)

Figure 3.4 Social encounters cue alternative color and behavioral morphs in the cichlid fish *Astatotilapia burtoni.* Male individuals can rapidly and reversibly switch between a cryptic, nonreproductive "subordinate" phenotype (a) and a conspicuously colored, aggressively territorial "dominant" phenotype (b), depending on their success in social interactions with other males (details in Burmeister et al. 2005). Photos courtesy of Sabrina Burmeister. For the color images, see Plate 7.

Environmentally induced epigenetic changes also profoundly shape gene activity: epigenetic effects clearly comprise a major mechanism of plasticity in plants, animals, and microorganisms, although one that is only beginning to be understood (see Chapter 1, Section 1.2.1; further references in Kalisz and Purugganan 2004; Chinnusamy and Zhu 2009; Fagiolini et al. 2009; S. Gilbert and Epel 2009; Duncan et al. 2014; Herman et al. 2014). For instance, epigenetic modifications can influence hormonal pathways, altering the regulation of developmental trajectories and other aspects of phenotypic expression (Latzel et al. 2012; Lema and Kitano 2013). However, clarifying the role of various epigenetic mechanisms in environmental response systems will require more integrative data than are currently available: to date, few studies have been designed to link naturally occurring environmental cues to particular epigenetic changes and resulting phenotypic adjustments in ecologically relevant traits. One innovative ecological epigenetics study investigated the role of methylation in mediating the interaction between a floral fungus and its host plant community. In a lab experiment based on field nectar measurements, researchers found that variations in sugar composition and concentration designed to mimic the different nectars of co-occurring plant taxa induced specific methylation changes in the flower-living yeast *Metschnikowia reukaufii* (Herrera et al. 2012). These nectar-induced "methylation polymorphisms" allowed *Metschnikowia reukaufii* genotypes to exploit a broader range of sugar conditions (Herrera et al. 2012; discussed in Chapter 7, Section 7.3).

Gene expression patterns influence phenotypic outcomes through direct effects, as well as through effects on hormone production and on biochemical interactions. "Systemic" signals such as hormones may interact with local conditions such as resource or osmotic status, to finely calibrate phenotypic responses at the level of individual tissues or organs (de Kroon et al. 2009). For example, the extent to which a plant's root system will preferentially proliferate into nutrient-rich soil patches is reduced if root cells have high tissue nutrient concentration as a result of earlier exposure to favorable nutrient conditions (references in de Kroon et al. 2009). Like external cues, these internal pathways constitute highly complex signaling networks (H. Knight and Knight 2001) that are characterized by "crosstalk" between different transduction pathways. Via these pathway interactions, the flow of information is allowed to "diverge, branch, converge . . . synergize, and integrate" (Trewavas 2005, 414). Such regulatory convergences, synergies, or antagonisms can occur at many points along a transduction cascade (e.g., Roelofs et al. 2008). Specific biochemical components—and indeed entire sequences or "modules" of regulatory pathways—can participate in several different aspects of response.

For example, the same hormonal machinery coordinates a host of developmental responses in insects, including facultative diapause, molting, alternative morph types, and reproductive onset and allocation (Nijhout 1994). In plants, the jasmonate signaling pathway regulates both light foraging and defensive responses to herbivory, via direct and interactive effects with other hormones (Ballaré 2009). Certain plant genes (such as *RD29A*) are up-regulated in response to both drought and cold, because these stresses induce different, specific transcription factors that bind to the same element in the gene's promoter region. This element, *DRE*, exemplifies a crosstalk "integration point" for cold and drought signaling (H. Knight and Knight 2001). There is also a separate, parallel regulatory pathway in response to drought alone, mediated by the broad-spectrum phytohormone previously mentioned, abscisic acid (Urano et al. 2010).

In some cases, the same components can be activated by different cues and initial transduction events to elicit a common plastic response (Mopper and Agrawal 2004). Two distinct types of environmental stress plants may encounter are shading, when neighbors intercept solar rays (see Section 3.4.1), and submergence, due to soil flooding. Many plants respond to both of these environmental challenges by rapidly elongating stems, leaf petioles, or both, a developmental response that allows the individual to either avoid shade cast by its neighbors or extend photosynthetic tissues above the water surface into a functionally suitable aerial environment (Schmitt et al. 2003; Voesenek et al. 2006). The elongation response is mediated by shared hormonal pathways that regulate (a) growth-restraining proteins in the *DELLA* gene family, and (b) *expansin* genes governing cell-wall elongation (Vreeburg et al. 2005; K. Franklin 2008; Pierik et al. 2009). However, the initial cues that converge on this shared regulatory module are entirely separate: shade is perceived, largely by photoconvertible pigments and other chemical receptors, as a set of spectral cues, and submergence is perceived as an internal buildup (due to limited diffusion under water) of the endogenous gaseous hormone ethylene (references in Sultan 2010). Maladaptive plastic effects of chemically very different environmental triggers can likewise result in the same phenotypic response. In one such case, the *DDT* metabolite *dichlorodiphenyldichloroethylene* (an organochloride) and the organometallic toxin *methylmercury* both lead to the production of thinner eggshells in several bird taxa by causing reduced prostaglandin synthesis and consequently lower calcium transport to the eggshell gland lumen (Atkinson and Thorndyke 2001). In other cases, multiple environmental cues can induce antagonistic rather than similar effects on gene expression; there is no "rule" for cue interactions (Roelofs et al. 2008).

Such network rather than linear topologies allow individual organisms to integrate their phenotypic responses to the complex, multifactorial environments they encounter in nature. In the model plant *Arabidopsis*, for instance, the *Leafy* gene is one of several that integrate photoperiod and vernalization (cold-period) cues to control the transition to flowering (Parcy 2005). These integrative cue and response networks can be particularly difficult to elucidate. Recent studies have revealed crosstalk between plant light and temperature signals that results in integrated effects on shoot development and flowering time; however, neither the precise hormonal signals nor even the temperature sensors in plants have been identified (K. Franklin 2008; Penfield 2008). Transduction networks for plant temperature responses are themselves, surprisingly complex (reviewed by Penfield 2008): instead of passive effects of temperature on metabolic rates, these signal pathways involve temperature effects on protein stability and membrane fluidity; interactions with abscisic acid, salicylic acid, and fatty acids that shape tolerance of temperature extremes; and mediation by *gibberellin* hormones of temperature effects on growth. The transcription factor gene *Flowering Locus C* (*FLC*) appears to be a key node in several temperature signaling pathways that shape life history and cold tolerance (Penfield 2008); interestingly, *FLC* is regulated in part by epigenetic resetting of the vernalization signal via methylation (Sheldon et al. 2000).

An even more elusive aspect of plant developmental response to multiple environmental signals is the way roots are able to grow generally *downward* to anchor plants and access soil moisture, as well as *horizontally*, as needed to avoid soil obstacles. As originally hypothesized by Charles and Francis Darwin (1880) and only recently confirmed (Massa and Gilroy 2003), the cells of the root cap are able to sense both gravity and contact with obstacles: touch sensitivity is localized in the tip cells of the cap, while graviperception occurs in the more basal columella cells because of their precipitating starch plastids. Touch-induced changes to the actin cytoskeleton of these gravity-perceiving cells have been proposed as the mechanism by which roots integrate their responses to both types of information (Massa and Gilroy 2003).

The signaling role of internal feedbacks can make transduction networks remarkably multilayered. An example that raises the issue of human-mediated environmental inputs is the effect of elevated atmospheric carbon dioxide on plant development. Carbon dioxide concentration affects photoreceptors and, consequently, the perception and transduction of light signals (Song et al. 2009). Carbon dioxide also directly influences photosynthetic activity to cause changes in the within-plant concentration of sugars such as sucrose, glucose, and fructose. These sugars participate in a broad array of signaling cascades that shape development, life history, defense, and stress responses via effects on enzymatic activity and gene expression (Eveland and Jackson 2012; additional references in Smeekens and Hellmann 2014). In addition, sugars directly affect light reception and other physiological events that feed back to influence photosynthesis (M. Paul et al. 2001). Finally, atmospheric carbon dioxide concentrations interact with mineral availability, temperature, and light conditions to shape plant growth and phenology (Johnston and Reekie 2008; Song et al. 2009). To date, little is known about how elevated carbon dioxide, and other aspects of future environments, will alter the complex cue and transduction pathways that inform phenotypic outcomes in natural systems. Empirical studies in this area will provide critical information about the immediate and evolutionary impact of anthropogenic changes on organisms and biological communities.

3.4 Cue and response systems: plant and animal case studies

Elucidating the perception and transduction steps in environmental cue and response systems is a major challenge, especially in the context of real-world conditions. Below, four particularly well-characterized plant and animal case studies are presented in detail, to illustrate the mechanistic complexity and ecological nuances of these systems (see Murata and Suzuki 2006 for a bacterial case study). These examples also point to the many potential avenues for ecological developmental investigation, from biochemical, molecular, and "-omics" approaches to ecological studies of multitrophic natural communities.

3.4.1 Light cues and plant developmental responses to shade from neighbors

Plants perceive the effects of shading by neighbors on both the spectral quality and the quantity of light they receive, through a series of distinct, partly redundant cues that integrate external and internal information (reviewed by Ballaré 2009; Sultan 2010; Pierik and de Wit 2013). Light *quality* cues make use of the "unique property of chlorophyll-containing tissues" in strongly absorbing red light wavelengths while reflecting far-red wavelengths (H. Smith 1982, 1995); indeed humans also exploit these spectral effects to detect the presence of plant canopies, for example, in aerial vegetation surveys. For a plant, a reduced ratio of red to far-red wavelengths in light that is reflected horizontally from adjacent leaves serves as an early cue for the presence of near neighbors (Ballaré et al. 1990). Once a neighbor's tissues actually begin to shade the plant, light transmitted through those tissues is also far-red enriched, prolonging and intensifying the initial signal. This signal is picked up, at both stages, by specialized phytochrome pigments located throughout the recipient plant's shoot.

These pigment molecules serve as highly sensitive detectors by switching between two photoconvertible forms, in response to small changes in the red to far-red ratio of incident light (reviewed in H. Smith 2000). This molecular switch can initiate fine-tuned, reversible growth responses within minutes of a spectral cue (H. Smith and Whitelam 1997). Interestingly, several distinct phytochrome proteins have been identified that have partly overlapping light-sensing and signaling functions; as is often the case in such "gene families," the coexistence of evolutionarily conserved and variable functional domains allows for phenotypic regulation that is at once robust and extremely precise (Schlichting and Smith 2002; Heschel et al. 2008). Phytochromes in the active molecular form are sent to the nucleus, where they bind to a group of dedicated transcription factors (*phytochrome-interacting factors* or *PIFs*) that regulate genes involved in stem and petiole elongation (Lorrain et al. 2008). Along with the red to far-red light ratio, the transcription levels of certain *PIF* genes are regulated by diurnal light/dark cycles and internal circadian rhythms (Salter et al. 2003; Nozue et al. 2007).

A second suite of cues signals the reduced *quantity* of usable light (photosynthetically active radiation) due to shade from neighboring plants. Because blue light is absorbed by plant tissues, its quantity is sharply reduced by vegetative shade, a condition perceived by specialized blue-light receptor molecules known as *cryptochromes* and *phototropins* (K. Franklin 2008). Phototropins sense the precise directionality of blue light to direct stem movements toward light and to regulate chloroplast movement within leaf cells so as to shape the spatial distribution of plastids in ways that maximize their exposure (and hence light-harvesting capacity) in low light (Ballaré 2009). (As in the phytochrome family described above, phototropins encoded by homologous genes have somewhat different but partially overlapping sensitivities to blue light [Galen et al. 2004].) Along with their primary role in receiving spectral information, phytochromes may also sense reductions in the quantity of both red and far-red components of incident sunlight (Pierik and de Wit 2013). In addition to these chemical signaling mechanisms, reduced light quantity is perceived through a complex network of external and internal cues. Lower light intensity directly changes the excitation level of the Photosystem II units within leaf cells and thus influences the expression of genes involved in leaf morphogenesis (Ballaré 2009). A lower flux of photosynthetically active radiation also results in a reduced concentration of carbohydrates in plant tissues; this reduction influences subsequent growth through effects on carbon storage, photosynthetic activity, and the expression of genes that influence shoot morphology (Eveland and Jackson 2012).

These light quality and quantity cues jointly initiate a well-studied complex of "shade-avoidance" growth responses that include rapid elongation of stems and petioles, more erect leaf angles, and suppressed branching (Casal and Smith 1989; Schmitt et al. 2003). In addition to PIFs, stem and petiole elongation are regulated by the *DELLA* family of growth-restraining proteins, which integrate several hormonal pathways. A reduced red to far-red spectral ratio leads to (1) increased biosynthesis of auxin (a key phytohormone for stem elongation) via interactions between *PIF* and *DELLA* proteins, and (2) increased ethylene synthesis, which affects *DELLA* stability; reduced light intensity causes an increase in gibberellins (another plant hormone with widespread effects), which promote elongation by degrading *DELLA*s (K. Franklin 2008 and references therein). The elongation response is also influenced by separate interactions between both ethylene and auxin with other, non-*DELLA* targets (Pierik et al. 2009).

This cue and response system has been exceptionally well studied in ecological context, as a primary way that plants mediate competition for light. Experimental manipulations of these signaling pathways in mutant and transgenic *Arabidopsis* plants grown under natural shade have confirmed that the presence and density of neighboring plants gives rise to these distinct red to far-red–ratio, blue-light, and gibberellin signals and their specific downstream effects on *DELLA* protein abundance and breakdown, which in turn directly regulate shoot elongation responses (Djakovic-Petrovic et al. 2007). Phenotypic manipulations (reviewed by Schmitt et al. 2003) have demonstrated that the specific developmental responses cued by neighbor shade or, alternatively, by *lack* of shade (and resulting light quality/quantity signals) enhance plant fitness in natural habitats and therefore constitute adaptive plasticity in response to alternative light environments (Dudley and Schmitt 1996; Donohue et al. 2000).

3.4.2 Habitat transience and risk cues for amphibian metamorphic timing

Amphibians such as frogs and toads begin life as free-swimming larvae or "tadpoles" which must undergo metamorphosis under aquatic conditions to become terrestrial (or semiterrestrial) juveniles and, ultimately, reproductive adults. The timing of metamorphosis represents a trade-off between prolonged larval growth, which results in greater adult size and fitness, and the need to quickly complete development when the aquatic environment poses high mortality risks (R. Newman 1992; C.-S. Wu et al. 2012). Once a tadpole has achieved a species-specific size threshold, the onset of its metamorphosis can be slowed or hastened in response to specific cues about food availability, predator and pathogen presence, and other aspects of the larval environment that indicate the prospects for continued survival (Denver et al. 1998; Rose 2004; Grözinger et al. 2014; and references therein). (Curiously, although metamorphic plasticity in these organisms is very well known, the three amphibian "model systems" that are chiefly used to study development [*Xenopus laevis, Rana catesbeiana*, and *Ambystoma mexicanum*] all have exceptionally canalized rather than environmentally responsive developmental trajectories [Rose 2004].) Environmental threats such as thermal or osmotic stress are sensed directly at the tissue level; less immediate risks such as predator presence or high conspecific density are perceived at higher brain centers through the animal's sensory systems and then transduced via the neuroendocrine system to shape developmental responses (Denver 2013). These fascinating cases provide detailed information about hormonally mediated developmental plasticity in vertebrates, illustrating how deeply conserved transduction mechanisms can be utilized in the context of a species' unique ecological scenario.

A particularly dramatic case of metamorphic plasticity is that of desert amphibians and other taxa that breed in temporary ponds, since these animals must complete metamorphosis before their aquatic environment disappears. Individuals in these species plastically accelerate development in response to pond drying, along a graded continuum corresponding to the rate of water loss (Denver et al. 1998 and references therein). Although the duration of the aquatic habitat is perhaps the most critical of all environmental variables for tadpole survival (Denver 2013), it is a difficult one for an animal to read, particularly so because pond desiccation must be perceived well in advance to provide a useful metamorphic cue. A classic ecological developmental study of the spadefoot toad *Scaphiopus hammondii* by Robert Denver and colleagues tested a suite of possible environmental covariates of pond drying as potential cues to initiate accelerated metamorphosis (Denver et al. 1998). Controlled experiments showed that several covariates of decreasing pond volume could be excluded as cues: metamorphic timing was not influenced by water temperature, chemical concentration (e.g., of mineral salts, waste products, hormonal secretions, or dissolved oxygen), or physical interactions among very crowded tadpoles. Instead, tadpoles in this species perceived the imminent risk of their aquatic habitat drying out through two distinct environmental signals: a visual cue indicating the animal's closer proximity to the water surface, and the animal's own reduced swimming activity due to lower pond volume (an example of the kind of internal feedback effect that can serve as an indirect environmental cue). Even closely related species may have evolved to use different correlates of falling water volume as plasticity cues: for instance, in *Scaphiopus couchii*, another spadefoot toad of ephemeral ponds, increased tadpole density is the metamorphic signal, perceived through more frequent physical interactions among increasingly crowded individuals (Figure 3.5; R. Newman 1994).

These sensory inputs are transduced into developmental effects through the neuroendocrine system, which in vertebrates provides a signaling interface between the nervous system and the endocrine pathways, which coordinate developmental processes including metamorphosis (reviewed by Denver 2013). The visual, reduced-activity, and physical-contact cues that signify a drying pond, perceived by the tadpole as sensory stimuli, are sent via neural pathways to the brain's hypothalamus, where they initiate an increase in the neurohormone *corticotropin-releasing factor* (CRF). This CRF increase stimulates the adjoining pituitary gland to release *thyrotropin* and *corticotropin*, chemicals that activate two distinct endocrine systems: the thyroid axis and the interrenal or "stress" axis (Denver 1997a, 1998). Because the intensity of the environmental stimulus determines the amount of CRF and consequent hormone secretion, the hypothalamus serves as "a continuously varying control switch" that allows for a specifically calibrated (and reversible) response to the rate of water loss (Rose 2004). This hypothalamus switch can be activated by distinct environmental cues to set in motion the same set of hormonal and therefore developmental events in amphibian species that perceive a risky larval situation in different ways (Sultan and Stearns 2005). (Indeed, the stress neurohormone CRF may mediate developmental timing responses in vertebrates more broadly, possibly including, e.g., early parturition in mammals due to fetal stress [Denver 1997b].)

(a)

(b)

Figure 3.5 Tadpoles of *Scaphiopus couchii*, a species that inhabits ephemeral ponds, accelerate metamorphosis in response to falling water level. In this species, the plastic life-history response to pond drying is cued, indirectly, by more frequent physical contact among the increasingly crowded tadpoles. The timing of metamorphosis represents a trade-off between continued larval growth and the survival risk of pond drying. (a) *Scaphiopus couchii* tadpole (photo credit and copyright Gary Nafis). (b) *Scaphiopus couchii* adult (photo courtesy of Michael Spencer).

Thyrotropin release stimulates the thyroid gland to release *thyroid hormone*, the primary regulatory molecule that coordinates the remarkable suite of morphogenetic changes that comprise amphibian metamorphosis (Brown and Cai 2007). These changes range from remodeling of the gut, to tail, gill, and mouthpart resorption, to the production of jaws and limbs (Shi 1994). This key role is consistent with the major regulatory influence of thyroid hormone on early development in birds and mammals, and on metamorphosis in other vertebrates such as lampreys and flatfish (Rose 2004; Brown and Cai 2007; Laudet 2011). Interestingly, in a third spadefoot species, *Scaphiopus multiplicatus*, the environmental signal for pond drying is a directly hormonal one: when the volume of water falls during pond drying, the density of brine shrimp that cohabit these ponds increases, and tadpoles ingest more of these prey items, which contain very high levels of thyroid hormone (D. Pfennig 1992). The circulating thyroid hormone is taken up by cells, in which it is bound to proteins that form complexes with specialized thyroid receptors and receptor coactivators to induce tissue- and stage-specific effects on gene expression; these gene expression effects lead directly to the morphogenetic proliferation, targeted death, differentiation, and migration of cells, all of which are required for metamorphic changes to occur (Buchholz et al. 2003; B. Paul et al. 2007; Grimaldi et al. 2013).

Along the second endocrine axis activated by CRF, the release of *corticotropin* stimulates the interrenal glands to synthesize and release stress hormones, primarily (in amphibians) *corticosterone* (Kulkarni and Buchholz 2014). These hormone molecules diffuse into cells, where they bind to specialized cytosolic receptors bound to heat-shock proteins. Hormone binding changes the conformation of these receptors so that they release the heat-shock proteins, translocate to the nucleus, and bind to DNA, to either promote or repress the expression of specific target genes (Denver 2013 and references therein). Although thyroid gland activity is the key regulator of metamorphosis and its onset, the acceleration response is governed by a synergistic interaction within target cells between thyroid hormone and corticosterone (Maher et al. 2013). There also appears to be a feedback effect whereby thyroid receptors are autoinduced in many tissues after the start of metamorphosis, so as to increase tissue sensitivity to the thyroid hormone signal and further accelerate metamorphosis; however, this mechanism is not yet well understood (Denver 2013).

Tadpoles can also alter developmental timing in response to deteriorating resource conditions, again through the neuroendocrine transduction system. However, these cases are less fully understood. In *Rana pipiens*, the response to either very high densities of conspecifics or restricted food availability depends on the individual's developmental stage. Prior to the onset of metamorphosis, tadpoles respond to unfavorable resource conditions by increasing their whole-body content of corticosterone, an endocrine change that slows growth and development (Glennemeier and Denver 2002). This response could be an adaptive one, since it allows tadpoles in circumstances where food is hard to find, due to low supply or intense competition, to spend longer building body mass before metamorphosis. However, in tadpoles that have already initiated metamorphosis, exposure to food restriction or crowding results in *accelerated* development, again mediated by changes in circulating corticosterone (Glennemeier and Denver 2002). In this case, it may be advantageous to more quickly exit a resource-poor environment, since the metamorphic size threshold related to later fitness has already been reached.

Other aspects of plasticity in larval amphibians can be cued by a perceived risk of predation. Predator presence is signaled via two chemicals (secreted by the skins of cohabiting tadpoles) that are perceived in combination as an alarm pheromone (Fraker et al. 2009). Response to this chemical cue is mediated (via the hypothalamus–pituitary–interrenal system described above) by corticosterone (Maher et al. 2013). In many vertebrates, activation of this system increases the amount of circulating corticosteroids so as to trigger a rapid escape or "flight" response. Tadpoles of *Rana* species show a more ecologically complex, bimodal response to the predator cue. The immediate, short-term response to predator presence (within one to four hours) is to *decrease* whole-body levels of corticosterone, to a degree that matches the dose intensity of the chemical signal. This hormonal change suppresses the stress response, bringing about reduced locomotion and foraging activity, a behavioral shift which reduces mortality by limiting exposure to the predators (Fraker et al. 2009). However, longer-term exposure to the predator cue initiates the opposite endocrine response: prolonged exposure causes *increased* corticosterone levels (within four days), leading to morphological changes such as a longer tail and decreased trunk length (by Day 14) that facilitate escape from predators (Maher et al. 2013 and references therein). Manipulation experiments using exogenous corticosterone application and a corticosteroid synthesis blocker confirmed this stress hormone as the proximate mechanism of both the endocrine-response suppression and its activation (Maher et al. 2013).

3.4.3 Plant (and some animal) defensive responses to predation cues

Predation cue and response systems provide a way for individual organisms to mount defensive biosynthetic and morphological responses and hence incur the materials and fitness costs of those responses, only at times when those individuals are under definite threat. The benefits of such "induced defenses" have been well established in cases where (a) predation risks vary and (b) defense traits can be expressed rapidly enough, following perception of a threat, to be effective (Tollrian and Harvell 1999; Petrusek et al. 2009). The responses of plants to animal predation (*herbivory*) comprise a particularly fascinating range of ecologically and mechanistically rich defensive systems.

The long coevolutionary history between land plants and their animal herbivores (e.g., 350 million years in the case of insects) has resulted in an extraordinary diversity of defensive chemicals (over 500,000) and structures, many of which are produced as rapidly induced responses of individual plants, or parts of plants, to herbivore attack (Karban and Baldwin 1997). The following is by no means a comprehensive review of this impressive literature.

3.4.3.1 Herbivory cues

Although the signaling pathways for these biosynthetic and morphological responses are not yet fully understood (J. Wu and Baldwin 2010), it is clear that plants use precise cues to identify attacking herbivores, as they typically respond quite differently to simulated herbivory (such as mechanical leaf damage) than to actually being chewed or pierced by a particular animal (references in Howe and Jander 2008; de Vos and Jander 2010). For example, tomato plants up-regulate and accumulate different defensive proteins depending on whether they are under attack by lepidopteran larvae (caterpillars), coleopteran leaf miners, or mites (references in Gatehouse 2002).

One source of highly specific, direct herbivory cues is the oral secretions (saliva and regurgitant) of feeding animals onto plant tissues contacted by their mouthparts. The first herbivore-derived signaling molecule to be chemically characterized was a fatty acid–amino acid conjugate (FAC) called *volicitin*, found in oral secretions of the beet armyworm, *Spodoptera exigua* (discussed by J. Wu and Baldwin 2010). Plants evidently perceive volicitin and other FACs by means of specific receptors. In laboratory tests, application of volicitin altered gene expression in maize seedlings so as to increase the biosynthesis of indole and terpene volatiles, chemicals that combat the attack by attracting parasitoid enemies of *Spodoptera* (Alborn et al. 1997; see below, Section 3.4.3.2). Interestingly, FACs such as volicitin are formed in the midgut of a feeding caterpillar when a plant-derived fatty acid and an insect-derived amino acid are actively combined. The fact that the insects themselves manufacture FACs suggests that these signaling compounds may have originated as digestive factors in the insect gut and been co-opted by plants to serve as cues for mounting a counterattack (Gatehouse 2002).

Other FACs that serve as herbivory cues occur in the oral secretions of tobacco hornworm larvae, *Manduca sexta*, which are specialist feeders on plants in the chemically rich Solanaceae family, such as the wild tobacco *Nicotiana attenuata* (Bonaventure et al. 2011). Application of *Manduca* FACs to damaged tobacco leaves quickly activates key regulatory elements in plant defense-chemical pathways, such as mitogen-activated protein kinases (MAPKs), jasmonic acid (JA), and ethylene. These internal signals trigger elevated synthesis of defensive compounds as well as FAC receptors, the latter of which raises the plant's sensitivity to subsequent herbivory cues (Halitschke et al. 2001; Giri et al. 2006). Another defense cue found in oral secretions of insect herbivores (e.g., the fall armyworm, *Spodoptera frugiperda*) is a set of molecules known as *inceptins*. These compounds are formed in the midgut of the insect and are derived from the proteolytic products of chloroplasts ingested in chewed plant tissue. Plants such as cowpea, *Vigna unguiculata*, are so highly sensitive to this plant-sourced biochemical cue that they respond to even 1 fmol (10^{-15} moles) of inceptin per leaf by strongly activating hormonal signals such as JA, salicylic acid, and ethylene, which in turn lead to increased defense-chemical production and emission (J. Wu and Baldwin 2009 and references therein).

Oral-secretion cues can be remarkably precise, as in the following example. Two species of corn earworms (*Heliothis virescens* and *Helicoverpa* [formerly *Heliothis*] *zea*) have different midgut enzyme activities and thus different concentrations of FACs in their oral secretions (Gatehouse 2002 and references therein). Apparently due to this rather subtle difference, plants recognize the two herbivore species as distinct, releasing different sets of volatile defense chemicals depending on which species chews their leaves. At the next trophic level, the parasitic wasp *Cardiochiles nigriceps* then "reads" these differing plant emissions to distinguish plants infested by its potential host, *Heliothis virescens*, from plants infested by the non-host insect *Helicoverpa* (De Moraes et al. 1998).

Placement of insect eggs on, or inside, tissues of a host plant signals future herbivory by growing larvae, even if immediate damage is negligible. Accordingly, oviposition fluids can also contain specific chemical cues that induce targeted plant defensive responses, such as the production of ovicidal substances (Seino et al. 1996) or the induction of necrosis in leaf regions to which eggs are attached (Balbyshev and Lorenzen 1997). The speed and frequency of feeding activity by a particular herbivore can provide an entirely different but equally precise type of cue, as demonstrated by studies that mechanically mimic characteristic temporal leaf-injury patterns (e.g., Mithofer et al. 2005). However, these mechanical cues, and the ways through which plants perceive mechanical damage, are as yet poorly understood (Howe and Jander 2008). The ability of plants to identify their herbivores with such remarkable precision suggests that, at least in certain cases, multiple cues rather than a single chemical signal could be involved (Gatehouse 2002).

3.4.3.2 Responses to herbivory

A plant's perception of herbivore attack sets in motion a host of both local and systemic responses. Signaling pathways for these myriad responses can be highly complex. In one microarray study in *Arabidopsis*, herbivory led to changed expression of over 700 mRNAs (P. Schenk et al. 2000; see Rehrig et al. 2011 for further examples). Although all of the signaling events triggered by the perception of herbivory are not yet known, the key players in these transduction networks have been identified (see overview in Howe and Jander 2008).

The primary hormonal signal for transducing herbivory cues into defensive responses is the JA pathway. The initial step in this pathway occurs when the perception of herbivore wounding leads (by a still unknown cellular signal) to the rapid synthesis and buildup of JA in the wounded tissue (Stork et al. 2009). The JA conjugates with isoleucine to form a bioactive molecule which binds to a specific JA-receptor protein (COI1) to form a compound that promotes the degradation of jasmonate *zim*-domain (JAZ) repressor proteins bound to JA-responsive transcription factors (Howe and Jander 2008; Sheard et al. 2010). The breakdown of JAZ repressors releases these transcription factors, thus leading to the activation of many JA-responsive genes (Verhaage et al. 2011 and references therein). The mobile signaling molecule that sends information from damaged leaves to other parts of an attacked plant is not yet known, although in some cases JA produced by damaged leaves is part of systemic as well as local signaling; electrical signals within the plant may also be involved (J. Wu and Baldwin 2010). Remarkably, the precise number and spatial location of herbivore-elicited JA bursts within an individual plant create a "memory" that modulates the downstream effects of JA and consequently shapes subsequent defense responses (Stork et al. 2009). In addition to its central role in herbivory response, the JA pathway is involved in "crosstalk" with numerous other pathways in plants through which both abiotic and biotic environmental cues are transduced (reviewed by Fujita et al. 2006).

A second key signaling element consists of intracellular changes in the concentration of calcium ions (and the corresponding effects on cell membrane potentials); these changes are mediated by proteins such as calcium-dependent protein kinases and calmodulins. In keeping with their diverse regulatory functions in eukaryotes, these calcium sensors also play signaling roles in plant responses to pathogenic infection, drought resistance, and other environmental stresses (J. Wu and Baldwin 2009). The diverse family of MAPK proteins, another group of highly conserved cell-regulatory elements in eukaryotes, also plays a central role in defense-related gene expression via both the direct effects of MAPK signals, and the interactions of MAPKs with calcium flux and JA (Howe and Jander 2008). Finally, reactive oxygen species (such as hydrogen peroxide and hydroxyl radicals) are produced as a result of herbivore (as well as pathogen) attack and may play a signaling role in defense responses (e.g., Maffei et al. 2006).

Herbivory-induced cellular signals lead to myriad effects on plant gene expression (reviewed by Gatehouse 2002; Rehrig et al. 2011; see also Karban and Baldwin 1997). One dimension of response is the up-regulation of genes that directly code for defensive compounds. These extraordinarily diverse compounds (discussed by J. Wu and Baldwin 2010) include *toxins* such as glucosinolates, cyanogenic glucosides, and alkaloids; phenolic and monoterpene *repellents* (such as those found in pine tree resin, chili peppers, and clove oil), and *digestive blockers* (e.g., protease inhibitors that inhibit protein breakdown in herbivore guts so as to reduce the availability of amino acids and consequently growth and eventual fertility [Koiwa et al. 1997; e.g., De Leo and Gallerani 2002]). Activation of these genes results in enhanced amounts and concentrations of these defense chemicals in the tissues of a plant under attack, often specifically at the attack site. Up-regulation also occurs for genes that encode biosynthetic enzymes for secondary metabolites, as well as for those that participate in signaling pathways leading to indirect defenses (discussed below). Gene regulatory changes also take place to systemically adjust metabolic function, for instance, to increase defense biosynthesis over photosynthetic activity, or to maintain the elevated levels of protein production and turnover needed to make nitrogen-rich defense compounds (Hermsmeier et al. 2001).

One particularly interesting class of defense chemicals produced in response to herbivory (as mediated by a JA signal) are *volatile organic compounds* such as terpenoids, aromatics, and fatty acid derivatives, all of which are synthesized primarily in leaf epidermal cells and glandular trichomes (Dalin et al. 2008; Figure 3.6). Some of these molecules are phylogenetically widespread, such as the "green leaf volatiles" produced as one branch of the JA biosynthetic pathway, while others are species-specific compounds produced along diverse secondary metabolic routes (Gatehouse 2002). Plant volatiles are typically released in tissue-specific combinations as highly complex chemical "bouquets." Changes in the composition and abundance of these bouquets are induced by a broad range of insect herbivores on leaf, stem, seed, or root tissues, as well as by insect oviposition (A. Kessler and Heil 2011). These lipophilic molecules can readily cross cell membranes to disperse into the air when leaf tissue is damaged; release of volatiles produced by cells within shoot tissues is regulated by stomatal opening (Baldwin 2010). In some cases, tissue disruption causes previously separate enzymes and substrates to combine, forming toxic volatile "bombs" such as cyanide from stored nontoxic molecules (Baldwin 2010). Volatile organic compounds are typically of low molecular weight and quantity and so are likely to entail very low costs to the plant; indeed no fitness costs to their production have yet been identified (references in A. Kessler and Heil 2011). Reflecting both their informational specificity and their potential as long-range airborne cues, plant volatile emissions serve three distinct defensive roles, as *repellents, tri-trophic attractants*, and *plant-to-plant signals*.

First, volatile chemical blends can serve a directly defensive function as local repellents produced in response to tissue damage. For example, herbivore-damaged plants of the tobacco *Nicotiana attenuata* emit volatiles that deter oviposition by herbivorous *Manduca* moths, either because of the toxicity of the chemicals or because of the insect's preference to lay eggs in tissue that has not previously been damaged (A. Kessler and Baldwin 2001).

Second, herbivore-induced volatiles can function as indirect defense substances by generating a three-way (*tri-trophic*) interaction between plants, their herbivores, and natural enemies of the herbivores (references in Hare 2011). In these cases, plant volatile emissions provide information used by specific predators and parasitoids to locate their own insect prey or hosts (Baldwin 2010; de Vos and Jander 2010). Such signals are used by diverse taxa, including insect-eating nematodes, predaceous mites, ants, beetles, parasitoid hymenopterans, and even birds (A. Kessler and Heil 2011; Quintero

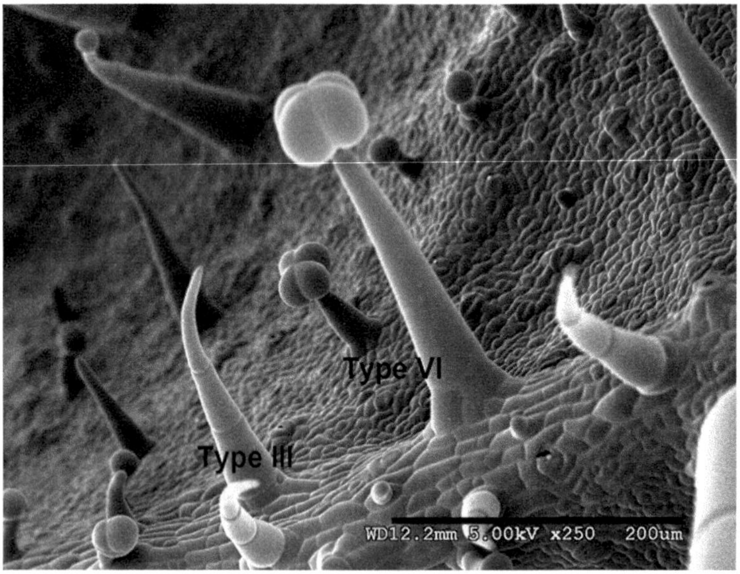

Figure 3.6 Plant responses to herbivore attack include producing volatile defensive compounds in glandular structures on the leaf surfaces. This cryoelectron micrograph of a tomato leaf, *Solanum lycopersium*, shows the glandular trichomes (labeled as *Type VI*) that produce and release sesquiterpenes, which are airborne toxins that defend directly against herbivorous insects and mites. The image shows several types of epidermal outgrowth that occur on tomato leaves, including hairlike trichomes that lack glands (labeled as *Type III*). Scale bar shown = 200 μm. Image and caption information courtesy of M. David Marks.

et al. 2013). *Brassica* plants under attack by cabbage white butterfly larvae produce volatiles that act as long-range signals to attract the parasitoid wasp *Cotesia rubecula* to the attack site. Since parasitism paralyzes the cabbage white caterpillars, the presence of *Cotesia* wasps decreases both the feeding damage to plants by current larvae and the growth of pest populations in the plant's locality (Geervliet et al. 1994).

Highly specific tri-trophic interactions are also initiated in response to aphids, a major group of herbivores that feed by piercing plant vascular systems and sucking out nutrient-rich phloem sap. These tiny, largely immobile insects have numerous natural enemies, including hoverflies, parasitoid wasps, and ladybug beetles, family Coccinellidae (Figure 3.7). Fava bean plants, *Vicia faba*, respond to attack by pea aphids, *Acyrthosiphon pisum*, by producing more of the volatile chemical *6-methyl-5-hepten-2-one* (Du et al. 1998). Electrophysiological studies of antenna activity showed that females of the pea aphid–parasitizing wasp *Aphidius ervi* are attracted to this plant signaling compound but not to volatiles released by the aphids themselves (references in de Vos and Jander 2010). Possibly host and non-host aphid species produce similar volatiles, which are therefore poorer cues for their parasitoids than the highly specific volatile emissions of plants under attack. For instance, bean plants attacked by the black bean aphid, *Aphis fabae*, which is not a potential host for *Aphidius ervi* wasps, do not release the above wasp-attracting compound (Du et al. 1998).

The adaptive benefits of indirect plant defenses can be substantial: for example, the three volatile compounds released by *Nicotiana attenuata* tobacco plants as direct defenses (see above) also attract a generalist predator on insect eggs, leading to a total reduction in each plant's herbivore load of 92%–94% from the combined direct and indirect impact of volatile emission (A. Kessler and Baldwin 2001). However, few fitness data have been collected that quantify the benefits to plants of indirect defenses (Hare 2011).

Along with their direct and indirect defensive functions, plant volatiles induced by herbivory can serve as plant-to-plant alarm signals that induce expression of defense-related genes in undamaged plant tissues (Gatehouse 2002). In this role, volatile compounds inform undamaged conspecific and heterospecific plant neighbors that herbivores are nearby, allowing them to mount a preemptive defense response (A. Kessler et al. 2006). For example, volatile chemicals released by chewed sagebrush plants are used by neighbor plants up to 60

Figure 3.7 In response to herbivory, some plants release volatile compounds that serve as indirect plant defenses by attracting the herbivore's natural enemies. Here, a grain aphid, *Sitobion avenae*, feeding on an unidentified grass leaf is being parasitized by a female *Aphidius rhopalosiphi* wasp. These wasps, which are specialist parasitoids of aphids, are attracted to volatile chemicals released by wheat and other grasses on which the aphids feed (Wickremasinghe and Emden 1992). Photo copyright Sonia Dourlot. For the color image, see Plate 8.

cm away as cues to increase their own defense-chemical production (Karban et al. 2006). The distance at which volatiles provide effective plant-to-plant signals depends on the stomatal behavior of receiving plants (and hence their water and carbon dioxide status), on atmospheric conditions such as humidity, wind, and temperature, and on the size and weight of volatile molecules, and hence their rate of diffusion and dilution in air (Baldwin et al. 2006). Some repellant volatiles can remain on leaf surfaces of neighboring plants, whence they are passively adsorbed and later re-released to provide defense from herbivores without any production cost to the recipient plant (Himanen et al. 2010), a rather subtle airborne mode of ecological facilitation.

In some cases, volatiles released by plant neighbors do not cause a plant to immediately upregulate its own defense chemicals but instead may "prime" it to respond more quickly to a direct attack when such an attack occurs (Engelberth et al. 2004). In a study by A. Kessler et al. (2006), tobacco plants showed such a response to volatile compounds released by experimentally damaged sagebrush neighbors. The tobacco receivers evidently perceived the volatile signal, which triggered numerous transcriptional changes in defense-related genes (e.g., significant accumulation of transcripts for trypsin proteinase inhibitor digestive blockers [TPIs]). However, these plants did not immediately increase TPI gene expression or the production of other defense chemicals. Instead, they expressed a response only when directly attacked by *Manduca* caterpillars: in the event, plants that had been previously exposed to the volatile signal increased production of TPIs and other secondary metabolites significantly faster than those without prior exposure. Consequently, the "primed" tobacco plants suffered less damage from the *Manduca* herbivores and inflicted greater mortality upon them (A. Kessler et al. 2006). This sophisticated two-part cueing system enhances the plants ability to mount an immediate, effective defense response while avoiding the cost of producing nitrogen-rich defense compounds until absolutely necessary.

Responses to herbivore attack can entail increased production of structural as well as chemical defenses. Several species of tropical acacia trees increase the length and density of thorns in response to grazing by giraffes, so as to significantly reduce shoot destruction (Milewski et al. 1991). This response is induced within a matter of weeks at the level of individual branches, the level at which grazing varies within a tree, because of limits on browsing height (Young et al. 2003). In a number of herbaceous plants, leaves produced following foliar herbivory have greater densities of leaf trichomes or stinging hairs, or increased toughness, making them less acceptable as food sources to both insects and mammals (references in Milewski et al. 1991; Dalin et al. 2008). The lag time built into such morphological responses by insect host plants may be a key part of their adaptive value. For example, gray willow shrubs, *Salix cinerea*, subjected to grazing by adult willow leaf beetles, *Phratora vulgatissima*, subsequently produced new leaves with nearly double the density of leaf trichomes. As a result, when larval offspring of the beetles developed on these plants, they consumed less leaf tissue, and did so in a more dispersed feeding pattern (due to repeatedly testing each leaf's palatability) than beetle larvae raised on plants not previously exposed to herbivory. In other words, the plants' structural response did not deter leaf herbivory by adults, but it did reduce subsequent larval feeding, which generally causes far heavier damage (Dalin and Björkman 2003).

3.4.3.3 Animal developmental responses to predation

Similarly, in many animal taxa, predation induces the expression of defensive morphologies, although the cues and underlying mechanisms for these changes are generally not well understood (references in Brookes and Rochette 2007; Bourdeau and Johansson 2012). The well-studied acorn barnacle, *Chthamalus anisopoma*, responds to the presence of the predatory snail *Acanthina angelica* by developing a "bent" shell morphology (instead of the typical conical shell) that impedes access to the aperture and consequently predation by this specialized carnivore (Lively 1986; community impact discussed in Chapter 6, Section 6.1). Marine snails exposed to chemical cues from snail-fed crabs enhance their own rate of calcification to produce thicker shells that resist piercing by crab predators (Brookes and Rochette 2007). Perhaps the most well-studied predator-induced morphological defenses in invertebrates are the various structures produced by individuals in several species of the planktonic freshwater crustacean *Daphnia* in response to specific water-borne predator kairomones (chemical cues). These structures include helmets, long spines, sharp protruding "neckteeth", and even a "crown of thorns" consisting of spiked cuticular shields that form a sort of "armor" shown to effectively protect the most vulnerable parts of the *Daphnia* body (Petrusek et al. 2009). The various *Daphnia* structures result in significantly reduced vulnerability to both insect and vertebrate predators (references in Sell 2000; Riessen and Trevett-Smith 2009; Juracka et al. 2011). The "crown of thorns" is a particularly striking case because this trait was formerly considered to characterize a distinct species; rather, this phenotype is diagnostic of a particular aspect of the Daphnia *habitat*, the presence of an extremely ancient type of predatory crustacean known as "tadpole shrimp," order Notostraca (Petrusek et al. 2009).

In vertebrates as well, highly specific chemical cues can induce morphological defense responses. In addition to amphibians (see Section 3.4.2, last paragraph), in the presence of predators, fish such as three-spined sticklebacks, *Gasterosteus aculeatus*, and crucian carp, *Carassius carassius*, develop different morphologies that reduce predation by gape-limited species (Brönmark and Petteersson 1994; Andersson et al. 2006; Frommen et al. 2011). Some fish, when attacked by piscivores, also release alarm pheromones (analogous to plant-to-plant volatile emissions following herbivory) that lead to predator avoidance behaviors by nearby individuals (references in Brönmark and Pettersson 1994).

CHAPTER 4

Ecological development as niche construction: How plasticity shapes the environment an organism experiences

By means of complex perception and transduction networks, environmental cues and conditions shape an organism's phenotypic expression according to evolved norms of reaction (see Chapter 1, Sections 1.2.2 and 1.3; also see Chapter 3). Eco-devo adjustments to the structure, morphology, size, and deployment of organs, as well as to the individual's physiology, life history, and behavior, in turn mediate the organism's experience of its external environment. In this sense, plasticity (i.e., aspects of trait expression that change from one environment to another [Chapter 3, Section 3.1]) can be seen as a mode of niche construction that allows the organism to *experience* a different and often more favorable physical and biotic environment, even without altering external conditions as such (see definitions in Chapter 2, Sections 2.3.2 and 2.3.4; niche-constructing effects that change the external environment per se are termed *habitat construction* [Chapter 2, Section 2.3.3; Chapter 5]). From this close perspective, adaptation arises in large part by virtue of ecological development, from the mutually defining interaction of organism and environment.

This chapter explores the multifaceted ways that animals, plants, and microorganisms modify and enhance the environment that they (and their offspring) experience, by means of plastic eco-devo responses. Following a brief introduction that starts with the simplest case, habitat choice, several detailed case studies are presented. These case studies begin with the relatively familiar example of thermoregulatory adjustment and then go on to provide examples of more unexpected, but equally critical, dimensions of individual environmental mediation. The goal is not to provide a comprehensive scheme for cataloging plasticity but rather to illustrate how the phenotypes organisms express in response to environmental conditions can specifically shape their experience of those environments. More broadly, these case studies of *experiential niche construction* show how adaptive insights can be gained from an eco-devo research perspective that focuses on the intimate organism–environment relationship.

4.1 Habitat choice and other modifications to the experienced environment

A brief look at several familiar adaptive phenomena will exemplify how phenotypic changes can be viewed as a form of niche construction. Perhaps the most straightforward way that organisms alter their environmental experience is simply to move to a favorable location. This is called *habitat choice*, a term which refers to the outcome and does not imply conscious deliberation (aka *relocation niche construction*; Odling-Smee et al. 2003). Habitat choice in this sense has been intensively studied in highly mobile organisms such as migratory birds, which choose

Organism and Environment. First Edition. Sonia E. Sultan.
© Sonia E. Sultan. 2015. Published 2015 by Oxford University Press.

repeatedly among possible sites for foraging, breeding, and overwintering (Cody 1985 and references therein). These "choices" primarily involve behavioral responses to environmental cues.

In ordinarily sedentary taxa, moving to more favorable conditions can require developmental plasticity. In many insect taxa with limited mobility, local resource limitation can lead to the production of dispersing (e.g., winged) morphs able to relocate to (metaphorically or literally) greener pastures. Even when they are produced by the same genetic clone, dispersing morphs can differ dramatically from the typical nondispersers with respect to sensory systems, physiology, and behavior, as well as in the production of wings and flight muscles (Braendle et al. 2006). One such example is the ordinarily slow-moving desert locust, *Schistocerca gregaria*, in which increased crowding of mothers, juveniles, or both induces a rapid transition to the conspicuously colored "gregarious" morphs that migrate as flying swarms to new, resource-rich sites (Maeno et al. 2013; Rogers et al. 2014; also see Chapter 3, Section 3.2 and Figure 3.2). In species of aphid (Aphididae), an even less motile type of insect, winged morphs can be plastically induced by several cues for poor local habitat quality. These cues can include increased contact with con- or hetero-specific individuals competing for resources, advancing age or lower quality of a host plant, or the presence of predators; conversely, production of dispersing morphs can be inhibited in the presence of protective ants that signal a favorable spot (references in Braendle et al. 2006). The individual or group movement of larval, juvenile, and adult animals to resource-rich habitat patches of various spatial scales has been studied in vertebrates and marine invertebrates as well as in many other insects (references in Huey 1991; D. Morris 2003; Van Beest et al. 2010). In recent decades, a substantial empirical and theoretical literature has grown up addressing the implications of habitat choice for the distribution of animal taxa, especially in relation to biodiversity conservation (e.g., Basille et al. 2013).

Habitat choice can also occur by means of phenological eco-devo adjustments that allow movement in *time* to favorable conditions. For instance, amphibian species of unpredictable habitats can hasten metamorphosis so as to remain in an aquatic environment throughout development rather than the mud of a drying pond (Denver et al. 1998; also see Chapter 3, Section 3.4.2). For sessile organisms unable to migrate in space, including of course terrestrial plants, such phenological adjustments can be a particularly important means of niche construction (Donohue 2005). The plant life cycle encompasses several ecologically critical phenological points (here described in angiosperm terms, although analogs exist for other land plants): time of flowering determines conditions during seed maturation and dispersal, release of dormant buds determines the conditions encountered by photosynthetically active leaves, and timing of seed germination determines the environment experienced by the young plant (Donohue 2005; Penfield 2008). In many tropical trees, the seasonal timing of new leaf production shapes the biotic environment with respect to local risk factors. Some species shift their production of new leaves to the time of year when the fewest pathogens and herbivores are present (generally the dry season), which has been shown experimentally to reduce tissue damage (Coley and Barone 1996 and references therein). Alternatively, co-occurring tropical trees may produce their new leaves in mass, synchronous "flushes" of which herbivores can attack only a modest proportion (Coley and Barone 1996), creating a shared temporal habitat with less individual risk.

With respect to habitat choice via germination timing, plants have evolved to await specific cues that signal suitable conditions for a seedling's initial emergence from the soil and for subsequent early development (Figure 4.1; Evans and Cabin 1995; Baskin and Baskin 1998; additional references in Leck et al. 2008). In temperate taxa, such cues often include a prolonged period of cold that ensures germination will occur only after winter's end, rather than in response to a mild midwinter day (Fenner and Thompson 2005 and references therein). In addition to insuring a suitable germination climate, germination cues can direct seedling emergence into favorable, open microsites (see Heschel et al. 2007). In many species, the low red:far-red wavelength ratio of sunlight that has passed through leaf tissue inhibits seed germination; such seeds germinate only in the absence of neighbors that would shade young seedlings (Baskin and Baskin 1998).

Figure 4.1 Plants have evolved to await specific cues that signal suitable conditions for a seedling's emergence and early growth—a temporal form of habitat choice. Heat or smoke cue germination in many tree species of fire-prone habitats, such as the pitch pine, *Pinus rigida*. This *Pinus rigida* seedling has emerged following a burn in the New Jersey Pine Barrens. Photo courtesy of Mary A. Leck. For the color image, see Plate 9.

Many plants in Californian chaparral, Australian bushland, and other fire-prone habitats require exposure to smoke, extreme heat, or both to release seeds from dormancy; such germination cues ensure seedlings will encounter the open conditions that follow a fire (Keeley and Fotheringham 2000).

Germination signals that lead to adaptive phenologies can themselves be tailored by maternal plants growing in alternative habitats. Such transgenerational plasticity can lead to adaptive life-history variation in species with limited dispersal, when the offspring is likely to develop in conditions similar to those of its maternal parent (Galloway 2005; see Section 4.4). In the gravity-dispersed woodland herb *Campanulastrum americanum*, maternal plants that grow in sunlit forest gaps produce seeds that germinate in autumn and develop as annuals, while plants in understory shade produce seeds that instead germinate the following spring and carry out a biennial life cycle; these alternative, plastically induced phenologies are associated with higher fitness (survival and population growth) in each habitat (Galloway and Etterson 2007).

Beyond spatial or temporal movement to patches or periods of suitable habitat, organisms shape the way they experience a given environment through more subtle adjustments to their morphology, physiology, and behavior. A familiar example is thermoregulation by individual animals to maintain functionally appropriate body temperatures despite the pronounced heterogeneity and occasional extremes of ambient conditions. Similarly, developmental adjustments that either enhance access to limiting resources or reduce resource requirements can be interpreted as ways that individual organisms change their experience of a given resource environment from severely limiting to more plentiful. Eco-devo adjustments that accomplish these kinds of changes (discussed in detail in Sections 4.2 and 4.3) allow organisms to experience a narrower, as well as a more favorable, range of conditions than the absolute variation that occurs in natural habitats.

Ecological developmental responses can also ameliorate the individual's experience of its biotic environment. For example, the production of pointed defensive helmets by *Daphnia* individuals alters their phenotypes (see Chapter 3, Section 3.4.3.3), but it also changes the environment they experience to one with many fewer potential predators (i.e., only those able to overcome this potent defense). In general, induced defenses allow an organism to experience a given biotic environment as less threatening, by reducing the activity or impact of its herbivores, predators, or pathogens (see Chapter 3, Section 3.4.3, and Chapter 6, Section 6.4.2). Individual organisms also alter the competitive environment they experience, often by movement and other aspects of behavior (in animals) or by deployment of resource-collecting organs (in plants), both of which direct foraging activities away from sites of intense competition (Huey 1991; Hodge 2009). Plant developmental responses to competition can be viewed as a type of microhabitat choice. For instance, split-container studies show that individual plant root systems proliferate preferentially in soil microlocations that are free of competing neighbors (Gersani et al. 1998; discussed by Hodge 2004), effectively occupying a low-competition soil environment. These within-individual deployment "decisions" are made in part through an unknown mechanism of discriminating between self and nonself roots in the soil (Gruntman and Novoplansky 2004; H. Schenk 2006).

This chapter examines some of the many types of phenotypic adjustments individual organisms make that mediate external conditions, to determine their environmental experience and that of their progeny. In contrast to the detailed case studies presented in Chapter 3, in many cases the precise mechanisms of these behavioral, morphological, anatomical, ultrastructural, and physiological adjustments have not yet been fully elucidated, although their ecological impact is generally well understood. Because these ameliorations of environmental experience affect the organism's functional success, and hence its metabolic and ecological presence, they ultimately feed back to alter external environmental conditions in absolute terms as well. The following three broad-based case studies illustrate the level of individual adaptation at which the organism and its environment most intimately and powerfully define each other. In this view, adaptation is not the playing out of a selectively scripted encounter between an internally determined phenotype and an impervious external environment. Certainly, selection and other evolutionary forces shape and delimit the repertoire of possible interactions (i.e., the genotypic norms of reaction that underlie eco-devo responses to given conditions). From an eco-devo perspective, adaptation emerges in the moment, from the two-way organism–environment interaction.

4.2 How animals and plants modify their experienced temperature conditions

Although organisms obviously cannot control the temperatures of their external (terrestrial or aquatic) environments, to maintain physiological function, it is essential that they keep their body temperatures within a relatively narrow range. This is achieved largely by means of behavioral, morphological, and physiological adjustments that (in conjunction with constantly expressed species-specific features) influence heat gain and loss so as to modulate the thermal heterogeneity of natural habitats (Huey 1991; Seebacher and Franklin 2005). Indeed, an individual's phenotype "can be viewed as a filter that transduces the environmental thermal regime into a particular body temperature" (Huey 1991, S93).

The temperature an animal experiences is determined first of all by its "habitat choice" behaviors in seeking particular climatic conditions (e.g., via seasonal migration; Huey 1991) or microsites. For example, solitary terrestrial species (including many mammals) nest in tree holes and other shelters that offer a less variable temperature and humidity microclimate (Terrien et al. 2011). Both vertebrates and invertebrates also make quite short-term temporal or spatial microclimatic adjustments. In extremely hot weather, anteaters switch their foraging activities to nighttime (de Sampaio et al. 2006). Fish thermoregulate by seeking warmer or cooler areas in water, where thermal gradients occur due to variation in depth and insolation; individual fish can detect and respond behaviorally to thermal differences as fine as 0.03°C (Van Dijk et al. 2002). Similarly, many reptiles bask in the sun to raise body temperature during cold times of day or, conversely, seek protective shade at midday (Huey 1982; Seebacher and Franklin 2005). Shorebirds are another familiar example: under intense, direct insolation, individuals often stand in shallow water, unloading excess body heat to the water via leg tissues.

A more esoteric example of microclimate tracking is the tropical flower mite *Proctolaelaps certator*, a tiny arachnid which inhabits the hummingbird-pollinated inflorescences of the New World tropical understory plant *Heliconia wagneriana* (O'Connor et al. 1991). The structural complexity of these inflorescences, together with the shifting patterns of sunfleck and shade they receive in the forest understory, leads to substantial microclimatic variation: within seconds to minutes, temperatures within a single flower can range from 26.5°C to 37.5°C (Dobkin 1985; Figure 4.2). Flower mites track this fine-scale variation by continually moving to cooler locations within inflorescences, actively avoiding hot, directly insolated microsites where their

Figure 4.2 The structurally complex inflorescences of the rainforest understory plant *Heliconia wagneriana* present a shifting range of microclimatic zones that are actively tracked by resident flower mites. As a result of this fine-scale spatial habitat choice, individual mites experience only favorable, cool conditions. *Heliconia wagneriana*, Gamboa, Panama. Image courtesy of Spencer C. H. Barrett.

reproduction and survival are reduced (Dobkin 1985). In general, insects may have access to a surprisingly wide range of microclimatic conditions within a very short distance (references in Coggan et al. 2011).

Adjustments in body position and orientation can also allow animals to experience more favorable temperature conditions in a given spot, including when solar radiation is excessive. Intertidal mollusks such as the periwinkle *Echinolittorina peruviana* have few behavioral options for avoiding excess heat, as they are unable to relocate to cool microenvironments during low tides. Instead, individuals thermoregulate by orienting frontally, which minimizes the surface area they expose to incident sunlight (Muñoz et al. 2005). Mammals such as wildebeests also alter body orientation to regulate heat gain in direct sunlight (references in Norris and Kunz 2012). Insects as well adjust basking posture as part of thermoregulatory behavior (Kemp and Krockenberger 2004; Coggan et al. 2011). Some animals even use body position to create their own shade: for example, giant Seychelles tortoises, *Testudo* spp., orient their bodies so that the animal's own carapace casts a shadow over its head and neck (Coe 2004), and Cape ground squirrels, *Xerus inauris*, shade their bodies with their tails (Bennett et al. 1984). In birds, flapping wings or raising back feathers are behaviors that increase convective cooling (Norris and Kunz 2012).

Body and behavior adjustments also allow animals to mitigate cold conditions. Many mammals can reduce their surface-to-volume ratio for heat loss by adopting a hunched or ball-like posture (Terrien et al. 2011), while birds maintain body temperature in cold air by either standing on one leg or sitting, to minimize heat loss from uninsulated leg tissues. Birds can also fluff out feathers in cold weather to reduce heat loss by increasing boundary layer thickness or extend wings to present a large surface for absorbing incident sunlight. In some cases, animals behave so as to thermoregulate collectively rather than individually. Penguins as well as mammals such as mice, bats, marmots, and pigs aggregate to reduce their collective surface-to-volume ratio, creating a group microclimate that markedly decreases energy expenditure in cold conditions (Schradin et al. 2006; C. Gilbert et al. 2007; Figure 4.3). Even a usually asocial, territorial mammal such as the American red squirrel, *Tamiasciurus hudsonicus*, may conserve heat by nesting communally, a behavior that becomes increasingly frequent with colder temperatures (C. Williams et al. 2013).

Individual animals also thermoregulate through short-term changes in activity level and physiology. When external temperatures are low, some mammals hibernate or enter torpor to minimize energy expenditure, whereas others produce heat through higher metabolic rates and changes in muscular activity (Terrien et al. 2011). Similar adjustments occur

Figure 4.3 Many animals reduce their collective surface-to-volume ratio by aggregating. Emperor penguins, *Aptenodytes forsteri*, huddle together in open air to create a group microclimate that prevents hypothermia and permits energy conservation during the Antarctic winter. These birds are huddling tightly during a blizzard. From C. Gilbert et al. 2007, reproduced with permission from *American Journal of Physiology.*

in several orders of large insects able to maintain stable body temperatures in cold conditions by means of added muscle activity before and during flight; one study reported that thorax temperatures in the honeybee *Apis mellifera* remained stable even when the ambient temperature ranged from 37°C to 19°C (Woods et al. 2005). Bees further regulate heat loss by varying the flow of hemolymph to the head versus the abdomen (Woods et al. 2005). Conversely, in high ambient temperatures, mammals may dilate blood vessels close to the skin surface to increase blood flow to areas of active heat loss, or they may pant or sweat to unload excessive heat by means of fluid evaporation (Norris and Kunz 2012). Reptiles also elevate heart rate and consequently peripheral blood circulation in hot conditions, preferentially sending blood to the limbs to enhance heat transfer to the air; also like mammals, in cool conditions they decrease cardiac rate (Seebacher and Franklin 2005).

Morphological traits such as size, surface-to-volume ratio, color, and skin outgrowths like feathers and hair also influence an individual's thermal properties. Animals that inhabit consistently cold environments may possess constitutive heat-conserving features, such as the layer of subdermal fat that insulates penguins against heat loss in cold water. Penguin bodies show several such constantly expressed traits, including tightly overlapping feathers that insulate the epidermis, body shape with a low surface-to-volume ratio, and a specialized vascular anatomy of the wing called the "humeral arterial plexus" that warms the body core by exchanging heat from outgoing arterial blood to incoming venous blood (Thomas and Fordyce 2012). Developmental plasticity can also shape certain aspects of the thermoregulatory phenotype. In several species of placental mammals, individuals reared at cold temperatures developed shorter extremities (i.e., lower surface-to-volume body ratios) than conspecifics reared at higher temperatures. This experimental result reproduces the biogeographic observation known as "Allen's Rule" regarding the relative size of extremities, including ears, tails, and limbs, in generally cold versus warm climatic regions (discussed by Riek and Geiser 2012).

In animals, behavior, physiology, and morphology jointly contribute to an individual's thermoregulatory phenotype. This has been studied in the fat-tailed dunnart, *Sminthopsis crassicaudata*, a mouse-like Australian marsupial that is born into the mother's pouch only 12 days after fertilization. Because dunnarts are exposed to external temperature conditions at this very early stage, they provide a useful system in which to study the integrated effects of developmental adjustments to temperature (Riek and Geiser 2012). In a split-brood experiment, individuals reared at 16°C developed a longer head and body then those raised at 22°C, an increase in surface area predicted to result in more efficient heat uptake through the animal's unusual, reptile-like basking behavior (Riek and Geiser 2012). When compared under cold (16°C) conditions, cold-reared dunnarts also employed longer bouts of torpor and had lower resting metabolic rates than warm-reared individuals, resulting in substantially reduced energy use. This combination of developmental, behavioral, and physiological adjustments to cold versus warm conditions may explain the extremely broad distribution of this species compared with its congeners across contrasting climatic regimes in Australia (Riek and Geiser 2012; cf. Tomlinson et al. 2012).

The thermoregulatory repertoires of both vertebrates and invertebrates can also include manipulation of their environment, for example, by making burrows or shelters. Eastern tent caterpillars, *Malacosoma americanum*, build and inhabit communal silk tents that serve as "greenhouses" which increase ambient daytime temperatures by at least 4°C while shielding the animals from direct insolation (Joos et al. 1988). In conjunction with (a) aggregating behavior and (b) the spatial orientation of these aggregations to the direction of incident solar radiation, these communal tents allow caterpillars to maintain body temperatures within a narrow, favorable range (Joos et al. 1988). Similarly, in ants, the combination of nest building, nest orientation, and active behaviors mediates the temperatures individuals experience within the nest (references in Norris and Kunz 2012).

Like animals, terrestrial plants adjust their behavior, morphology, and physiology in specific ways that maintain functionally favorable body temperatures that can be quite different from the surrounding air (Lambers et al. 2008). Although the environmental temperature sensors used by plants are not yet understood, plant responses are known to be regulated via multiple hormone signaling networks (Patel and Franklin 2009). Most familiar are the two physical processes that allow plants in hot, sunny conditions to either avoid or unload excess heat (Mahan and Upchurch 1988). The first of these is *convective heat transfer* from leaves and other shoot surfaces to the surrounding air. This transfer is maximized by a high surface-to-volume ratio, as in the many desert species characterized by constitutively very small leaves (Lambers et al. 2008). A second, behaviorally mediated aspect of thermoregulation is *evaporative cooling* of leaves, due to transpiration from open stomates into the aerial atmosphere. This process allows the plant to unload the considerable energy used in turning liquid water taken up from the soil into water vapor; it is remarkably effective in air with low relative humidity (Gates 1968). Evaporative cooling is critical for the leaves of plants exposed to direct sun, which receive an intense energy load that they must rapidly dissipate–indeed, without such cooling, leaf temperatures would be predicted to reach 100°C within less than one minute (Lambers et al. 2008). Because stomatal aperture controls the loss of water vapor for cooling, stomatal behavior over short timescales integrates a plant's energy and water budgets so as to avoid both heat and drought stress (Körner 2013).

Plants also orient their bodies so as to thermoregulate, much as animals do. Individuals in many plant taxa adjust leaf angles in response to temperature load, holding leaves parallel to incident solar radiation (rather than perpendicular to it) when temperatures are high, so as to minimize photon absorption (Gamon and Pearcy 1990; Figure 4.4a). This orientation response is especially pronounced when soil water, and hence the opportunity for transpirational cooling, is limited (Kao and Forseth 1992). Some plants also reduce radiation load by rolling leaves and, in more extreme conditions, may close stomates and allow leaves to wilt, a short-term stress response which likewise reduces photon absorption by changing leaf orientation, but at the cost of photosynthetic activity.

In addition to these positional and stomatal "behaviors," plastic aspects of plant morphology and anatomy can contribute to thermoregulation in hot conditions. *Arabidopsis* individuals grown at a high temperature (28°C) expressed a suite of developmental adjustments that resulted in greater transpiration rates than plants raised in cooler (22°C) conditions (Crawford et al. 2012). By virtue of this increased transpiration, plants raised at the high temperature maintained leaf temperatures that were 1°cooler on average than those of plants grown at the low temperature, when both were transferred to a high-temperature treatment

Figure 4.4 Plants can lower or raise the temperature they experience through organ position, metabolic activity, or solar tracking. (a) The jojoba shrub, *Simmondsia chinensis*, which is native to the American desert Southwest, orients leaves vertically to reduce midday heat load. Photo credit and copyright John R. Seiler. For the color image, see Plate 10. (b) The skunk cabbage, *Symplocarpus foetidus*, emerges in very early spring in Eastern North American wetlands. At low ambient temperatures, flowers enclosed within the emerging spadix (flowering shoot) raise oxygen production, actively generating heat that maintains favorable tissue temperatures. Photo credit and copyright Michael Lynch. For the color image, see Plate 11. (c) Flowers of the alpine plant *Ranunculus montana* show heliotropic orientation movements that create a warm microsite within the flower despite cold air, maintaining favorable temperatures for ovary development and for attracting insect pollinators. Photo credit Andrea Moro (Univ. of Trieste), CC-BY-SA. For the color image, see Plate 12.

(Crawford et al. 2012). This enhanced cooling capacity evidently reflected a plastic adjustment to shoot architecture: plants grown at the high temperature produced longer and more steeply angled petioles that raised leaves well away from the soil surface, increasing diffusion from open stomates into air currents. The engineering consequences of this high-temperature-induced shoot architecture were confirmed in a simulation model (Bridge et al. 2013). The less dense stomates on these high-temperature plants also may have increased water-vapor diffusion rates out of the leaf by reducing the humidity of the air just outside each stomate, an external niche-constructing effect that feeds back to influence function (Crawford et al. 2012).

The plastic production of reflective hairs or layers of wax on leaf surfaces may also alter a plant's energy load. For example, in the desert shrub *Encelia farinosa*, leaves produced in winter bear sparse hairs, while those that develop in high summer temperatures are covered with dense hairs that dramatically reduce absorption of incident sunlight. The combination of dense hairs with stomatal behavior that promotes transpirational cooling results in leaf temperatures measurably below that of the surrounding air (Ehleringer and Mooney 1978).

Plants under high heat load further increase their heat tolerance by producing heat-shock proteins—as do animals. Indeed, some of these plant heat-shock proteins are coded by genes that are homologous to those of animals (Lambers et al. 2008). (Along with animal and plant cells, mesophilic microorganisms such as bacteria can adjust to abrupt temperature changes by rapidly accumulating heat- or cold-shock proteins which maintain correct protein dynamic and membrane properties; Smirnova et al. 2001 and references therein.) Cold stress also leads to the up-regulation of plant genes believed to contribute to cold tolerance, such as the "antifreeze proteins" produced by certain frost-tolerant grasses (M. Griffith and Yaish 2004).

Other physiological mechanisms of plant thermoregulation to cold are even more surprising. In certain phylogenetically basal clades, plants are capable of an astonishingly animal-like response to low ambient temperatures: the internal, metabolic production of heat. The most well-studied cases to date are the thermogenic flowers of the sacred lotus, *Nelumbo nucifera*, and skunk cabbage, *Symplocarpus foetidus* (Figure 4.4b). At cold ambient temperatures, these flowers increase oxygen production, generating heat that maintains favorable, consistent floral temperatures across a broad range of external conditions. As a result, within-flower temperatures in these taxa have been shown to remain between 30°C and 36°C despite ambient temperatures as low as 10°C (R. Seymour 2001). Plants also express positional responses to cold conditions. Leaves as well as flowers may orient perpendicular to incident radiation, following the sun's diurnal path to maintain this orientation and maximize energy absorption. Flowers of many arctic and alpine species express this *heliotropic* behavior, which causes solar radiation to be reflected toward the ovary (Figure 4.4c). This creates a warm microclimate within the flower that contrasts with the cold external temperature, attracting insect pollinators and enhancing the rate of seed development (e.g., Luzar and Gottsberger 2001). In this case, the plant's alteration of its own body temperature also offers a favorable microclimate to ecological interactors.

4.3 Plastic responses that mediate the experienced resource environment

Although the levels of key resources such as food and oxygen are generally viewed as a given property of any ecological setting, organisms in fact make behavioral, developmental, and physiological adjustments that often allow them to experience a given resource environment as more plentiful. This fundamental aspect of experiential niche construction can be carried out by means of a variety of responses in very different types of organisms. In this case, as with temperature, the organism's environment is determined not purely externally but through close interaction with the phenotype.

4.3.1 Microbial niche construction for food and oxygen availability

It has become increasingly well recognized that microbial habitats comprise tremendous spatial and temporal complexity. Microorganisms show a range of behavioral responses to this heterogeneity, just as they contribute to its creation. Microscale habitat choices allow microbes to experience these shifting, heterogeneous environments as consistently rich in nutrients.

Microorganisms are largely motile, using amoeboid movement to glide across surfaces, or flagella to swim in liquid media (Fenchel 2002 and references therein). These motile behaviors are informed by diverse chemosensory signals, allowing bacteria to accumulate at high-nutrient patches within their habitats (J. Seymour et al. 2010). This type of habitat choice or "relocational" niche construction is most well studied in marine microbes, including phytoplankton, heterotrophic bacteria, and the microzooplankton that consume them. Although ocean water is generally low in nutrients, tiny patches or "microzones" of dissolved organic matter (e.g., phytoplankton photosynthates, lysed algal cells, and zooplanktonic wastes) occur at a spatial scale of micrometers to centimeters (Stocker et al. 2008). These high-nutrient microzones occur only briefly before being dissipated by diffusion and water movement (Fenchel 2002). Marine microorganisms rely on "infochemicals" (such as dimethylsulfoniopropionate released by grazed plankton cells) to locate nutrient patches as well as prey organisms (J. Seymour et al. 2010).

Experiments with the marine bacterium *Pseudoalteromonas haloplanktis* showed that rapid chemotactic swimming responses allowed the bacteria to exploit transient resource patches: in response to fine-scale nutrient pulses such as those generated by lysed algal cells, bacteria clustered within tens of seconds in foraging "hot spots" containing three times more bacterial cells than other microsites. As a result of these rapid foraging behaviors, the fastest-moving 20% of the population experienced a tenfold higher nutrient environment than nonmotile cells (Stocker et al. 2008). Motile bacteria were also able to quickly colonize nutrient plumes of sinking organic particles with environmentally realistic dimensions and dynamics, resulting in fourfold enhanced nutrient availability compared with that obtained by nonmotile cells (Stocker et al. 2008). In nature, such rapid, targeted movements guided by fine-scale "chemical landscapes" (J. Seymour et al. 2010) allow microorganisms to experience their generally nutrient-poor surroundings as nutrient rich. Similarly, chemosensory movements also allow these tiny organisms to seek out microsites that have very specific, preferred oxygen levels (Fenchel 2002).

4.3.2 Plant niche construction for available nutrients, water, light, and oxygen

Although plants (unlike motile microorganisms and foraging animals) lack the option of moving to resource-rich locations, they do have an impressive developmental repertoire that can serve to enhance their experience of the below- and above-ground resource conditions they encounter as sessile organisms (Huey et al. 2002). The concentration of essential plant nutrients varies widely in soils of natural habitats at within- and among-individual scales, both spatially and temporally (Bazzaz 1996; Hodge 2010; and references therein). Individual plants are able to experience an environment that is consistently high in nutrients despite this resource patchiness because of the developmental and physiological plasticity of root systems. This resource enhancement is accomplished through the following specific changes to plant "foraging behavior" (sensu M. Hutchings and de Kroon 1994): (a) adjustments in total surface area of root systems in response to soil nutrient levels; (b) targeted proliferation of new roots in nutrient-rich locations; and (c) increased systemic rates of mineral ion uptake in those locations (Hodge 2009 and references therein).

One well-studied developmental response to nutrient-poor soil is increased biomass allocation to root tissue relative to shoots (Fitter and Hay 2002). This allocational adjustment is often accompanied by a morphological shift to thinner roots that have a higher surface-to-volume ratio (Fitter 1994; Bell and Sultan 1999; and references therein). Together, these responses result in a much greater root-system length and surface area per gram of tissue to be supplied, allowing the plant to take in more mineral ions from its rhizosphere and hence experience it as a less nutrient-poor environment (Sultan and Bazzaz 1993c).

Plastic, targeted root proliferation was first investigated in a series of classic lab experiments during the 1970s (e.g., Drew et al. 1973; further references in Hodge 2004). In these studies, individual barley plants, *Hordeum vulgare*, were grown in containers that were divided into compartments containing different concentrations of a given macronutrient (nitrate, ammonium, potassium, or phosphate). Plants grown in containers that had a high concentration of the nutrient in all three compartments showed consistent root proliferation throughout the container. In contrast, individuals grown in containers with two low-nutrient compartments and a single, very high-nutrient compartment showed intense proliferation of lateral roots in this nutrient-rich zone, and very little lateral root production in the nutrient-poor compartments (Drew 1975; confirmed in several species by Robinson 1994). Localized proliferation based on relative resource richness reflects the number of new primary and especially lateral roots initiated, as well as the elongation and further branching of these fine, physiologically active roots (Hodge 2009). The proliferation response, which is highly dynamic, is apparently governed by a combination of local sensors (which recognize the immediate concentration of soil resources) and systemic signals (which reflect the plant's previous resource acquisition); even in the presence of very high nitrate levels, for instance, the response occurs only when the plant is nutrient stressed (de Kroon et al. 2009 and references therein).

The physiological component of plant nutrient-foraging plasticity is equally dramatic: in most species, roots of nutrient-deprived plants transiently increase uptake rates for mineral ions from 2-fold to 11-fold in response to a local increase in nutrient supply (Hodge 2004). This short-term increase in nutrient uptake may provide an internal signal for the plant to initiate new root investment at this nutrient-rich location.

Two other key soil resources for plants are water and oxygen. Individuals express the same kind of foraging adjustments to enhance the availability of these resources: increased whole-plant allocation to root tissues in dry soil or to oxygen-collecting organs such as pneumatophores in flooded conditions; rapid, targeted root proliferation in moist or aerated soil locations in the event of water or oxygen stress; and changes to uptake thresholds and utilization (conductance or respiration) rates (references in Bell and Sultan 1999; Fitter and Hay 2002; Sultan 2003).

Developmental and physiological plasticity also allow plants to experience favorable levels of aboveground resources such as light, oxygen, and carbon dioxide. In low-light conditions, plants maximize the total surface area for catching photons through increased leaf biomass allocation and morphological changes to leaves (Chapter 2, Section 2.3.4; Ryser and Eek 2000; Fitter and Hay 2002). "Shade" leaves produced in low light are anatomically altered so as to most efficiently utilize a less dense photon flux (i.e., with a single layer of chloroplast-rich palisade parenchyma; references in Sultan 2003). Access to solar radiation at the whole-plant level is also determined by crown architecture, that is, by the deployment of individual leaves and branches in space (reviewed by M. Hutchings and de Kroon 1994; Valladares and Ninimets 2007). Within phylogenetically constrained parameters, plants adjust leaf angle and placement through developmental shifts in internode and petiole length, as well as through diurnal movements (Fleck et al. 2003; Huber et al. 2008). For example, leaves may be held more vertically in the upper part of an individual's canopy, and more horizontally at lower levels, allowing more light to penetrate to the latter and hence maximizing whole-plant insolation (Valladares and Ninimets 2007). The production of broad, thin "shade leaves" lower in the canopy helps to maximize light interception as well. Increased production of vegetative buds on branches in high-light positions can also result in effective light foraging, analogous to root proliferation in resource-rich soil patches. Because the exposure of shoot organs to direct sun also influences the temperature of leaf tissues and their water status, these adjustments simultaneously shape several aspects of the plant's experienced environment (Valladares and Pearcy 1997; Valladares and Ninimets 2007). For example, leaf- and branch-level adjustments can also allow a plant to avoid the negative effects of photoinhibition and heat stress in conditions of high insolation and moisture deficit (Valladares and Pearcy 1997).

Several fascinating aspects of plant experiential niche construction take place within the leaves, where the supply of photosynthetically active radiation is shaped by dynamic ultrastructural adjustments. One such adjustment consists of orientation movements that redistribute chloroplasts within leaf mesophyll cells, first described in 1908 (reviewed by Wada et al. 2003). These light-processing organelles move continuously inside cells, in response to the flux density, direction, and spectral quality of light (Haupt and Scheuerlein 1990; W. Williams et al. 2003). In weak light, chloroplasts accumulate across cell surfaces that are perpendicular to the direction of incident light, maximizing the cell's photon harvest (Kasahara et al. 2004; Davis and Hangarter 2012). By contrast, under strong irradiance, chloroplasts line up along the side walls of the cell, parallel to the angle of light (Figure 4.5a). This chloroplast "avoidance" response allows more light to be transmitted through the leaf tissue, protecting the photosynthetic machinery of the cell from photodamage due to excess energy (Kasahara et al. 2002; Gabryś 2012). Oddly enough, it also makes it possible to create images within living leaf tissues, by "drawing" with direct irradiance (Figure 4.5b). Although their regulatory mechanisms are not yet fully understood (Wada et al. 2003; Davis and Hangarter 2012), these niche-constructing intracellular movements, and the phototropin blue-light receptors that mediate them, are evidently conserved across algae, mosses, ferns, and angiosperms (references in DeBlasio et al. 2003).

At an even finer level, plants modulate their experienced light environment by means of plastic ultrastructural changes to photon-processing components of chloroplasts themselves (Kirchhoff 2013). In *Arabidopsis*, for instance, plants grown in low light alter the morphology and conformation of thylakoid membranes and structurally rearrange Photosystem II protein supercomplexes, so as to more efficiently transfer energy between photosystems (Kouril et al. 2013). Analogous ultrastructural changes take place in a distantly related photosynthetic organism, the purple bacterium *Rhodopseudomonas palustris*: in this proteobacterium, the number, protein composition, and size of light-harvesting units change in response to contrasting light levels, resulting in changed cellular light absorbance spectra (Brotosudarmo et al. 2011). It appears that plants, algae, and photosynthetic bacteria may shape their experienced light environment even at the atomic level, by means of a biophysical mode of plasticity known as *quantum coherence* (discussed by Anna et al. 2014). Through coherence, a photon-excited electron within a chlorophyll molecule interacts with its protein scaffold so as to realize a wave-like transfer of energy that moves across the light-harvesting system, unerringly finding the most efficient path (Engel et al. 2007). It is believed that quantum coherence could explain the remarkable, near-perfect efficiency at which solar energy is captured for conversion to chemical energy during photosynthesis, a biophysical behavior that maximizes the availability of light as a resource.

Developmental responses at several levels also allow plants to experience an enhanced oxygen environment in the event of flooding. Plants of flood-prone habitats confront a severe drop in atmospheric oxygen if shoots become submerged. Individuals respond to this challenge with species-specific morphological and ultrastructural changes that allow them to experience the environment as one with sufficient oxygen. In some taxa, submerged shoots produce leaves with enlarged surface area, thinner cuticles, and thinner epidermal cell walls, changes that jointly enhance the plant's internal supply of carbon dioxide (Mommer et al. 2005). Together with a reorientation of chloroplasts toward the epidermis, these changes lead to higher underwater photosynthetic rates that result in elevated tissue oxygen concentration despite the poor solubility and diffusion of oxygen in water (Mommer and Visser 2005; Mommer et al. 2006). In other species, stems or petioles on submerged shoots rapidly elongate, or orient more vertically, so that leaves maintain direct contact with the oxygen-rich aerial environment (Voesenek et al. 2006 and references therein). One signal for this elongation response is an internal one, the buildup of the gaseous hormone ethylene due to reduced gas diffusion out of plant tissues under water (Bailey-Serres and Voesenek 2010). Interestingly, subsequent to the initiating cues (ethylene rather than phytochrome signals), elongation responses to flooding and shade-avoidance shoot responses are mediated by largely shared hormonal pathways (Sultan 2010 and references therein). Physiological shifts to alternative metabolic pathways can also help plants to avoid experiencing an oxygen deficit (Bailey-Serres and Voesenek 2010). Response to flooding is one of the few cases in which species differences in ecological developmental response capacities have

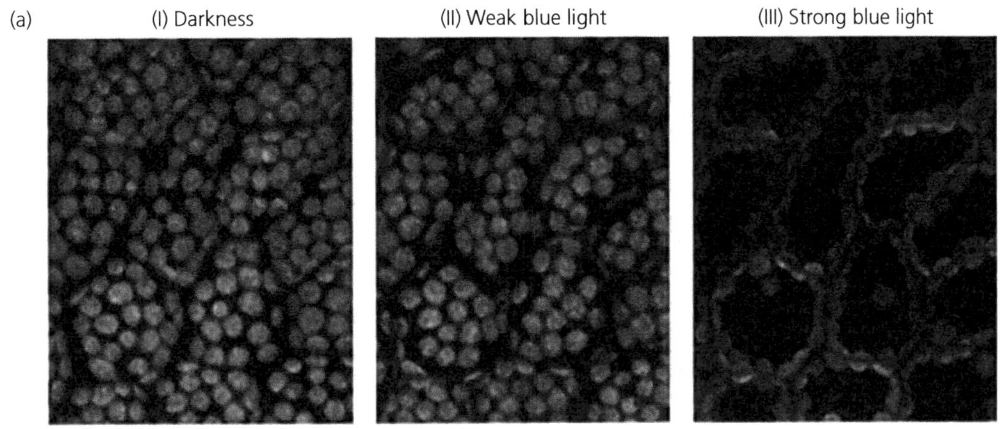

Figure 4.5 Chloroplast movements within plant cells mediate the supply and intensity of photosynthetically active radiation. (a) In darkness (I) or weak light (II), chloroplasts cover cell surfaces, maximizing photon interception; in strong light (III), they move to cell margins, thus allowing solar energy to pass through the cell. Image courtesy of Mark Fricker. For the color image, see Plate 13. (b) Because light-induced chloroplast movements alter light transmission, they can be used to reproduce a high-contrast image (such as a photograph) within leaf tissue. After approximately 45 minutes of blue-light illumination, the areas where light reached the leaf will be more transparent than those covered by dark parts of the image. A portrait of the pioneering plant physiologist Julius von Sachs is here reproduced in a live coleus leaf. In his 1887 *Lectures on the Physiology of Plants*, Sachs was the first to observe that a "light-picture" could be created in living leaf tissue. Leaf portrait courtesy of Roger Hangarter. For the color image, see Plate 14.

been studied both mechanistically and with respect to realized ecological consequences in the field (Voesenek et al. 2004; Benschop et al. 2005).

4.3.3 How animals modify their experienced food and oxygen supplies

Active foraging to seek out suitable foods is a familiar way by which animals maximize their resource supply despite the spatial and temporal variability of real habitats. Foraging behavior may result not just in an adequate supply of food but also in a favorable balance of specific nutrients, regardless of the relative abundance of various local foods. Individuals of many invertebrate taxa (including herbivores, carnivores, and omnivores) choose foods so as to take in a balance of required nutrients despite the heterogeneity of their nutritional environments (Mayntz et al. 2005). In one study, predatory ground beetles and spiders individually chose to consume either more protein-rich or more lipid-rich foods, depending on which nutrient their previous food supply had lacked (Mayntz et al. 2005). In addition to prey choice, dietary balancing can be accomplished by selectively extracting more of a specific nutrient from a given prey item. In the latter study, for example, desert spiders changed the ratio of nitrogen to carbon they extracted from each of their fly prey, through an unspecified physiological mechanism (Mayntz et al. 2005).

Along with foraging behavior, animal morphology is a key mediator of resource availability. Animal biologists from Georges Cuvier to Charles Elton have noted that traits such as jaw and tooth morphology equip animals to best utilize certain foods. Indeed, traits such as bill shape in birds and jaw features in fish have served as standard examples of species-level ecological adaptations. More recently, researchers have recognized that diet and feeding structures can show reciprocal causation: not only does morphology shape diet, but diet can also affect morphology, by means of developmental plasticity (West-Eberhard 2003 and references therein). This aspect of plasticity can allow individuals to develop the appropriate feeding apparatus to process whatever foods are locally plentiful, thereby maintaining ample access to resources despite site variation in the availability of particular prey items. An animal expressing such developmental responses thus experiences its environment as one with plenty of food resources.

Diet-based developmental plasticity was initially recognized in fishes: as early as the 1960s, it was noted that jaws developed differently in several fish taxa depending on the type of prey available during juvenile growth (references in Wimberger 1992). This was subsequently confirmed through a number of controlled experiments. For example, the neotropical cichlid *Cichlasoma managuense* was found to develop substantially different, functionally appropriate cranial features when its diet consisted of soft foods such as nematode worms, versus harder planktonic foods such as larval crustaceans (Meyer 1987). Individuals of the congener *Cichlasoma citrinellum* likewise developed contrasting jaw morphologies and pharyngeal apparatuses in response to diet: either a gracile jaw with narrow, pointed teeth, efficient for feeding on soft prey, or a robust jaw with strong, rounded teeth, well adapted to cracking and feeding on snails (Meyer 1990). Remarkably, the latter morph can be induced reversibly in response to seasonal or site differences in the abundance of snails as a food source (Meyer 1990). Plastic responses can even allow for dramatically different modes of feeding: in another cichlid, the rock-dwelling species *Neochromis greenwoodi*, individuals raised on zooplankton developed jaw anatomical traits suitable for suction feeding, while those fed algae substitute (which requires biting feeding) developed traits that resulted in the appropriate biting force (Bouton et al. 2002). Further experiments have confirmed that early food type can affect several key feeding traits in fish, such as mouth size, biting force, and handling efficiency for specific sorts of prey (e.g., Hegrenes 2001).

These effects of different food types on fish development likely arise from the distinct physical stresses on the musculoskeletal system that result from processing different types of food (Wimberger 1992). The result is a causal cycle between phenotype and resource environment: behavioral flexibility that allows a juvenile to feed on available foods can lead to developmental consequences that shape its subsequent resource use (Wimberger 1992). Such plasticity is now considered to explain many fish "polymorphisms" with respect to musculoskeletal features of jaws, tooth morphology, and head and body shape (discussed by Meyer 1987) and may explain the existence of foraging diversification within so many fish species (Hegrenes 2001 and references therein).

This food type–developmental plasticity–food type cycle has also been documented in other vertebrates. For example, juvenile oystercatcher birds that feed on different foods develop distinct bill types that lead them to specialize on those foods as adults (references in T. Smith and Skúlason 1996). Similarly, the developmental trajectory in northern water snakes, *Nerodia sipedon*, that shapes jaw length, head proportions, and body size is influenced by the size of prey items during the juvenile stage (Queral-Regil and King 1998). Although this phenomenon has been less well studied in insects, insects can also show jaw plasticity similar to that seen in vertebrates: for example, *Melanoplus femurrubrum* grasshoppers fed fibrous, poor-quality food developed larger mandibles and mandibular muscles than those receiving more nutrient-dense food (Thompson 1992).

In other animals, diets consisting of low-quality food result in a different type of adaptive developmental response: changes in the internal food-processing organs, such as greater gut capacity and an increased intestinal area for nutrient uptake (Sassi et al. 2007 and references therein). These changes (which have been studied in many small mammals) allow individuals to compensate for a poor resource environment by processing a greater volume of less nutrient-dense food (Sassi et al. 2007). For example, in the herbivorous arid-habitat rodent *Microcavia australis*, individuals found in sites (or seasons) with the lowest quality, highest-fiber foods had the largest cecum, the organ in which fiber digestion occurs. The greater volume of cellulose fermentation that took place within these enlarged organs released additional nutrients from the fibrous plant foods (Sassi et al. 2007). These individuals also showed functionally appropriate differences in the mass and area of small and large intestines (Sassi et al. 2007). Interestingly, certain grasshoppers also developed a larger gut when fed fibrous, low-nutrient foods, compared with individuals given nutrient-rich food (Yang and Joern 1994; discussed by Whitman and Agrawal 2009).

Developmental responses to food conditions can occur in various structures associated with resource uptake, depending on the animal. For example, larvae of the blackfly *Simulium lundstromi* capture food particles from their aquatic habitats using a pair of "labral fans" that filter out bits of food from moving water. In sites with fast-flowing water, the supply of food particles that reaches these structures is more plentiful. However, individual blackfly larvae influence their supply of food particles through developmental plasticity. In a controlled experiment, larvae developing in slow-moving water produced large labral fans with widely spaced rays that provide a larger surface area for trapping particles; under high flow rates, larvae developed stronger, more compact fans that create less drag. These alternative morphologies are considered an adaptive plastic response that balances the rate at which food particles are encountered with the fluid drag in slow versus fast currents (Y. Zhang and Malmqvist 1997).

Some vertebrates make analogous developmental adjustments to maintain an adequate supply of a second critical resource, oxygen. Again, this type of plasticity is most well established from controlled rearing experiments with fish. In the African cichlid *Pseudocrenilabrus multicolor victoriae*, individuals raised under hypoxic conditions developed longer total gill filaments and larger gill surface area than those raised at normal oxygen levels (L. Chapman et al. 2000, 2008). These changes allowed for an increased oxygen uptake capacity that would maintain a physiologically favorable supply despite the external limits the species encounters when conditions vary in their natural habitats. The adaptive change in gill size and morphology was accommodated by developmental changes to the surrounding structural elements, resulting in an expanded body shape that may cause a locomotory trade-off.

An even more dramatic case of "adaptive remodeling" of fish gills was found in the crucian carp, *Carassius carassius*, a species that inhabits small ponds which can become very low in dissolved oxygen when temperatures drop in winter. After several days in hypoxic water, individuals in this species showed a sevenfold increase in respiratory surface area, a developmental change that led to a 50% reduction in the threshold oxygen concentration needed to support an aerobic resting metabolism (Sollid and Nilsson 2006). In this case, the change in surface area was achieved by means of cell death in the cell mass that ordinarily imbeds the gill lamellae where most gas exchange occurs (Figure 4.6). Interestingly, this change in cell mass is reversible within several days if normal oxygen conditions return (Sollid and Nilsson 2006). Similarly, small mammals that develop at high altitudes, where oxygen concentration is reduced, develop lungs with larger alveolar surface area and show physiological adjustment in the form of higher blood hematocrit (Hammond et al. 2001; Hsia et al. 2005). Indeed all vertebrates exposed to hypoxia adjust respiratory and cardiovascular function so as to maintain an adequate oxygen supply to metabolically active tissues (Hicks and Wang 2004).

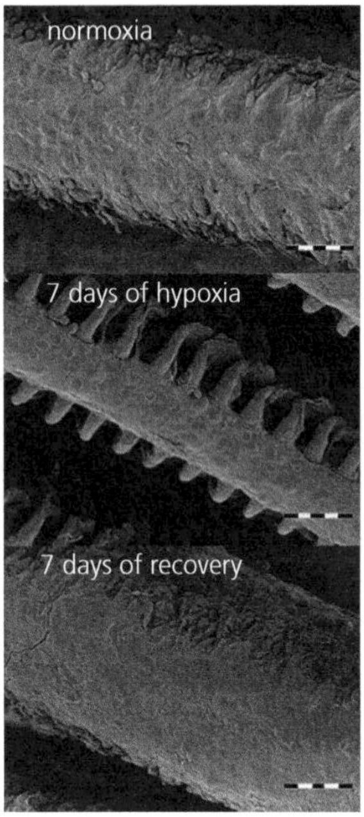

Figure 4.6 Fish in certain taxa modulate their oxygen supply by means of gill remodeling. Gills of the crucian carp, *Carassius carassius*, undergo reversible remodeling in response to variation in aquatic oxygen levels, as shown in this experimental sequence (scale bars = 50 μm). In this species, under normal oxygen conditions the gill lamellae, where gas exchange occurs, are embedded in a cell mass; however, in hypoxic water, cell death takes place, and the lamellae consequently protrude so as to maximize oxygen uptake. Caption modified from Sollid et al. 2003. Image courtesy of Jørund Sollid, reproduced from J. Sollid, P. de Angelis, K. Gunderson, and G. E. Nilsson. 2003. Hypoxia induces adaptive and reversible gross morphological changes in crucian carp gills. *Journal of Experimental Biology* 206: 3667–73.

Some reptiles and fish also reduce their oxygen needs by down-regulating aerobic metabolism, in part by changing their target body temperature through behavioral modifications such as seeking colder microsites (Hicks and Wang 2004). Generally speaking, reducing resource needs is another way that many organisms can experience a limited-resource environment as being sufficient. Such reductions can be effected through a diversity of phenotypic changes. For example, in ectotherms such as reptiles, fishes, and insects, individuals at higher body temperatures require more food resources to maintain maximum rates of metabolism and growth (Coggan et al. 2011). When nutrient availability is low, many such animals conserve chemical energy by maintaining a lower body temperature and reducing levels of activity (Coggan et al. 2011). Individual *Rutilus rutilus* fish made both of these changes when subjected to a naturalistic three-week starvation period: at night, they migrated to cooler water (in an aquarium with a horizontal thermal gradient); during the day, they remained in warmer water (with potentially higher food density) to forage but decreased swimming activity by approximately 50% (Van Dijk et al. 2002). These behavioral shifts decreased the fish's metabolic rates and hence their energy expenditure. By contrast, *Rutilus rutilus* individuals that were provided with unlimited food at all times preferred warmer water (Van Dijk et al. 2002). In Australia, migratory locusts, *Locusta migratoria*, also adjust thermoregulatory behavior depending on food availability. Individuals given a reduced supply and quality of food resources selected the lowest available temperatures within an experimental temperature gradient, while those given an excess of high-quality food chose the higher temperatures (Coggan et al. 2011). Because nutrient assimilation is most efficient at lower temperatures, such responses maximize resource availability but reduce growth rates.

Resource needs can also be adjusted directly, and in some cases reversibly, through changes in growth rate and hence body size. This has been studied in populations of the long-lived Galapagos marine iguana, *Amblyrhynchus cristatus*. These herbivorous lizards experience drastic fluctuations in their main food resources, red and green algae, as a result of the El Niño weather events that occur repeatedly during each individual's life span. Surprisingly, adult marine iguanas can adjust body length depending on food availability, by changing bone absorption and regrowth so as to switch between growth and shrinkage (Wikelski and Thom 2000). These changes in bone metabolism were found to result in body length reductions of up to 20% within two years, in response to reduced nutrient availability. Because this shrinkage both reduces energy expenditure and increases foraging efficiency, this plastic response allows individual iguanas to experience a less severe food reduction during the El Niño algal population drops and to survive better during these harsh periods (Wikelski and Thom 2000). Many small temperate-region mammals also reduce body size to decrease energetic requirements during the cold and resource-poor winter (Heldmaier and Werner 2003).

4.4 How organisms shape the environments their offspring experience

A critically important way that individual organisms influence adaptive success is to appropriately shape the environment experienced by their offspring. These niche-constructing processes, and the behaviors parents employ, have been extensively studied in animals as well as, more recently, in plants. These include parental influences on the site and timing of seed or egg dispersal, pre- and postbirth provisioning and protection to offspring, transgenerational plastic effects on functional offspring traits, and longer-term environmental mediation in the form of "parental care" (reviewed in Clutton-Brock 1991; Fenner and Thompson 2005; Royle et al. 2012). These effects on offspring environmental experience comprise a rich set of biological phenomena with important ecological and evolutionary implications for natural systems (see Mousseau and Fox 1998; Bateson et al. 2004; Donohue 2005; Badyaev and Uller 2009; Donohue 2009; Wolf and Wade 2009; Herman and Sultan 2011). In humans, recent studies have revealed how the prenatal environment that results from maternal nutritional status and toxin exposure can alter progeny phenotype in ways that can positively (or negatively) affect the offspring's experience of its postbirth resource environment (Bateson et al. 2004; Gluckman, Lillycrop, et al. 2007; S. Gilbert 2011). The resource costs to parents of these developmental and behavioral mediations has traditionally been interpreted as a conflict of interest between parents and offspring (Royle et al. 2012), but a reassessment of this theory indicates that this conflict has generally been overestimated (Bossan et al. 2013 and references therein). Indeed, the fitness benefits and costs can be very complex to evaluate, as in cases where parent organisms enlist unrelated mutualists to assist with specific aspects of offspring care.

To begin with, parent animals and plants can implement spatial and temporal *habitat choice* for developing offspring. Most birds as well as many reptiles, mammals, and insects (such as ants, termites, wasps, and beetles) create favorable physical conditions for their young by building nests or burrows (see below). In other cases, maternal individuals implement a favorable habitat for their offspring by means of sophisticated oviposition decisions (reviewed by Resetarits 1996) such as those made by insect mothers to insure larval access to host plant tissues of suitable type and quality. Although this kind of direct construction of the offspring's environment is unusual in plants, it does occur. Maternal plants in a number of (primarily tropical and semidesert) taxa actively place their offspring in suitable germination sites in the soil, a phenomenon known as *geocarpy* (Barker 2005 and references; e.g., Popovkin et al. 2011; Figure 4.7). The most familiar of such taxa is the common peanut or groundnut, *Arachis hypogaea*, which is indeed named for this trait. After fertilization, the small stalk on which the peanut's flower is borne elongates into a specialized structure called a *gynophore* which uses light, gravity, and touch cues to grow downward, penetrate the soil, and bury the immature ovary underground (Moctezuma 2003). There, the developing seeds are protected from both desiccation and herbivores (Van Rheede van Oudtshoorn and Van Rooyen 1999).

Both animal and plant parents also effect temporal habitat choices for their offspring. In vertebrates such as birds and mammals, reproductive phenology is timed according to seasonal schedules

Figure 4.7 Maternal plants in some species practice parental care by burying their developing progeny. Fruit stalks of the newly discovered geocarpic annual *Spigelia genuflexa* curve downward and elongate to deposit developing fruits in the moist mossy layer on the soil surface. The plant is known only from the sandy Atlantic forest habitat of northeastern Bahia, Brazil. Photograph by Alex V. Popovkin, CC BY 3.0. For the color image, see Plate 15.

that insure ample availability of resources for developing offspring (references in Visser et al. 2004). Indeed, disruption of such coordinated schedules due to global warming is a major concern for the persistence of natural populations of all kinds (Stenseth and Mysterud 2002; Visser and Both 2005). Depending on the species' ranges of phenological plasticity, individual mothers may fine-tune these temporal habitat choices to a greater or lesser degree (e.g., Visser et al. 2004). In some cases, prey organisms participate in this fine-tuning. For example, in certain generalist leafhopper insects (Membracidae), mothers synchronize the life history of their progeny with specific deciduous host plants by inserting dormant eggs into the plant's vascular tissue in spring, before the leaves have expanded. When the host plant's sap ascends (which occurs at a different time in different host taxa), the eggs absorb water from the sap, which triggers egg hatching and the onset of juvenile development (Wood et al. 1990). The resulting plant-induced variation in insect life history is hypothesized to be a source of assortative mating and hence sympatric speciation in these insects (Wood 1993).

Maternal plants also implement temporal niche construction that determines the conditions into which offspring will emerge (see Section 4.1). The time of flowering in the maternal plant (which is itself strongly influenced by environment) determines what seasonal conditions seeds will encounter when they are released and eventually germinate (Donohue 2003; Galloway and Etterson 2009). In a number of cases, this causal chain determines not only the timing of seedling emergence but also which of several alternative life histories offspring will express (Donohue 2005). Interestingly, the different possible life histories initiated through maternal timing effects can be specifically adapted to the particular maternal conditions that bring them about (e.g., Galloway and Etterson 2007; discussed in Section 4.1). Maternal plants also influence the habitat their offspring will encounter through environmentally conditioned adjustments to fruit and seed traits that affect the timing of emergence and development (e.g., Sultan 1996; Herman et al. 2012). Such traits may include the structure and thickness of the fruit wall and of the seed's protective coat (both of which are actually maternal tissues), as well as the seed's ratio of water-imbibing surface to its volume (references in Herman and Sultan 2011). These maternally regulated traits play a key role in seed dormancy, germination behavior, and rate of emergence from soil and thus allow the mother plant to indirectly affect offspring habitat choice in time (Donohue 2009 and references therein).

Animal and plant parents may also manipulate specific aspects of the environment their developing offspring will experience, such as temperature. As noted above, many insects (e.g., ants, termites, and wasps) as well as reptiles, mammals, and most birds provide developing offspring with temperatures above ambient by rearing them in nests or burrows; in some cases, adults move eggs among different depths to more finely regulate developmental temperature (Clutton-Brock 1991). Adults of certain lizard and snake taxa collect heat by basking in the sun and then return to the nest where they transfer this heat energy to the young, while birds, mammals, and insects may cluster together to create favorable incubation temperatures (references in Farmer 2000).

For example, both male and female parents in the monogamous California mouse, *Peromyscus californicus*, engage in huddling over pups to keep them warm (Gubernick and Teferi 2000). Interestingly, birds and bumblebees have evolved a very similar specialization for warming eggs: an uninsulated, highly vascularized spot on the abdomen, termed the *incubation patch*, that efficiently transfers the adult's body heat to eggs. Endothermy itself may have evolved convergently in mammals and birds as a result of selection for parental care, since thermogenesis allows parents to keep temperature-sensitive vertebrate embryos within the relatively narrow range of temperatures required for proper development (Farmer 2000).

Maternal plants can also alter the temperature experienced by their developing young, for instance, through the solar tracking that occurs in many alpine flowers (Section 4.2; see Figure 4.4c). Individual plants may also show thermoregulatory "parental care" behavior by adjusting the color and reflectance of reproductive tissues, as has been studied in the widespread herbaceous perennial *Plantago lanceolata* (Lacey and Herr 2005). Plants of this species produce their seeds on long inflorescences or "spikes" held vertically above the foliage. At high ambient temperatures, maternal plants produce light-colored reproductive spikes that better reflect incident radiation (especially in the visible and near-infrared regions of the spectrum). In contrast, mother plants growing in low temperature conditions produce dark-colored spikes that reach a higher internal temperature when insolated (Lacey and Herr 2005).

More generally, maternal plants can alter the physical conditions experienced by their young offspring simply through their close proximity, because their larger bodies shelter seedlings from excessive ultraviolet radiation, drought, and temperature extremes. In a study of alpine Colorado field populations, the close proximity of rosette plants to their seedling offspring resulted in significantly reduced evaporative water loss from soil compared with exposed microsites, leading to lower seedling water stress and greater establishment and survival (Wied and Galen 1998). This "nurse plant" effect results in a characteristic distribution pattern in alpine populations, where seedlings emerge from beneath maternal plants. Seedlings of other parent individuals or even other species can benefit from this kind of protective facilitation, which evidently outweighs any competitive disadvantages in these physically exposed habitats (references in Wied and Galen 1998).

Parents may also enhance the resource environment experienced by their offspring. This is directly effected in both animals and plants by means of maternal provisioning of nutrients to eggs, seeds, and developing embryos (Roach and Wulff 1987; Bateson et al. 2004). Parents in many animal taxa continue to provide food to their young after birth, including lactation in mammals (Royle et al. 2012). Although in mammals offspring provisioning is predominantly undertaken by the maternal parent, in some birds and many fish, the male broods, feeds, and even incubates the young (Gubernick and Teferi 2000). In social mammals and birds, parents may indirectly maintain adequate access of their juvenile young to food resources by attacking competing individuals (Clutton-Brock 1991). Although complex parental provisioning behaviors are often studied in such groups as mammals and social insects, close attention may reveal such behaviors in other animal taxa as well. In the dung beetle *Onthophagus taurus*, adult females (and sometimes males) dig tunnels into which they drag bits of dung, which they then form into compact "brood balls." On each brood ball, an egg chamber containing a single egg is placed. The larval beetle completes development and metamorphosis within its brood ball, which comprises its entire food supply (Moczek and Emlen 1999 and references therein). In building the brood balls, parent beetles judge the nutritional quality of the available dung and adjust ball size accordingly, such that each larva is provided with a sufficient supply of nutrients (Moczek 1998).

In plants, the quantity and composition of resources provided to developing embryos is determined by the maternal individual on whose body fertilized ovules are borne. This provisioning is highly plastic and can be strongly influenced by several aspects of the mother plant's environment (Roach and Wulff 1987; Donohue and Schmitt 1998). Maternal provisioning includes nutrients as well as a variable cocktail of growth hormones that together shape offspring gene expression and metabolic rates (references in Donohue 2009 and Herman and Sultan 2011). Environmentally induced epigenetic changes may also be implicated in transgenerational effects of both maternal and paternal plant environment on offspring phenotypes (Verhoeven and van Gurp 2012).

In some cases, maternally or paternally mediated effects (or both) result in adaptive developmental adjustments to offspring phenotypes that enhance their success in environments similar to those encountered by the parent, for example, when those environments entail specific resource limits. Experimentally drought-stressed parents of the generalist annual *Polygonum persicaria* produced seedling offspring that developed large, deep root systems more rapidly than offspring of genetically similar but well-watered parent plants did (Sultan et al. 2009). Presumably due to their enhanced early water-collecting capacity, these seedlings had significantly greater survival in dry soil (Herman et al. 2012). Nutrient-deprived *Polygonum persicaria* parents also produced seedling offspring that allocated proportionately more biomass to roots, which collect soil nutrients as well as water (Sultan 1996). Similarly, seedling offspring of shaded parents in this species and in the congeneric *Polygonum hydropiper* allocated proportionally more biomass to shoot tissue, and more quickly produced true leaves, than offspring of genetically similar plants that were given full sun (Sultan 1996; Lundgren and Sultan 2005).

Such adaptive transgenerational effects of parental conditions on offspring phenotype can evolve by natural selection if the parental environment accurately predicts the environment its progeny will encounter (Donohue and Schmitt 1998; Galloway 2005). In the *Polygonum* system, for instance, propagules are dispersed simply by gravity—that is, by just falling off—and hence are likely to reach a microsite close by, and similar to, that of the mother plant. In species that have evolved this type of specific transgenerational effects on resource-collecting tissues, parent plants can enhance the supply of resources that offspring experience in a given microsite. Note that the offspring's own, immediate, developmental responses to growth conditions will also influence resource availability and other aspects of its environmental experience. For example, all *Polygonum persicaria* seedlings growing in dry rather than moist soil increased their allocation to root tissues, whether parents had been drought-stressed or not (Sultan et al. 2009). However, the parental effect provides a key "head start" to seedling response: in this case, the joint phenotypic impact of parental and offspring drought stress created a significantly earlier and more pronounced enhancement of root-system development than the immediate response alone.

In addition, animal and plant parents exposed to predators may change offspring phenotypes so that their young experience a less threatening environment. For example, female water fleas, *Daphnia cucullata*, that are exposed to a chemical predator cue produce offspring that are more well defended by virtue of a large, sharply pointed defensive "helmet" that makes it difficult for predators to ingest them (Agrawal et al. 1999). Indeed, neonates produce this helmet as a result of maternal exposure to the cue even if they are born into a predator-free environment (Agrawal et al. 1999). Like the *Polygonum* plants discussed above, this parentally mediated developmental effect provides an early advantage that avoids any lag time associated with the offspring's own plastic response to the predator cue. Similarly, wild radish plants, *Raphanus raphanistrum* and *Raphanus sativus*, that were damaged by herbivorous caterpillars not only raised their own levels of defense chemicals but also produced seedling offspring with greater concentrations of defensive glucosinolates and denser leaf trichomes (Agrawal et al. 1999). Yellow monkey flower plants, *Mimulus guttatus*, subjected to simulated herbivore attack also produce offspring that have greater leaf trichome density, an adaptive effect that is epigenetically mediated by both maternal and paternal parents (Scoville et al. 2011).

In other cases, parent animals use habitat choice to insure that offspring experience relatively safe conditions early in life. Monarch butterflies, *Danaus plexippus*, infected with protozoan parasites use oviposition behavior to insure that their offspring experience an environment in which the parasites are not a severe threat. If they are infected, mothers preferentially oviposit on host plants that produce high levels of cardenolides that are toxic to the parasites (Lefevre et al. 2010). Since the emerging larvae feed on these highly toxic leaves, they experience lower parasite burdens and hence a more favorable developmental environment (Lefevre et al. 2010). In many animals, parents directly create a safer offspring environment through their own protective behaviors (e.g., many invertebrates, fish, reptiles, amphibians, and birds; Clutton-Brock 1991). For instance, maternal care in many leafhopper insects includes not just guarding eggs until they hatch but also protecting newly hatched young with the mothers' comparatively large, spiny bodies (Wood 1993).

Membracid insects that cannot carry out prolonged maternal care enlist ants to protect their emerging broods by means of oviposition behavior (Wood 1993). Females of *Entylia bactriana* initially deposit just a small number of eggs on a suitable host plant patch. If these eggs are not found by ants, the mothers desert them and oviposit in a different location. But if ants do locate the initial group of eggs, the mothers then deposit a large egg mass in the same location and guard it until they die, at which point the ants take over the protection of the young nymphs, in return for a chemical reward the juvenile leafhoppers produce (Wood 1993).

Clearly, ecological mutualists may participate in shaping conditions for offspring. An obligate example is the tiny dipteran gall midge *Asteromyia carbonifera* and its fungal symbiont *Botryosphaeria dothidea* (Heath and Stireman 2010). In this complex relationship, female midges collect fungal spores from the environment and store them in specialized pockets on their abdominal segments (*mycangia*). When the females oviposit on the leaves of nearby goldenrod plants, *Solidago* spp., they deposit the spores along with their eggs. As the eggs hatch and larvae begin to develop, the spores germinate and proliferate, forming a protective gall structure filled densely with fungal hyphae. The fungus both provides the developing midge insects with food and serves to protect them against parasitoids. Interestingly, the growth of the fungal hyphae requires the presence of developing midge larvae (Heath and Stireman 2010).

Many herbaceous plants create a favorable offspring site through a surprisingly common mutualism with ants; this mutualism is known as *myrmecochory* and makes use of the ants' own parental care behaviors (reviewed by Ness et al. 2009). In these taxa, maternal plants produce seeds that bear a fleshy, lipid-rich appendage called an *elaiosome* that nutritionally mimics a dead insect (L. Hughes et al. 1994). Such seeds attract carnivorous (rather than seed-eating) ants, which take the seeds to their nests and feed the elaiosomes to their young. With the elaiosomes removed, the unwanted seeds are discarded in underground chambers that provide them with a germination site that is both rich in nutrients and protected from seed-eating mammals (Lengyel et al. 2009). Since certain species of ants preferentially build nests in light gaps, plants that attract ant dispersers may also effectively place their offspring in favorable light conditions (Ness et al. 2009). Furthermore, plants that engage in myrmecochory often release their seeds asynchronously. As a result, seeds will be collected by several different ant species and dispersed to a variety of nests, rather than to a single nest, where seedlings would more strongly compete with each other (Ness et al. 2009).

Like other types of functionally adaptive plasticity, parental effects that mitigate the offspring's experience of environmental stresses may provide resilience in the face of rapid, anthropogenic environmental changes (see the Epilogue for a full discussion). For example, there is great concern about the impact of globally increasing carbon dioxide in seawater, since high carbon dioxide can disrupt key physiological processes such as oxygen transport and energy metabolism in many marine organisms (references in Miller et al. 2012). When juveniles of the anemonefish *Amphiprion melanopus* were subjected to the elevated carbon dioxide concentrations that are predicted to occur by the end of the current century, they showed decreased growth and survival, as expected. But, when parent fish were exposed as adults to these same very high levels of carbon dioxide and then their offspring were also raised in that environment, the juveniles showed none of these negative effects, apparently because of parentally transmitted epigenetic changes in the function of enzymes that influence mitochondrial metabolism or acid–base regulation (Miller et al. 2012).

CHAPTER 5

Habitat construction and functional feedbacks: How organisms modify their external conditions

5.1 An overview of organismic impacts on the environment

The effects of organisms on their environments are ubiquitous, both individually and collectively, in the short term and over eons of time. Every plant, animal, bacterium, yeast cell, or other organism takes in resources and expels waste products, occupies space, absorbs or radiates energy, and so on. Whether positive or negative with respect to the organism itself, these external impacts inevitably shape the conditions of subsequent function and evolution in that habitat. This chapter explores the nature and variety of habitat-constructing effects of organisms, identifies conceptual and management issues, and presents detailed case studies of animals, plants, and bacteria. Tracing these effects in diverse systems reveals and reiterates an essential feedback cycle from individual organismic activities (which are themselves influenced by environmental conditions) to external environments, and back to the organisms. This perspective on familiar ecological scenarios focuses attention on the reciprocal influence of phenotypic expression on environmental conditions, opening new questions framed explicitly around these fine-scale physiological, behavioral, and developmental effects and providing insights to higher-level downstream consequences.

5.1.1 An elementary example: from photosynthesis to air

Well over two billion years ago, the Earth went through the most marked environmental change in its history, a change that transformed it from a planet resembling Saturn's moon Titan in its methane-rich atmospheric haze to one with a clear, breathable, oxygen-rich envelope (Kasting and Siefert 2002; Johnson et al. 2013; and references therein). This dramatic rise in atmospheric oxygen concentration was brought about by some of the earliest-evolved, unicellular organisms, the cyanobacteria that inhabited the Earth's oceans (Beerling and Berner 2005; Taylor and McElwain 2010). Their metabolic repertoire included the complex process of photosynthesis, which fixes solar energy in a chemically usable form while generating molecular oxygen as a byproduct. The collective photosynthetic activity of these countless microscopic organisms gradually and irreversibly transformed the chemistry of the aerial environment (Figure 5.1a), such that subsequent evolution and ecology have unfolded in the context of adaptation to an oxygenated atmosphere (Lyons et al. 2014). For instance, the symbiotic origin of the eukaryotic cell, with its new capacity to carry out the far more efficient oxidative (rather than anaerobic) metabolism, is believed to have led to the evolution of multicellular organisms (Taylor and McElwain 2010 and references therein).

The ongoing activities of photosynthetic organisms continue to determine the composition of the Earth's atmosphere in the present day (although very recently human activities have also taken a major role). Because the ancestor of the biotically dominant clade that includes green algae and land plants had incorporated cyanobacteria symbiotically, both unicellular marine plankton "pastures"

Organism and Environment. First Edition. Sonia E. Sultan.
© Sonia E. Sultan. 2015. Published 2015 by Oxford University Press.

(a)

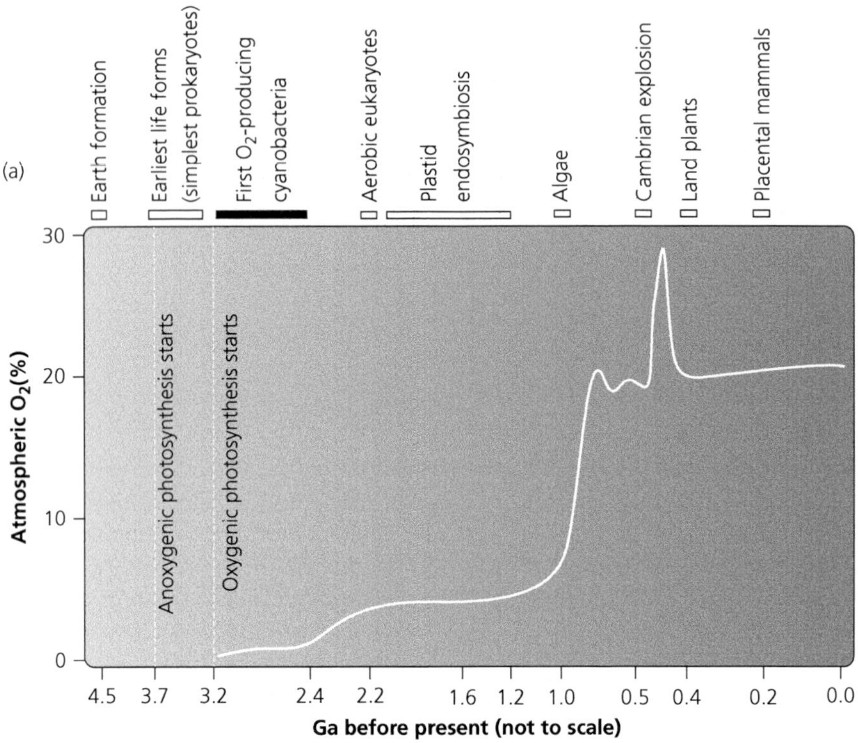

(b)

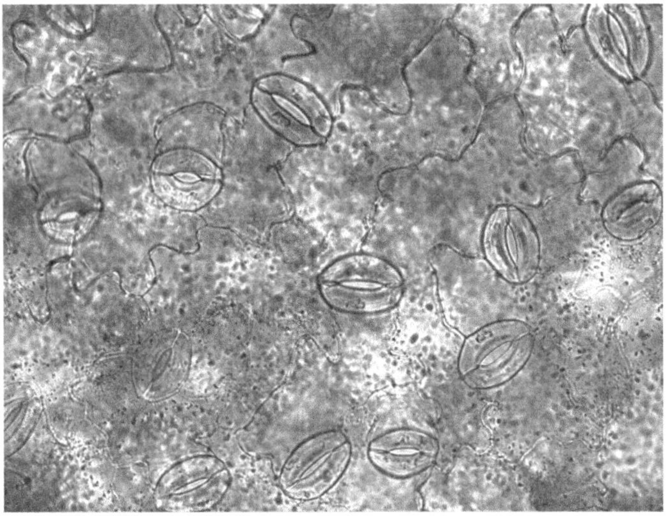

Figure 5.1 The photosynthetic activity of organisms has collectively transformed the Earth's aerial environment to one rich in oxygen. (a) A schematic showing the change in atmospheric oxygen concentration over geological time. The general association between estimated oxygen concentration changes and major evolutionary events is shown based on currently available data, with approximate dates in billions of years (Ga). Figure reproduced (with updated information) from D. Shevela, R. Y. Pishchalinikov, L. A. Eichacker and Govindjee (2013) "Oxygenic Photosynthesis in Cyanobacteria", in "Stress Biology of Cyanobacteria," edited by A. K. Srivastava, A. N. Rai, and B. A. Neilan, CRC Publishers. pp. 3–40. Amended version provided courtesy of the corresponding authors D. Shevela and Govindjee. (b) The release of oxygen from plants into the atmosphere is regulated at the cellular level by the opening and closing of stomates. Photomicrograph of a live *Brassica rapa* leaf shows several open stomates on the lower epidermal surface (400× magnification; field of view = 0.037 mm^2). Image courtesy of Dana L. Royer and Peter Martin.

and later-evolved multicellular taxa in this clade are able to carry out photosynthetic carbon fixation and produce oxygen (Kasting and Siefert 2002). Indeed, it is estimated that these organismic activities draw approximately 120×10^{15} g of carbon annually from the atmosphere's 730×10^{15} g into the primary production of plant tissues (Hetherington and Woodward 2003), leading to a profound feedback on the function and evolutionary diversification of plants and, through them, terrestrial faunas (Beerling and Berner 2005; Baldwin 2010). In addition, the release of water vapor via transpiration that is coupled to photosynthesis (an estimated $30–40 \times 10^{18}$ g of water vapor) makes an enormous contribution to atmospheric moisture content and hence to the global water cycle (Lake et al. 2002; Gerten et al. 2004; and references therein).

These vast global effects are governed by developmental and behavioral events that scale down spatially to the individual cell, and temporally to the microsecond. These events in turn are governed by environmental conditions which elicit responses at both immediate and selective timescales. Stomates (from the Greek *stoma*, or mouth; see Chapter 3, Section 3.2) are tiny epidermal structures distributed on plant shoots, primarily on leaf undersides and surfaces. Each stomate consists of a pair of guard cells that alternately swell shut or relax to reveal a central pore through which carbon dioxide enters the leaf tissue and water vapor exits (Figure 5.1b). The size and duration of stomatal opening thus regulates both the uptake of carbon dioxide for photosynthesis and the loss of water from the plant's tissues. This aperture is determined both by developmental "decisions" within the leaf epidermis regarding the size and density of stomates and by the rapid behavioral responses of the guard cells to immediate atmospheric conditions such as concentrations of carbon dioxide and water vapor.

Both these longer-term developmental effects and the moment-to-moment behavior of stomates are conditioned in highly complex ways by the plant's external and resulting internal environment (see Chapter 3, Section 3.2). The precise ways in which environmental factors enter into the developmental pathways that determine stomate size and density in the epidermis are not yet known, although both light and carbon dioxide appear to play roles (possibly integrated by MAP kinase signaling pathways; D. Bergmann 2006; Casson and Hetherington 2010). On a scale of seconds to hours, stomates open and close in response to a complex, interacting set of environmental inputs (Hetherington and Woodward 2003; also see Chapter 3, Section 3.2). Larger-scale patterns of environmental variation participate in the selective evolution of traits that affect photosynthetic function, including norms of reaction for biochemical, anatomical, allocational, and morphological responses to ambient carbon dioxide and humidity conditions, at the population and species levels. Clearly, the "intimate and bidirectional relationship" between living organisms and the composition of the planet's atmosphere (Taylor and McElwain 2010, 272) consists of feedback cycles at cellular, individual, and macro—or even mega—evolutionary and ecological scales, exemplifying the "vice versa" of environmental impact on organisms.

5.1.2 Environmental "engineering," community effects, and habitat heterogeneity

In recent years there has been a renewed interest in the measurable external impacts of organisms on natural systems (Stinchcombe and Schmitt 2006). Species whose presence and activities most dramatically modify their physical and chemical environment (i.e., its structural properties, resource levels, and flows of energy and mass) have been recognized as "ecosystem engineers" (C. Jones et al. 1994, 1997; Boogert et al. 2006; J. Wright and Jones 2006). In defining "the role that many organisms play in the creation, modification and maintenance of habitats" as *engineering* (C. Jones et al. 1994, 373; Pickett et al. 2001), ecologists have focused attention on the potentially critical feedback between these organisms and their own continuing functional context (Hastings et al. 2007). This ecological feedback is illustrated by the way photosynthetic organisms have created, and now utilize, the Earth's oxygenated environment (see Section 5.1.1). At the same time, changes that organisms make to their external environments can generate an evolutionary feedback by shaping selective pressures on both the "engineering" species and the cohabiting taxa, again as exemplified by the spectacular diversification of oxygen-exploiting organisms (Odling-Smee et al. 2003, 2013; also see Chapter 7, Section 7.4).

To some, the rather anthropomorphic term "engineering" implies, inappropriately, that organismic changes to the environment are both favorable to the organism and deliberate. This implied intentionality may stem from the fact that the founding example of ecosystem engineering was not unicellular carbon fixing but rather the habitat-building behavior of the North American beaver, *Castor canadensis*. By felling specific trees, these modest-sized mammals construct dams that transform streams into wetlands with changed patterns of sedimentation and nutrient cycling, changes which completely alter the immediate biological community and lead to greater species richness at the landscape scale (Naiman et al. 1988; J. Wright et al. 2002; discussed by C. Jones et al. 1994). Other often-cited examples of taxa that create complex new habitats likewise entail purposeful building behavior, as with the nests or structural colonies of many birds and social insects (Odling-Smee et al. 2003; Stinchcombe and Schmitt 2006). However, the critical feedback dynamics arise simply from the universal fact of organismic impacts on the environment, without either the conscious intent or the exceptional impact attributed to distinct "ecosystem engineers." A general term for these external effects, and one in keeping with a niche construction framework, is "habitat construction" (see Chapter 2, Sections 2.3.2 and 2.3.3; for a further discussion of "ecosystem engineers" as a special case, see Section 5.1.4). Note that "habitat construction" simply refers to the external consequences of an organism's presence and activities, whether favorable or unfavorable to that organism and others.

Apart from the industrious beaver, how widespread are transformative, persistent habitat-constructing effects in ecosystems? A widespread example of landscape transformation is the production of *Mima mounds*, which are uniformly spaced low hillocks from 2 m to 50 m wide that are found spread across diverse vegetation types on six continents (references in Gabet et al. 2014; Figure 5.2a, b). These geomorphological features shape soil drainage properties, the formation of vernal pools, and other aspects of terrestrial habitats. Despite their landscape prominence, longevity, and ecological importance, the origin of these mounds has been unknown for centuries; their formation often been attributed to abiotic forces such as glacial or seismic activity. However, it appears that Mima mounds are in fact the cumulative result of soil movement by small burrowing mammals such as the North American pocket gopher *Thomomys bottae* (Figure 5.2c). Computer simulations incorporating observed patterns of sediment transport by individual gophers over several generations produced strikingly similar dimensions and soil mound distribution patterns as actual Mima mounds, suggesting that these widespread topographical features originate with animals (Gabet et al. 2014). The mound-constructing activity of each animal, and consequently the size and distribution of these large-scale landscape features, is informed by individual territorial behavior, the soil drainage properties of the site, and soil moisture content at a given time. Because individual gophers are born into these sites, each generation adds to the habitat construction of the last, resulting in enormous effects over a period of several decades.

Another behaviorally mediated type of habitat construction was recently uncovered by researchers investigating the long-standing mystery of "devil's gardens" in the Amazon rainforest. In contrast to the typically diverse forest flora, these are large stands consisting only of the single species *Duroia hirsuta*, an understory tree which is a member of the coffee family and is involved mutualistically with a local species of leafcutter ant. The local Kichwa people have traditionally explained these anomalous *Duroia* stands as "gardens" cultivated by an evil forest spirit. A field manipulation experiment (Frederickson et al. 2005) showed that these distinctive forest stands are in fact created and maintained by *Myrmelachista schumanni* ants, which nest in the swollen, hollow "domatia" (from the Latin *domus*,"home") formed by *Duroia* stems and feed on the plant's leaves. *Myrmelachista schumanni* workers rapidly attack saplings of other species planted near *Duroia* trees, but never saplings of their host species. During these attacks, ants inject into sapling leaves formic acid, a broad-spectrum toxin that kills vegetative tissues. By actively maintaining and promoting "gardens" consisting entirely of *Duroia*, the ants

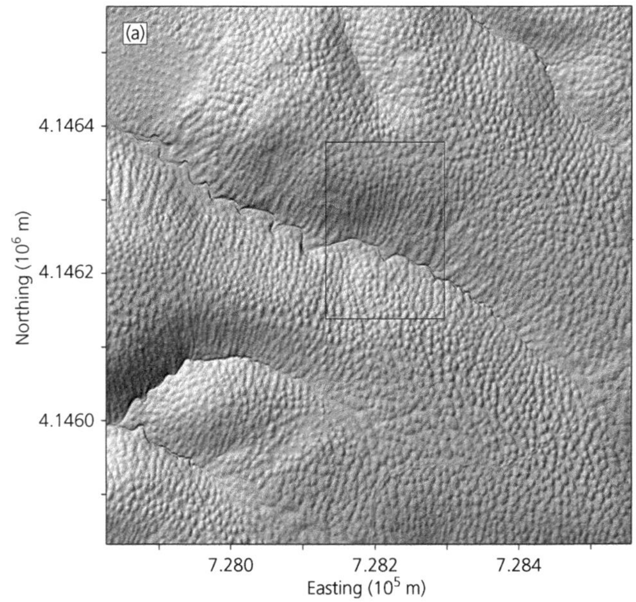

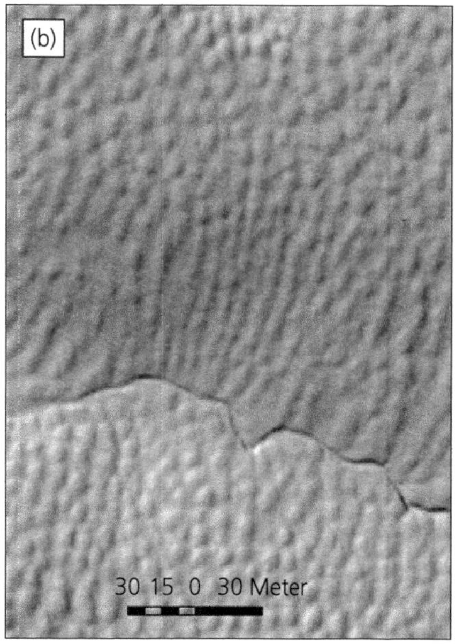

(c)

Figure 5.2 Organisms may alter many aspects of their external environments, including landscape features. A widespread example is the production of Mima mounds, uniformly spaced low soil mounds that have been found to occur on six continents. Hillshade images created from LiDAR data show (a) an area of Mima mounds near Merced, California, USA, and (b) a magnified view of the outlined portion of the area. Images reproduced by permission of Emmanuel J. Gabet, from E. J. Gabet, J. T. Perron, and D. L. Johnson, 2014, Biotic origin for Mima mounds supported by numerical modeling, *Geomorphology* 206: 58–66. (c) These large-scale topographic features appear to result from soil movement by small burrowing mammals such as the North American pocket gopher *Thomomys bottae*, shown here. Photo courtesy of John C. Avise.

provide themselves with plentiful nesting sites well into the future; these monospecific stands can persist for hundreds of years. At the same time, the *Duroia* trees gain relief from competing neighbors (at the cost of providing leaf tissue as food for the ants; Frederickson and Gordon 2007). Hence, this distinctive, ant-created type of habitat within the forest provides for the long-term success of ant colonies and creates novel conditions for other ecological interactions.

When a species alters its own environment, it changes the conditions encountered by co-occurring species (see Chapter 6, Sections 6.1, 6.2.2, and 6.3). Changes made by one species to the physical and chemical environment can have negative as well as positive ecological consequences for cohabiting species (C. Jones et al. 1997), as is all too convincingly demonstrated by the contemporary habitat-constructing activities of *Homo sapiens*. Nevertheless, many of the examples that have motivated studies of "ecosystem engineering" (i.e., habitat construction) are cases where the environmental impact of one or more ecologically dominant species creates favorable conditions for biotic neighbors, a type of ecological facilitation (sensu Bertness and Callaway 1994).

One such case is that of African bush elephants, *Loxodonta africana*, which browse on acacias and other trees in their savannah habitat (discussed by C. Jones et al. 1994; J. Wright and Jones 2006). By stripping bark, tearing and splintering branches, and creating openings within the canopy, browsing elephants cause long-lasting tree damage. Paradoxically (or at least unlike the deliberate construction of dams by beavers), this inadvertent effect of their feeding activity creates precisely the kind of arboreal crevices used as territorial nesting sites and refuges for the local lizard species, the arboreal *Lygodactylus keniensis*. A field manipulation experiment confirmed that individual lizards preferentially occupied trees with elephant damage and that sites with a greater number of elephant-damaged trees also had the highest local density of *Lygodactylus* lizards (Pringle 2008). However, since lizard nesting choices depend on several environmental factors, this positive impact on lizard individuals and populations varied depending on community context and on scale. In sites dominated by acacia trees with abandoned ant domatia on their branches, the effect of elephant damage on lizard habitat choice was weak because other potential refuges were plentiful. At the landscape scale, overall lizard densities were higher in areas where elephants and other large herbivores were *absent*, because the increased density of shrubs and leaf litter in these areas provided even more abundant lizard refuges (Pringle 2008). In general, the habitat-constructing and downstream community effects of large herbivores may be much greater than has been previously recognized (Pringle et al. 2007).

In some cases, the positive community impact of a species' habitat-altering effects can be greater in stressful environments (Crain and Bertness 2005). For example, certain perennial wetland plants such as the tussock sedge, *Carex stricta*, create raised hummocks as their shallow root systems capture sediments that promote further root proliferation. This feedback process builds up elevated masses or "tussocks" of aerated, nutrient-rich soil; these tussocks persist for decades, providing diverse plant taxa in these waterlogged habitats with a favorable long-term rooting environment (Crain and Bertness 2005). In a saline wetland, the formation of a biotic community may depend critically upon the enhanced rooting conditions provided by tussock formation, while in the more physiologically benign context of a freshwater tidal marsh, this environmental modification by sedges may have little or no effect on species distribution or diversity (demonstrated empirically by Crain and Bertness 2005; for theoretical confirmation of context dependence, see Kylafis and Loreau 2008).

In several ecological systems, the presence and activities of certain species result in physically complex environments that provide habitat for diverse taxa. Familiar examples include reef-building by marine corals and coralline algae; the growth of forests (of both woody terrestrial plants and large marine plants such as kelps); shellfish beds built by mollusks; and burrows created by marine animals such as crabs, amphipods, and fish as well as small terrestrial mammals (Coleman and Williams 2002; Odling-Smee et al. 2003). As in the case of Mima mounds, organismic activities can even shape the physical features of the landscape: plants of tidal shallows and estuaries, such as seagrasses and mangroves, can capture sediment to create stable, stratified intertidal habitats that buffer coasts from wind and wave action; beach grasses trap wind-blown sand particles so as to build dune ridges that can protect coastal terrain; dense vegetation such as forests stabilize soil systems; and populations of freshwater plants can alter the physical structure of riverbanks and modify channel dynamics so as to reduce erosion (Zarnetske et al. 2012 and references therein).

Habitat construction can also consist of changes to a species' external *biotic* environment, in some cases by engaging eco-devo pathways. A well-studied example is the effect of *Pheidole bicornis* ants on the *Piper* plants they inhabit in the shaded understory of Costa Rican rainforests. The relationship is mutualistic: the plants constitutively form tube-shaped petioles that serve as domatia to house the ants; the ants in return aggressively defend against any herbivorous insects, competing vines, and pathogenic fungi that attack their plant hosts (Risch and Rickson 1981; Fischer et al. 2002 and references therein). *Pheidole* ants also developmentally alter their *Piper* hosts so that the plants provide them with food as well as lodging. Through an unknown signaling compound, the presence of an ant colony induces epidermal cells along the margins of the host plant's leaves to enlarge up to eightfold and to fill their cytoplasm with proteins and large lipid droplets. As a result of this induced developmental and allocational change, the number of epidermal food bodies produced by a single *Piper* leaf typically increases from less than 20 to approximately 1,500 (Risch and Rickson 1981). These food bodies are a high-energy food source that is well matched to the nutritional needs of the actively patrolling ants that induce their production (Fischer et al. 2002).

In addition to specific chemical, physical, and biotic changes, organismic effects on the environment are also predicted to increase the heterogeneity or "complexity" of ecological systems, a quality that is key to system function and productivity (Pickett et al. 2001; Coleman and Williams 2002). According to this view, environmental changes made by organisms generate a more variable distribution of resources and physical structures in a habitat because of the inherent, environmentally mediated variability of individual effects. For example, in the short term, the digging activity of gophers generates spatially clustered burrows, each with an adjacent mound of loose earth. The pattern and extent of this heterogeneity will depend on a host of organismic factors such as the gopher's individual behavioral decisions, its size, its activity level, and its prior access to resources, as well as gopher population density, season, and so forth (Pickett et al. 2001). Due to such idiosyncratic factors, the spatial and temporal distributions of organism-based microsite "state changes" generate environmental heterogeneity; moreover, unlike abiotic sources of heterogeneity, such as storms or wind, these factors may touch on other complex chains of interaction with the environment (Pickett et al. 2001). Niche-constructing effects can increase heterogeneity at one spatial or temporal level but decrease it at another (Hastings et al. 2007). However, even if an organism-changed habitat patch supports fewer species than formerly, across broader scales habitat-constructing changes are predicted to generate greater environmental heterogeneity because *some* sites will be modified *some* of the time. As a result, the net effect will be increased habitat diversity and hence species richness at the regional scale, the long-term scale, or both (C. Jones et al. 1997), eventually leading to evolutionary diversification (Hastings et al. 2007).

Organisms can also enhance environmental heterogeneity through their specific responses to habitat patches or edges. For example, by concentrating their feeding activities on one side of a boundary between distinct vegetation patches, herbivorous animals can enlarge the differences between those habitat patches. Similarly, fast-growing plants might respond to gaps in a forest canopy in ways that increase the environmental differences between those gaps and the shaded understory (Pickett et al. 2001). Indeed, organisms themselves may generate as well as respond to such environmental patches and boundaries (Pickett et al. 2001). In theoretical terms, organismic effects on the environment, combined with the utilization of those environments by the organisms themselves, can create particularly complex dynamic patterns of heterogeneity, such as oscillations (Han et al. 2009). Whether or not organisms change their environments in ways that favor their own persistence, the environmental heterogeneity their activities generate can be seen as a kind of "self-organized ecological imprinting" (Han et al. 2009, p 449).

5.1.3 Implications for conservation and management

Because their activities shape the physical and resource conditions encountered by other taxa, the presence or absence of species with strong habitat-constructing effects can dramatically affect biological communities (references in Hastings et al. 2007; e.g., Karl et al. 1997; Coleman and Williams 2002; Odling-Smee et al. 2003; Crain and Bertness 2006). This insight suggests that, as a matter of environmental policy, conservation of these high-impact or "engineering" taxa should be accorded particularly high priority (Boogert et al. 2006). According to this view, effective ecosystem conservation will depend on correctly identifying the key habitat-constructing players in natural systems; in the past, only trophic "keystone species" such as high-level predators have received this kind of ecologically based special priority (Boogert et al. 2006).

A case in point is that of the tiny benthic diatoms that inhabit Canada's Bay of Fundy. In this environment, wave action is a major physical force against the sandy estuarine shores. Individual diatoms secrete filamentous carbohydrate polymers that bind sand particles, such that their cumulative impact is to greatly stabilize the substrate (Figure 5.3). As a result of this stabilization, other organisms colonize the shores, so the diatoms can be viewed as the major habitat constructors in the system (C. Jones et al. 1997). However, the ecological "keystone species" in this community is generally considered to be the migratory sandpiper *Calidris pusilla*, a bird that feeds on the amphipod *Corophium volutator*, which is the main grazing consumer of the diatoms

(a)

(b)

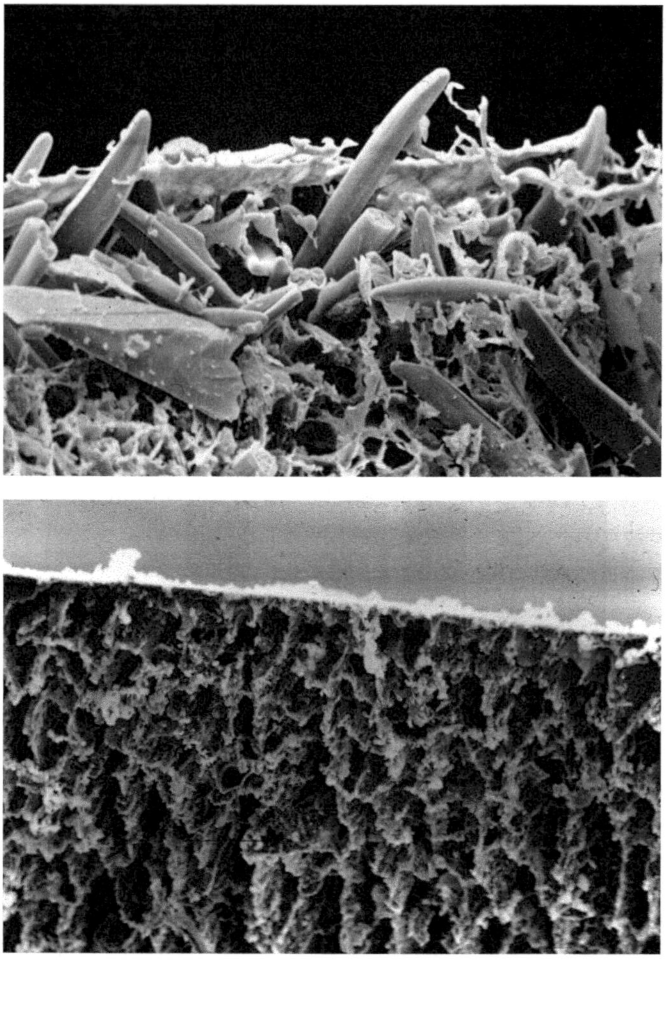

Figure 5.3 Inconspicuous organisms can play major roles in habitat construction, as in the case of benthic diatoms in Canada's Bay of Fundy. These tiny organisms produce polymeric filaments that bind together sand particles, creating a stable substrate that permits other organisms to colonize the shores. (a) With diatoms present, the mudflat structure is bound together by extracellular filaments. (b) With diatoms absent, the mudflat sediment is loose and unstable. Images courtesy of Kevin Laland and David Paterson, reproduced with permission from Boogert et al. 2006.

(Boogert et al. 2006 and references therein). Without the sandpipers, the amphipods become so incredibly abundant that diatom populations, and their substrate-stabilizing impact, are reduced; thus, the predatory birds are indeed critical to community structure. But the actual environmental modification that creates the habitat for this community is accomplished by the diatoms (C. Jones et al. 1997; Boogert et al. 2006), so conservation efforts must focus on protecting them as well.

The fact that habitat construction can arise from the collective effects of numerous individuals leads to a further conservation insight. The impact of organisms in suitably, and sufficiently, modifying external conditions to support biological communities may require a certain threshold population size. For example, the diatoms described above must be present at a certain abundance and density to stabilize the substrate sufficiently for their own species, and others, to inhabit the Bay of Fundy's shores. If a given population is reduced below this threshold size, perhaps due to habitat loss or fragmentation, such cumulative ameliorations of the environment may not occur. As a result, a small or disturbed-site population could go extinct, even in an environment that species ordinarily could tolerate, because it lacks the positive functional feedback that a larger population could create through more substantial environmental effects (Kylafis and Loreau 2008). Moreover, the loss of the main habitat-constructing population in an ecological community could destabilize other species that depend upon the environmental modifications. Clearly, the loss of one of these high-impact species from a community can also abruptly change the selection pressures operating on other organisms in the system (Boogert et al. 2006).

Along with these implications for ecosystem conservation, the potential habitat-constructing impact of exploited populations such as fisheries merits an important role in management decisions (Coleman and Williams 2002). At present, however, even ecologically informed management policies may overlook a species' impact on the environment and consequently on the larger system. Although it is recognized that overfishing of top predators such as the tilefish *Lopholatilus chamaeleonticeps* and the grouper *Epinephelus flavolimbatus* will result in trophic effects on marine communities, for instance, the far more wide-ranging ecological impact of these species as habitat builders has not been considered (Coleman and Williams 2002). These are burrowing fish, whose excavations directly provide habitat used by a wide range of invertebrate and fish taxa, such that tilefish and grouper populations correlate positively with the abundance of diverse marine animals (Coleman and Williams 2002). Moreover, their burrows also promote the breakdown of organic matter in sea-floor sediments, increasing microbial populations and facilitating the movement of mineralized nutrients into the water column. Due to their habitat-constructing impact as burrowers, reduced stocks of tilefish and grouper will alter the biogeochemistry of the entire system, as well as reduce its biological diversity (Coleman and Williams 2002).

The ecological impact of biological introductions will also depend in part on nonnative species' habitat-constructing effects. Just as the loss of a key environment-modifying species can alter a community, introducing into an existing community a novel species, with its distinctive effects on the environment, can result in dramatic changes to habitat structure, resource availability, and, ultimately, community composition (reviewed by Crooks 2002). Such a case was investigated by Zarnetske et al. (2012) in the coastal dunes of the Pacific Northwest of North America. In this system, the sparse native beach grass *Elymus mollis* has recently been replaced by more densely growing, nonnative, invasive beach grasses from the genus *Ammophila*. Because *Ammophila* plants have a different growth habit with greater tiller (shoot) density, they more effectively capture sand particles; as a result the beach-grass invasion has resulted in a transition from open, shifting sand dunes to more stable, continuous foredune ridges, changing the topography and biodiversity of these coastal areas (Zarnetske et al. 2012).

Conversely, when the environmental impact of an introduced organism is similar to that of a native species, its replacement of the native can be relatively "benign" (Odling-Smee et al. 2003), in the sense that it does not destabilize the rest of the ecosystem. An example is the introduction of an alien (Sonoran desert) species of pocket mouse, *Chaetodipus baileyi*, into Chihuahuan desert sites in southwestern North America. The introduced pocket mice are similar in size and diet to a seed-eating kangaroo rat, *Dipodomys merriami*, that had been experimentally removed two decades earlier, so the effect of the introduction has been to restore the system's energy flow through granivores (Ernest and Brown 2001; discussed by Boogert et al. 2006).

5.1.4 An inclusive view of environmental engineering

As noted by Hastings et al. (2007), the recognition that organisms can profoundly alter their physical and chemical environments is far from new. Several nineteenth-century works of natural history, including Darwin's charmingly titled *The Formation of Vegetable Mould through the Action of Worms, with Observations on their Habits* (1881), explicitly investigated how animals and plants modify the soil they inhabit (see Sections 5.2 and 5.3). Frederic Clements's pathbreaking 1916 book on ecological succession examined how plants change their immediate abiotic environments in ways that regulate both their own persistence and the composition of their local communities (references in Hastings et al. 2007).

If attention to organismic impacts echoes an earlier insight, that is not in itself a conceptual weakness or an objection to renewed investigation. In this case, as with ecological development and norm of reaction studies (see Chapter 1, Section 1.3), a return to more inclusive, earlier approaches that better accommodate current data can constitute important progress. What is problematic, however, is whether it is meaningful to distinguish only those taxa that most strongly alter their environments as "engineers," given that any organism inevitably alters its environment to some extent (Reichman and Seabloom 2002). This conceptual difficulty is exemplified in published lists of "ecosystem engineers" that offer a seemingly arbitrary selection of examples out of countless potential cases (e.g., C. Jones et al. 1994; Odling-Smee et al. 2003). Is this separate category useful? What is the threshold between subtle environmental impacts that can presumably be safely ignored and modifications that lead to significant functional and selective feedback effects? This connects to a key dilemma in coevolution studies: while pragmatism dictates focusing on the subsets of a community's taxa that are engaged in strong (often pairwise) interactions, even those direct interactions are shaped by more diffuse, multispecies effects that must therefore be taken into account (Inouye and Stinchcombe 2001). Yet, practically speaking, how can environmental impacts be studied if they emanate from all species? In light of these questions, what strategies can be employed to identify, and ultimately understand, the ecological and evolutionary feedbacks generated by the activities of organisms?

One, often implicit, approach to refining the concept of "engineers" is based on scale: for instance, to recognize as "engineering" only those organismic effects that occur at a greater spatial scale, or last longer than, direct biotic effects such as predation (C. Jones et al. 1994; Hastings et al. 2007). Other authors do include immediate trophic effects (Odling-Smee et al. 2003) but consider as "engineering" only those environmental impacts that lead specifically to increased organismic abundance and diversity (references in Pringle 2008; Odling-Smee et al. 2013).

A third approach is to consider as habitat construction *all* effects of organisms on their environments, from the stunning chemical impact of global photosynthesis and the creation of vast coastal mangrove habitats, to the breaking of a tree's branches and the resulting formation of usable lizard refuges. Within that unified context, the type, extent, and duration of species' effects can be quantified and compared in various ecological settings. This inclusive approach may promote recognition of both ecologically important but less apparent environmental effects and combinatorial and context-dependent impacts (of species or groups of species) that would be missed by focusing solely on recognized "engineers." Starting with one such well-known case and proceeding to the less obvious, the following detailed examples show that, whether strong or subtle, cumulative or short term, the effects of taxa on their external environments can be studied in ways that enrich our insight into the organism–environment relationship. To date, however, studies of these environmental effects have seldom extended to their functional or evolutionary consequences (e.g., see Chapter 7, Section 7.4). Precisely how to encompass this causal reciprocity in empirical work and models of ecological interactions, ecosystem processes, and selective evolution remains a considerable challenge.

5.2 Earthworms and soil quality

As first documented in Darwin's 1881 book, the earthworm is a classic example of an organism that profoundly alters its external environment, in this case, the "earth" or soil it inhabits (see also C. Jones et al. 1994). *Earthworm* is the common name for a number of species in the oligochaete subgroup of annelid worms. These species are unusual in being terrestrial, as the vast majority of annelid species inhabit either freshwater or marine environments. In fact, because they are phylogenetically constrained to be "physiologically aquatic" animals (J. Turner 2000; on which the following discussion is based), earthworms must effectively transform their terrestrial habitat in order to exist.

In functional terms, these small creatures are ill-suited to the central stress that characterizes life out of water—desiccation. Like freshwater oligochaetes, earthworms produce large quantities of relatively dilute urine, losing 60%–90% of their body weight in water daily (compared with less than 10% in terrestrial mammals such as humans). As a result, earthworms require an environment that is not too drying, and from which water (as well as oxygen and food) can easily be extracted. It is precisely this kind of moist, aerated, nutrient-rich soil environment that earthworms create in terrestrial habitats, through the humble but collectively powerful daily activities described below. In so doing, they shape the living conditions of countless other terrestrial organisms and microorganisms.

Earthworms tunnel through soil, compacting it and leaving a coating of polysaccharide-rich mucus that acts as an adhesive to aggregate soil particles and provides a ready carbon source that promotes microbial activity (Lavelle 1988; Bossuyt et al. 2005). Their production of this mucus is surprisingly copious: 1 g of earthworms can produce an average of 5.6 mg (dry weight) of these skin secretions per day (Pan et al. 2010 and references therein). As they tunnel, earthworms ingest bits of soil, digesting its constituent organic matter and bacteria and eliminating fecal pellets that are permeated with gut secretions and calcium carbonate (Daniel and Anderson 1992; J. Turner 2000). This complex of digested soil particles, calcium carbonate, and organic secretions dries in the form of durable macroaggregate "casts" that substantially enhance soil fertility (Syers and Springett 1984; Lavelle 1988; additional references in Bossuyt et al. 2005). For example, these fecal castings have higher moisture content, increased microbial activity, and greater concentrations of soluble organic carbon than the same soils that have not passed through a worm gut (Daniel and Anderson 1992). At the same time, worms incorporate into soils additional organic materials such as decomposing bits of leaf litter, gathering them from the soil surface and bringing them into their tunnels to use as an eventual food source (J. Turner 2000; Dempsey et al. 2013). Laboratory studies using radioactively labeled leaf material show that these activities result in the production, from rapidly incorporated organic residues, of large soil macroaggregates with elevated amounts of total carbon (Bossuyt et al. 2005). Interestingly, these macroaggregates within worm casts contain microaggregates in which much of this newly incorporated carbon is held in a protected form, possibly providing long-term stabilization of soil carbon (Bossuyt et al. 2005).

These activities enhance soil nutrient content and promote more rapid nutrient and carbohydrate cycling (Lavelle 1988; Dempsey et al. 2013). By building structurally aggregated soils, earthworm activities also transform the soil's physical and hence water-holding properties. If not for this aggregating action, soils would weather into increasingly fine particles, becoming dense matrices of clay to which water molecules would be very tightly bound and hence biologically unavailable (J. Turner 2000). One evident effect of earthworm activities is to increase the penetrability of soils to plant roots, arthropods, and fungal hyphae (Syers and Springett 1984). This structural effect also improves gas exchange between the soil and the aerial atmosphere (Bossuyt et al. 2005). In addition, worm burrowing activity enhances the moisture content of the soil by creating large structural macropores that make it easier for rain to infiltrate and allow a given soil volume to hold more water (J. Turner 2000). Another key physical effect of aggregation is the formation of larger soil micropores; these hold moisture less tightly, reducing the osmotic forces that could otherwise draw water out of worm bodies.

Together, earthworm activities result in the production of organically rich, structurally aggregated soil—Darwin's highly fertile "vegetable mould." Through their own transformative activities, these small animals "develop, maintain and expand" a soil zone that suits their aquatic physiology (and which can even be viewed as an external organ for water balance): a habitat where rain infiltrates readily; where plenty of water is held, but not so tightly as to be unavailable; and where the pores allow both high humidity and abundant oxygen (J. Turner 2000, 118). In the process, the earthworms create a fertile soil horizon for other soil inhabitants such as terrestrial plants and soil microorganisms. This understanding of how earthworm activities mediate soil conditions could perhaps be expanded to include functional and selective feedback effects, by testing individual and population consequences of soils with various levels of prior mediation by earthworms.

The prodigious soil-transforming capabilities of earthworms may provide urgently needed new benefits to terrestrial ecosystems in the future. Recent work has revealed that earthworm mucus has a high affinity for organic molecules such as certain insecticides and that earthworm activities can change the distribution and biological availability of toxic heavy metal ions (Pan et al. 2010 and references therein). These interactions with soil contaminants raise the intriguing possibility that earthworms may be employed as agents of bioremediation for soils that have been chemically altered by the habitat-constructing activities of human beings.

5.3 Habitat construction by terrestrial plants

It is well known that plants are highly plastic developmentally, adjusting the number, form, and structure of their organs in response to environmental conditions. The converse is equally true: although their habitat-constructing activities may be less conspicuous than those of beavers and elephants, plants substantially shape their external environments both above and below the ground, beyond the simple effects of resource uptake (and in addition to photosynthetic oxygen production; see Section 5.1.1). Through the processes detailed below, individual plants modify the soil surface microclimate, shape the physical, chemical, and biotic properties of the soil, and create spatial patterns of water and chemical deposition. Recently, evolutionary ecologists have begun to focus not only on these external changes but on how these changes in turn affect plants and cohabiting species (e.g., Ridenour and Callaway 2001; Stinchcombe and Schmitt 2006; Wolkovich 2010; see also the discussion of plant flammability and fire regimes in Chapter 2, Section 2.3.3). In view of the diverse abiotic and biotic effects noted below, such studies (under both field and controlled conditions) could provide a wealth of new insights into ecological and evolutionary dynamics in terrestrial systems.

These external impacts scale according to plant size and longevity: individual trees typically alter their surroundings across a 5–15 m spatial radius and at a temporal scale from decades to centuries or, in some cases, millennia (Binkley and Giardina 1998). At the level of plant communities, different types of vegetation create characteristic environmental impacts. For instance, the canopy formed collectively by forest trees (i) intercepts solar energy and radiant heat to maintain cooler daytime and warmer nighttime soil surface temperatures, (ii) leads to more intensive water cycling due to heavy evapotranspiration, and (iii) provides high inputs of organic matter to the soil and its microbial communities, thus supplying a sustained nutrient supply (Binkley and Giardina 1998).

Such collective effects create a generally positive feedback loop through which plants promote conditions favorable to their own growth. However, certain plant taxa or communities can create conditions that instead promote their replacement by other types of plants. A familiar example is the deep shade cast by pine trees of the north temperate region. This environment provides too little light for the growth of pine seedlings but just the right conditions for the shade-tolerant seedlings of several broad-leaf deciduous trees. It is this type of self-suppressing habitat modification that provides the motive force for ecological succession (Bazzaz 1979).

5.3.1 Above-ground habitat construction by plant shoots

By intercepting solar radiation, plant shoot systems determine both the quantity and spectral quality of light that reaches their less lofty neighbors and their own lower leaves (H. Smith 1982; Valladares and Niinemets 2008; and references therein). These effects on light conditions are determined by plant variation at several levels: species-specific aspects of shoot and leaf form, individual height, architecture, and age, as well as the size, angle, and anatomy of leaves (Falster and Westoby 2003; Bartemucci et al. 2006). The interception of solar energy also creates altered temperature and humidity conditions within and beneath plant shoot systems (Fitter and Hay 2002). The light attenuation and associated microclimatic changes below plant canopies modulate the germination and early growth environment for their own, and their neighbors', offspring (Bazzaz 1996). As a result of these effects, plant community composition and dynamics are a cause as well as a consequence of variation in light conditions (Valladares and Niinemets 2008). Living plant shoots also shape moisture conditions for themselves and their neighbors, both through evapotranspiration from leaf surfaces (as this process elevates ambient humidity) and by redistributing rainfall as it falls through the complex layers of the shoot system (Fitter and Hay 2002).

Individual plants further alter the microclimate of the nearby soil surface by seasonally depositing vegetative litter. This litter consists primarily of dead leaves but includes as well dried reproductive parts and broken bits of twig and bark (March and Watson 2007; Figure 5.4). This layer of plant detritus physically reduces the amount of direct light that reaches the soil surface zone and increases its humidity and temperature, influencing plant germination and seedling development (Stinchcombe and Schmitt 2006 and references therein). Even small amounts of plant litter can increase soil moisture and thus indirectly improve plant growth conditions (Foster and Gross 1998). By reducing light levels, the deposition of leaf litter can influence community composition as well, either directly or via the suppression of otherwise dominant competitors (Facelli 1994; Foster and Gross 1998).

Figure 5.4 Vegetative litter deposited by plants, such as dead leaves and broken twigs, strongly affects light, temperature, humidity, and nutrient cycling at and below the soil surface. The timing, amount, and tissue composition of litter deposits have consequences for seed germination, nutrient cycling, and soil biota that scale up to the ecosystem level. Photo of autumn leaf litter in a mixed deciduous forest (Four Marks, Hampshire, UK), courtesy of Chris Rose, http://fourmarksbirding.blogspot.com; for the color image, see Plate 16.

Beyond these straightforward physical effects, plant litter plays a surprisingly important role in shaping nutrient conditions below the ground. The decomposition of these discarded vegetative parts (whether or not hastened by earthworm burial as described above) creates nutrient inputs near the plant, including pulses of released carbon and nitrogen, that alter the structure, chemistry, and biota of the soil (Madritch and Hunter 2005). In conjunction with the physical impacts of increased temperature and soil moisture, these chemical inputs speed up rates of mineralization, promoting arthropod and bacterial populations and ultimately cycling back into vegetation (Hodge 2004; March and Watson 2007). Accordingly, the deposition of litter is a critical determinant of nutrient availability and cycling in natural communities, with profound effects on ecosystem productivity and diversity (Wolkovich 2010).

Since tissue decomposition rates are influenced by tannin concentration and other specific aspects of plant secondary chemistry, the chemical effects of vegetative litter vary depending on plant species, population, and genotype (Madritch and Hunter 2005). Indeed, one indirect consequence of reduced biodiversity in plant communities (and of altered atmospheric chemistry, soil conditions, and other factors that influence plant tissue composition) may be changes in mineral cycles resulting from altered leaf litter chemistry (discussed in Madritch and Hunter 2005). Conversely, the introduction of new plant species to terrestrial communities can result in either negative or facilitative effects via their leaf litter, a generally ignored aspect of biological invasion. For instance, the invasion of a shrub-dominated, semiarid California coastal area by a non-native annual grass increased soil moisture in the system, as formerly bare soil was covered by grass plants and their copious leaf litter (Wolkovich 2010). In this moisture-limited system, leaf litter deposits by an introduced plant enhanced both the growth of the native flora and that of the associated arthropod community of shrub herbivores and their spider predators (Wolkovich 2010).

In some cases, leaf litter can influence community composition by means of specific allelopathic effects. Chemicals that leach from the fallen leaves of the sub-Arctic heath plant *Empetrum hermaphroditum* have been experimentally shown to suppress the growth of pine seedlings, for instance (Gallet et al. 1999; additional references in Ridenour and Callaway 2001).

One well-studied example of the habitat-constructing effects of leaf litter is the Australian mistletoe plant *Amyema miquelii*, whose distinctive life history causes marked changes in ecosystem nutrient dynamics. This endemic hemiparasite and its congener *Amyema pendula* are commonly found growing on the branches of various dominant species of *Eucalyptus* tree across Australia, a region with generally nutrient-poor soils (Figure 5.5). Although individual *Amyema miquelii* plants are small relative to the host tree's canopy, their extremely rapid leaf turnover results in a very high production of leaf litter (March and Watson 2007). Indeed, much of the live leaf biomass of the mistletoe plants is deposited as litter every year, since their leaves live only 1.5 years on average (compared with over 4 years for *Eucalyptus* leaves; March and Watson 2007). As a result of this rapid turnover, the leaf litter below host trees consists disproportionately of the leaves of their mistletoe parasites, and the total amount of leaf litter beneath trees that host mistletoes is more than double that of mistletoe-free trees (March and Watson 2007).

Like other parasitic and hemiparasitic plants, mistletoes accumulate and concentrate the host's resources in their own tissues (Press 1998 and references therein). Consequently, mistletoe leaf litter contains higher concentrations of nitrogen, phosphorous, and other minerals than does litter from *Eucalyptus* trees (the amount and nutrient content of which is not significantly affected by mistletoe presence; March and Watson 2010). The greater quality and quantity of mistletoe leaf litter combine to create extremely nutrient-rich patches directly below host trees. In many such patches, twice the nitrogen, four times the phosphorous, and ten times the potassium are returned to the soil than is returned via litter from *Eucalyptus* trees alone; the mineral returns can be even higher depending on the size and number of mistletoe plants present (March and Watson 2010). These mistletoe litter inputs substantially alter the dynamics of nutrient cycling in this system, as minerals previously held in the biomass of long-lived woody host plants are rapidly

(a)

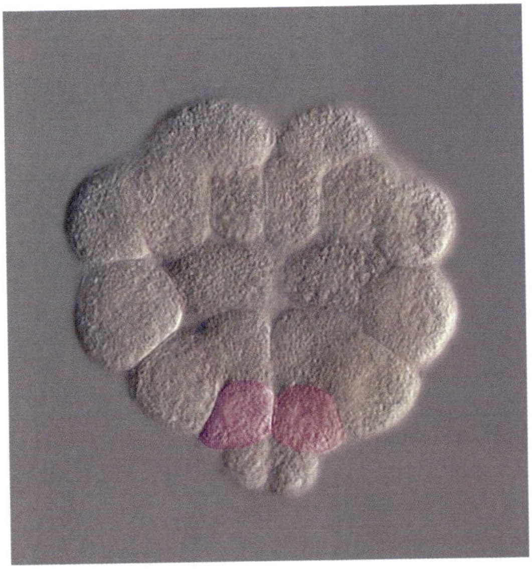

(b)

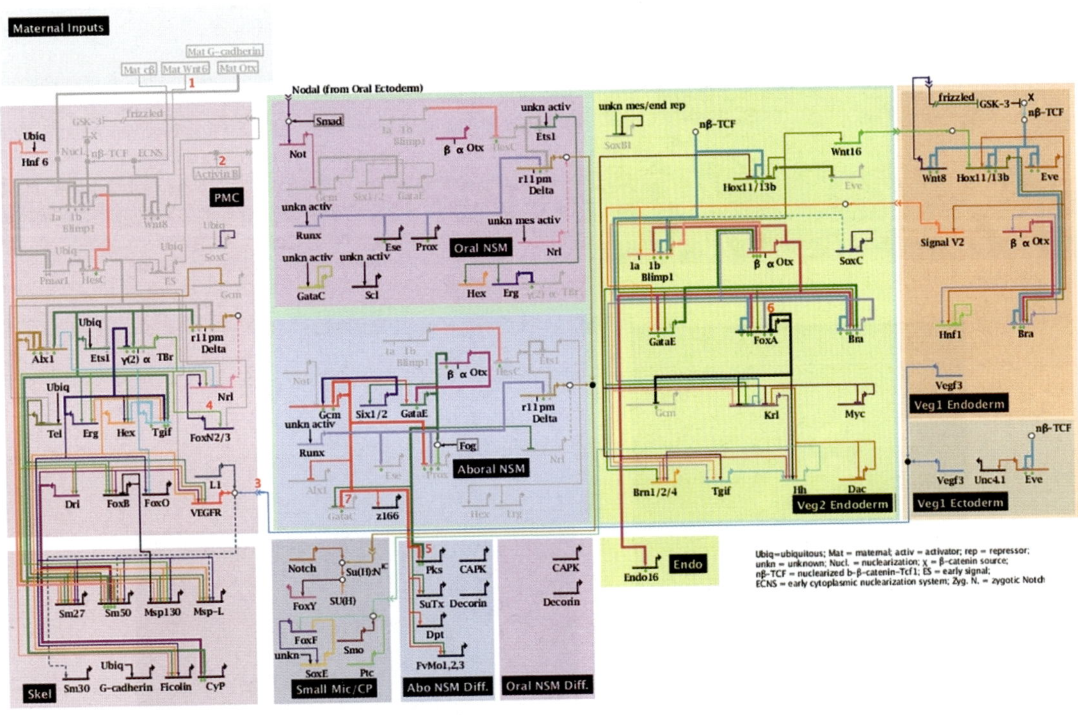

Plate 1 A sea urchin embryo (a) is shown at a very early (sixth-cleavage) stage of development. Four of the embryo's micromere cells (two of which are visible here, stained pink) give rise to the endomesoderm cell lineage that subsequently differentiates into the animal's gut lining and skeletal rods. Image courtesy of Andrew Ransick, California Institute of Technology. In (b), a data-based network model shows the complex *cis*-regulatory gene interactions that lead to the differentiation of these endomesoderm cells. Arrows and solid lines indicate gene activation or repression; rectangles show downstream differentiation genes. The signaling pathways that activate *endo-16* expression are shown in the yellow panel. Details are given by Davidson et al. (2002) and Peter and Davidson (2010, 2011). Image downloaded 2/22/2015 from the Davidson Lab website (http://sugp.caltech.edu/endomes/); reproduced (with modified caption) by permission of Eric Davidson, California Institute of Technology (copyright Hamid Bolouri and Eric Davidson).

Plate 2 The kleptoplastic (plastid-grabbing) green sea slug *Elysia chlorotica* is a brilliant green color, owing to chloroplasts it acquires as a juvenile by feeding on the alga *Vaucheria litorea*. These ingested chloroplasts are incorporated as functional organelles into the sea slug's own tissues, allowing it to support itself through active photosynthesis for the duration of its 10-month life span. Mature adults such as the one shown here are typically 2–3 cm in length. Patrick J. Krug, photo credit and copyright.

Plate 3 The desert rhubarb, *Rheum palaestinum*, modifies its environment to mitigate a key stress. Due to the unusual shape, thick cuticle, and channeled surface of the leaves, over the course of a year *R. palaestinum* plants (which are endemic to the central Negev desert) each collect and deposit over four liters of rainwater around their own roots. As a result, these "self-irrigating" plants inhabit soil moisture conditions equal to those found in climates that receive nearly six times as much rain as the Negev (details from Lev-Yadun et al. 2009). Photo courtesy of Gidi Ne'eman, University of Haifa-Oranim.

(a)

(b)

Plate 4 A parasitic fungus radically alters the body of its plant host in a biotic form of niche construction. (a) An uninfected individual of the alpine meadow plant *Boechera stricta* (Brassicaceae), with the species' characteristic pale, cross-shaped flowers. (b) A *B. stricta* plant that is infected with the parasitic rust fungus *Puccinia monoica*. The fungus causes the plant to produce elongated, bright yellow shoots instead of the plant's own reproductive structures. The surfaces of these insect-attracting "pseudoflowers" are covered with fungal spermatogonia. Photos courtesy of Tom Lesback.

Plate 5 Ecological developmental responses such as leaf plasticity can change the environment that an organism experiences. (a) Under high-light glasshouse conditions (100% of summer sun), *Polygonum persicaria* plants produce narrow, thick "sun" leaves that minimize transpirational water losses from the leaf surface. (b) When grown in reduced light (24% of summer sun), genetically identical plants instead produce broad, thin "shade" leaves. Because the expanded surface area of these "shade" leaves catches more photons, this developmental response increases the amount of available light. Images are shown at the same scale (leaves are approximately 8 cm long). Photo courtesy of Dan B. Sloan and S. E. Sultan.

Plate 6 Individuals of the desert locust, *Schistocerca gregaria*, develop into one of two alternative forms: the slow-moving, pale-green *solitarious* morph, at left, or the darkly pigmented, highly active *gregarious* morph, at right. The gregarious individuals form cohesive, migratory flying swarms. This dramatic morphological and behavioral change is induced by visual and tactile cues that indicate high conspecific density. Photo of laboratory-reared individuals courtesy of Gregory A. Sword.

(a)

(b)

Plate 7 Social encounters cue alternative color and behavioral morphs in the cichlid fish *Astatotilapia burtoni*. Male individuals can rapidly and reversibly switch between a cryptic, nonreproductive "subordinate" phenotype (a) and a conspicuously colored, aggressively territorial "dominant" phenotype (b), depending on their success in social interactions with other males (details in Burmeister et al. 2005). Photos courtesy of Sabrina Burmeister.

Plate 8 In response to herbivory, some plants release volatile compounds that serve as indirect plant defenses by attracting the herbivore's natural enemies. Here, a grain aphid, *Sitobion avenae*, feeding on an unidentified grass leaf is being parasitized by a female *Aphidius rhopalosiphi* wasp. These wasps, which are specialist parasitoids of aphids, are attracted to volatile chemicals released by wheat and other grasses on which the aphids feed (Wickremasinghe and Emden 1992). Photo copyright Sonia Dourlot.

Plate 9 Plants have evolved to await specific cues that signal suitable conditions for a seedling's emergence and early growth—a temporal form of habitat choice. Heat or smoke cue germination in many tree species of fire-prone habitats, such as the pitch pine, *Pinus rigida*. This *P. rigida* seedling has emerged following a burn in the New Jersey Pine Barrens. Photo courtesy of Mary A. Leck.

Plate 10 The jojoba shrub, *Simmondsia chinensis*, which is native to the American desert Southwest, orients leaves vertically to reduce midday heat load. Photo credit and copyright John R. Seiler.

Plate 11 The skunk cabbage, *Symplocarpus foetidus*, emerges in very early spring in Eastern North American wetlands. At low ambient temperatures, flowers enclosed within the emerging spadix (flowering shoot) raise oxygen production, actively generating heat that maintains favorable tissue temperatures. Photo credit and copyright Michael Lynch.

Plate 12 Flowers of the alpine plant *Ranunculus montana* show heliotropic orientation movements that create a warm microsite within the flower despite cold air, maintaining favorable temperatures for ovary development and for attracting insect pollinators. Photo credit Andrea Moro (Univ. of Trieste), CC-BY-SA.

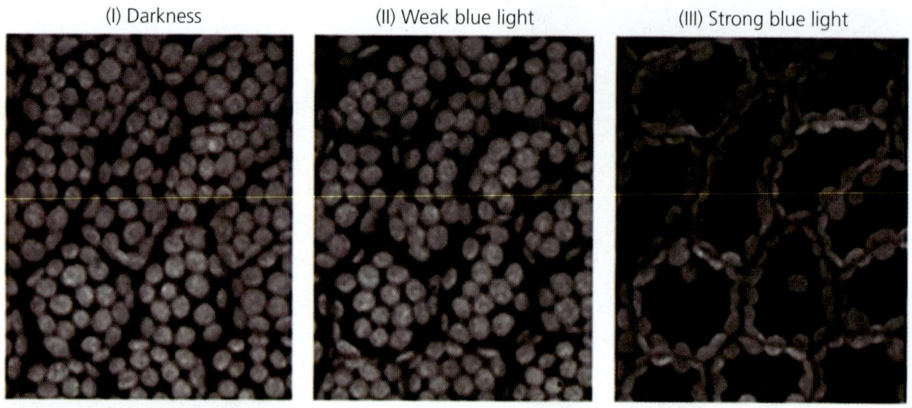

Plate 13 In darkness (I) or weak light (II), chloroplasts cover cell surfaces, maximizing photon interception; in strong light (III), they move to cell margins, thus allowing solar energy to pass through the cell. Image courtesy of Mark Fricker.

Plate 14 Because light-induced chloroplast movements alter light transmission, they can be used to reproduce a high-contrast image (such as a photograph) within leaf tissue. After approximately 45 minutes of blue-light illumination, the areas where light reached the leaf will be more transparent than those covered by dark parts of the image. A portrait of the pioneering plant physiologist Julius von Sachs is here reproduced in a live coleus leaf. In his 1887 *Lectures on the Physiology of Plants*, Sachs was the first to observe that a "light-picture" could be created in living leaf tissue. Leaf portrait courtesy of Roger Hangarter.

Plate 15 Maternal plants in some species practice parental care by burying their developing progeny. Fruit stalks of the newly discovered geocarpic annual *Spigelia genuflexa* curve downward and elongate to deposit developing fruits in the moist mossy layer on the soil surface. The plant is known only from the sandy Atlantic forest habitat of northeastern Bahia, Brazil. Photograph by Alex V. Popovkin, CC BY 3.0.

Plate 16 Vegetative litter deposited by plants, such as dead leaves and broken twigs, strongly affects light, temperature, humidity, and nutrient cycling at and below the soil surface. The timing, amount, and tissue composition of litter deposits have consequences for seed germination, nutrient cycling, and soil biota that scale up to the ecosystem level. Photo of autumn leaf litter in a mixed deciduous forest (Four Marks, Hampshire, UK), courtesy of Chris Rose, http://fourmarksbirding.blogspot.com.

Plate 17 At exposed montane sites such as this limestone cap (Cataract Mountain, Glacier National Park), closely spaced neighboring plants facilitate each others' survival and growth by mitigating the cold, dry conditions. Interactions among heterospecific individuals may be predominantly positive rather than competitive in such stressful habitats, while in more favorable conditions individuals may instead compete with close neighbors for resources. Photo courtesy of R. Callaway, Glacier National Park, Montana, USA.

Plate 18 Plastic adjustments by individual organisms may play important roles in structuring ecological communities. In the presence of a predatory hunting spider, the herbivorous grasshopper *Melanoplus femurrubrum* will switch from its preferred host plant, a grass, to the less nutritious goldenrod *Solidago rugosa*, a densely growing plant in which it is more well hidden (as shown). This change in foraging behavior diminishes the abundance of goldenrod and consequently that plant's soil-shading effect; this promotes a different community of herbaceous plants that produce different leaf litter, resulting ultimately in changes to nutrient cycling. Photo courtesy of Colin Donihue.

Plate 19 Reef-building corals collectively produce structurally complex underwater habitats that support diverse communities. The coral polyps themselves provide habitat for unicellular photosynthetic symbionts, zooxanthellae, which are major primary producers. Increased seawater temperature, pollutants, and other stresses can cause corals to expel these pigmented symbionts, causing the coral to appear "bleached" and, eventually, to die. This photograph of a reef in Hawai'i shows its three-dimensional complexity; the dark-colored corals retain living symbionts, while the "bleached" coral (top right) has lost its zooxanthellae. Photograph by Raphael Ritson-William, courtesy of Ruth Gates.

Figure 5.5 Australian mistletoe plants (*Amyema miquelii* and *Amyema pendula*) are hemiparasites of the native *Eucalyptus* trees. The mistletoes redistribute soil nutrients both spatially and temporally, acting as vegetative "Robin Hoods" by accumulating resources from their host trees and depositing them via leaf litter. Although individual mistletoe plants are small relative to the host tree's canopy, their extremely rapid leaf turnover results in very high leaf litter deposition rates. Photo of *Amyema pendula* on *Eucalyptus* sp. (Kimberly region, Western Australia), courtesy of Clare Morton.

returned to the soil. Moreover, the seasonal timing of mistletoe litterfall occurs primarily when their *Eucalyptus* host trees produce the least litter, so mistletoes also extend the period of nutrient inputs to soil (March and Watson 2007).

Hemiparasitic Australian mistletoes thus can be seen as vegetative "Robin Hoods" that raise soil productivity by spatially and temporally redistributing nutrients, creating locally rich soils that promote the growth of smaller organisms (including tree seedlings) and extending the seasonal availability of key minerals (Press 1998; March and Watson 2010). Root hemiparasites of nutrient-poor grasslands and sub-Arctic communities apparently play a similar role in increasing soil productivity through enhanced nutrient cycling (references in March and Watson 2007). It may be that this habitat-constructing role is a key to the ecology of plant–parasite interactions; the 3,000 to 4,000 known species of parasitic and hemiparasitic flowering plants are indeed most common in nutrient-limited ecosystems (Press 1998). More broadly, the spatial and temporal redistribution of mineral nutrients and organic matter from long-lived plants via their depositions of vegetative litter is one way that plants in general influence their environments.

5.3.2 Below-ground habitat construction by plant roots

The presence and activities of plant roots powerfully shape the physical, chemical, and biological processes that create and condition the roots' external environment, the soil (reviewed by Hinsinger et al. 2009; Philippot et al. 2013). In recognition of this organism–environment complex, the relatively shallow soil zone in which physiologically active plant roots occur is termed the "domain of roots" or *rhizosphere* (a term coined in 1904 by the German botanist and soil ecologist Lorenz Hiltner, as discussed by Hartmann et al. 2008). As Hiltner emphasized, this small volume of soil comprises a dynamic milieu which supports an extraordinary abundance and diversity of invertebrate, fungal, and bacterial soil inhabitants along with the plant roots with which they interact: a single gram of soil can contain 10^{10} bacteria, 10^4 unicellular eukaryotes, 10^4 nematode worms, and up to 25 km of fungal hyphae (references in Hinsinger et al. 2009). Recent work has revealed highly complex rhizosphere "signal traffic" between roots and soil microorganisms or fauna, between roots within a single plant's root system, and between the roots of one plant and those of others (Philippot et al. 2013). The functional key to this hidden ecosystem is the soil matrix, with its enormous surface area, complex pore structure, and transient patches of water, air, and minerals. These critical soil properties, which strongly influence the morphology, deployment, and physiological activities of roots (see Chapter 4, Section 4.3.2), are in turn strongly influenced by active root systems. In addition to simply taking up nutrients and water, plant roots alter their physical, chemical, and biotic soil environment in ways that feed back to affect their own function, both directly and via complex effects on the soil community.

First, roots physically stabilize and compact soils, reducing the size of the pores that hold air, water, and nutrient solutes (references in McCully 1999). These structural changes determine the physiological availability of these soil resources to the root system itself, as well as to other soil organisms. They also determine the mechanical pressure required for growing roots to penetrate the soil (Hinsinger et al. 2009). The physical soil environment is further shaped by the channels formed by now-dead roots of current and previously living plants in the same location, which often form sites of greater organic matter and microbial density (Bundt et al. 2001).

Second, active roots influence soil moisture patterns and, consequently, the distribution of dissolved mineral ions (Marschner 1995; Hodge 2009; and references therein). By taking up water at different rates in different microsites, plant roots generate heterogeneity in soil water and nutrient content (Fitter and Hay 2002; Hinsinger et al. 2009). In addition, roots of many species redistribute water from wetter to drier soil patches at night, when transpiration ceases and a steeply negative water potential gradient occurs that brings water from roots into their immediate soil environment (Caldwell et al. 1998).

Third, growing roots change soil chemistry in several ways. Roots regulate the uptake and efflux of protons so as to maximize the availability of scarce mineral ions and avoid heavy metal toxicity (Marschner 1995; Hinsinger et al. 2009). Active roots also change soil chemistry by depleting oxygen and releasing substantial amounts of carbon dioxide through respiration (Nye 1981; Philippot et al. 2013). Because microbial communities are powerfully influenced by soil pH, this increase in soil acidification feeds back to influence the soil biota and, consequently, the plants interacting with it (Fierer and Jackson 2006). Another habitat-constructing chemical effect of roots is the addition of atmospheric oxygen into waterlogged and consequently anaerobic soils. This often critical environmental remediation is effected by individuals of the many wetland species that produce an anatomically distinctive, axially continuous porous tissue known as *aerenchyma*. Aerenchyma channels conduct oxygen from shoot organs, which are in contact with the aerial environment, down to buried roots, from which some of the oxygen is emitted into the rhizosphere (Blossfeld and Gansert 2007 and references therein). Interestingly, the ability of aerenchymatous plants to oxygenate the soil by releasing oxygen from their roots varies, for instance, among different cultivars of rice (references in Marschner 1995).

Roots also continuously shape the physical, chemical, and biotic properties of their surrounding soil zone by synthesizing and releasing into it a broad array of compounds with important habitat-constructing properties (reviewed by Walker et al. 2003; Philippot et al. 2013). These *root exudates* consist largely of organic compounds such as sugars, amino acids, and phenolics; it is estimated that 5%–21% of all photosynthetically fixed carbon is transferred to the rhizosphere through root secretion (Marschner 1995). Plant roots also synthesize and secrete high molecular weight compounds such as proteins and the polysaccharides found in mucilage. The primary sites of root secretion, as well as of water and nutrient uptake, are the very fine, ephemeral "feeder roots" that (together with their even finer epidermal root hairs) constitute much of the enormous surface area through which plant root systems interface with soil (Badri and Vivanco 2009). Mucilage and other exudates are synthesized and secreted by cells at the tips and sides of the root caps of these tiny roots, as well as by the surface bacteria associated with the roots (Humphries et al. 2005).

These secretions lubricate the root as it passes between soil particles (McCully 1999); plants increase production of these secretions in response to dry or compacted soil (Badri and Vivanco 2009).

As a plant's roots push through the soil, they leave a trail of mucilage that contributes to soil structure by binding soil aggregates to each other (Read et al. 2003; Hinsinger et al. 2009); at the same time, the sloughed-off root cap cells with their mucilage provide food for soil microbial communities (Humphries et al. 2005). Mucilage also contains powerful phospholipid surfactants that, even in small amounts, alter the biophysical properties of the rhizosphere, making a given amount of soil moisture easier for roots to extract and increasing the availability to plants of dissolved nutrients such as phosphate (Read et al. 2003; Hinsinger et al. 2009). Mucilage and other root exudates also interact chemically with the soil to help regulate plant uptake of mineral ions and to prevent the entry of heavy metal toxins (McCully 1999). The production of specific root exudates can be closely regulated by the plant's immediate soil conditions. Specific types of nutrient deficiency increase the secretion of metabolites that increase the availability for uptake of the limiting nutrients; similarly, in response to aluminum stress, plants secrete organic acids that detoxify aluminum in soil (Badri and Vivanco 2009 and references therein).

For individuals of many grass species, mucilage secretion results in the formation of a distinctive plant–environment phenotype. The roots of these plants encase themselves in stable, close-fitting soil sheaths that are formed by the expansion and subsequent contraction of mucilages around soil particles during cycles of soil wetting and drying (McCully 1999). Interestingly, the mucilages involved in producing these *rhizosheaths* are secreted jointly by root cells and associated root-epidermal bacteria (Watt et al. 1994). Rhizosheaths create a moist zone immediately around the root and thus are believed to facilitate nutrient uptake in dry soils (Watt et al. 1994). Field observations support the view that sheath formation is a response to dry soil: the rhizosheaths that form during midsummer, when the soil is dry, are thicker and more strongly adherent to the root than sheaths that form in moist soil; plants in dry soil also produce mucilage that is more highly adhesive than that produced in moist soil (Watt et al. 1994). The roots of many southern hemisphere plants (including rushes in the Restoniaceae family and perennial desert grasses) form similar sand sheaths in dry soils, but by means of a different mechanism. These plants have long root hairs which develop lignin-encrusted walls that keep them in place; these root hairs trap sand particles close to the root in a thick sheath. This structural sheath comprises a transition zone that facilitates the uptake of water and nutrients from the soil and may also provide physical protection from herbivores (Shane et al. 2011).

The chemical impact of root exudates on the rhizosphere environment occurs largely through effects on soil cohabitants (possibly including competing plant individuals; see de Kroon 2007). Secreted compounds that contain proteins and readily available carbon support soil microorganisms that directly promote plant growth, such as symbiotic *mycorrhizae* (root-associated fungi; Philippot et al. 2013 and references therein; Figure 5.6). Other root exudates indirectly benefit plants by inducing spore germination of beneficial mycorrhizal fungi (Marschner 1995). Root exudates also promote rhizobacterial populations that suppress plant pathogens by outcompeting them, by producing specific antibiotic compounds, or by inducing systemic resistance in the plant itself (Humphries et al. 2005). It has also been suggested that roots secrete into the rhizosphere defensive compounds that mimic or block bacterial signals involved in collective pathogenic activity (Walker et al. 2003). These complex interactions are not yet well understood, although, intriguingly, the protein secretion profiles of plant roots and soil bacteria are known to change in response to each other's presence (Badri and Vivanco 2009). Indeed, the potential utility of manipulating root exudates as means of biological pathogen control is a subject of ongoing investigation (Humphries et al. 2005). Signaling molecules secreted by plant roots mediate other complex rhizosphere interactions. For example, attacks by larvae of the Western corn rootworm, *Diabrotica virgifera*, on the leaves of corn, *Zea maize*, induce the plants' roots to synthesize and secrete β-*caryophyllene*, which attracts a nematode that is pathogenic to the insects (Rasmann et al. 2005). The roots of plants in the legume family increase rhizosphere nitrogen content

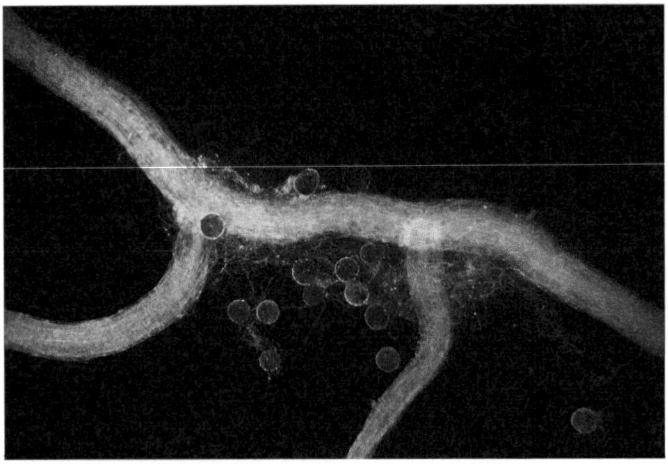

Figure 5.6 Plant root systems shape conditions in the rhizosphere in ways that feed back to affect plant function. Carbon-rich root exudates support symbiotic mycorrhizal fungi that enhance plant growth and induce fungal spores to germinate. This micrograph shows the intimate relationship between a beneficial mycorrhizal fungus and the corn root on which it is growing. The threadlike filaments around the root are fungal hyphae, and the round bodies are spores. Photo by Sara Wright, Courtesy of USDA_ARS.

by exuding compounds that initiate a symbiosis with nitrogen-fixing bacteria (references in Walker et al. 2003; see Section 5.4.2).

The release of habitat-constructing root exudates into the rhizosphere can play a tactical role in plant invasions, in part because the soil biota will lack coevolved responses (Broz et al. 2007). The roots of certain invasive plants release compounds that suppress the fungal symbionts of native species and thus allow the invaders to more easily outcompete their neighbors (references in Hodge 2009). A dramatic example of root-mediated habitat construction is that of the introduced allelopathic plant *Centaurea maculosa*. For over a century, this Eurasian member of the sunflower family has been aggressively displacing the native flora in western North America to form extensive monospecific stands. Early experiments using activated carbon to adsorb organic exudates showed that *Centaurea* roots released a substance that was toxic to other types of plant (references in Ridenour and Callaway 2001). Recent work has revealed this root-secreted toxin to be a particular molecular form of the plant secondary metabolite *catechin* (Bais et al. 2002; Walker et al. 2003). In addition, compounds that leach from *Centaurea* leaf litter suppress the germination and growth of neighboring plants (Ridenour and Callaway 2001). The chemical impact of both the leaves and the roots of this formidable invader creates a soil environment that native plants cannot inhabit. This altered rhizosphere is also inhospitable to native soil fungi: a molecular phylotype study showed that the presence of *Centaurea* plants both reduced the abundance of soil fungi and changed the composition of the soil fungus community (Broz et al. 2007).

5.4 How bacteria change the environment at several scales

5.4.1 Ecosystem impacts of marine bacteria

As in the case of ancient cyanobacteria and the Earth's atmosphere (see Section 5.1.1), the collective metabolic activities of even the smallest organisms can fundamentally shape environmental conditions within and beyond their own functional sphere. Present-day marine microorganisms exemplify these external impacts in several ways. One environmental effect arises from the recently discovered fact that marine phytoplankton release significant quantities of sulfur-containing metabolic products into seawater, thereby providing the major natural source of the volatile compound *dimethyl sulfide* (references in J. Seymour et al. 2010). This chemical oxidizes in the atmosphere, where it plays an important climate-shaping role by influencing the formation and albedo of clouds (J. Seymour et al. 2010). These effects, in turn, indirectly influence temperature and light conditions at various depths within and above the sea, including the phytoplankton zone.

A second surprising environmental impact of certain planktonic marine bacteria is the effect they have on the temperature of the seawater they inhabit. Herbert et al. (2008) investigated this in the South Andros Black Hole, which is a marine cave system that is located in the Bahamas and has very stable water stratification (i.e., little mixing with the open sea). In this system, sulfur-based (anaerobic) photosynthetic purple bacteria from the genera *Allochromatium* and *Thiocapsa* form a dense cell layer 1 m thick where the water becomes saline and anoxic, nearly 18 m below the surface. Their broad-spectrum light-harvesting pigments (primarily carotenoids) absorb virtually all incident light energy at this depth, but these pigments are rather inefficient at transducing the absorbed photons via photosynthetic pathways. With a photosynthetic energy efficiency of only approximately 30%, the massive bacterial population releases substantial excess energy into the water as heat. This energy transfer raises the water temperature at this depth from 29°C to 36°C, which is very close to the lab-determined growth optimum for these taxa; one meter below the bacterial layer, water temperature declines back to 29°C (Herbert et al. 2008). The bacteria also produce gelatinous extracellular polymers that trap heat within their massive cell layer, keeping water temperatures elevated at night as well as in daylight. Through these habitat-constructing activities, purple bacteria create a high-temperature zone that is hypothesized to explain their competitive dominance over co-occurring types of bacteria that are less heat tolerant (Herbert et al. 2008).

The warming of marine waters by dense layers of phytoplankton dissipating absorbed light energy as heat may occur more generally, if less dramatically, in other systems (references in Herbert et al. 2008). Although they do not participate in this collective energy transfer, populations of heterotrophic marine bacteria also influence the marine environment: their behavior in rapidly, intensively, and efficiently exploiting transient local concentrations of dissolved organic matter (see Chapter 4, Section 4.3.1) results in faster rates of biogeochemical cycling and greater net release of mineral nutrients (Fenchel 2002; Stocker et al. 2008). Directly testing the consequences of these energy and resource feedback effects on bacteria and their communities could provide valuable new insights to aquatic systems.

5.4.2 Shaping terrestrial and aquatic nitrogen supplies

A fundamental way that nutrient conditions in both aquatic and terrestrial habitats are shaped is by a small but taxonomically diverse group of microorganisms able to carry out a key biochemical transformation: converting the plentiful, but nearly inert, triple-bonded nitrogen gas in the atmosphere (N_2) into chemical forms that are usable by organisms (primarily ammonium [NH_4] and nitrate [NO_3] ions). These microorganisms include bacteria, cyanobacteria, and archaea species that, either as free-living cells or in symbiotic associations with multicellular organisms, produce biologically available forms of nitrogen (Karl et al. 1997; Zahran 1999). The primary mode of this conversion is *nitrogen fixation*, in which dinitrogen is reduced to ammonia in an energy-intensive process involving an enzyme complex known as *nitrogenase* (Zahran 1999). The fixed nitrogen enters biological food chains when plants metabolize this ammonia (after conversion to ammonium) into proteins and amino acids.

It is through this biochemical process, largely carried out in various symbioses, that soil bacteria fundamentally shape terrestrial habitats: total annual inputs of soil nitrogen from biological nitrogen fixation are estimated to reach 175 million tons, compared with approximately half that amount from industrially produced nitrogen fertilizers (references in Zahran 1999). One ancient symbiotic relationship in particular plays the dominant role in terrestrial nitrogen cycling: that between root-colonizing bacteria (*rhizobia*) in the genera *Rhizobium* and *Bradyrhizobium*, and their host plants in the large and diverse legume family (Fabaceae, or Leguminosae), which includes nearly 20,000 species worldwide of trees, shrubs, and herbs (Judd et al. 2002), as well as major crop and forage plants such as soybean, groundnut, clover, and alfalfa.

Because nitrogen fixation provides a critical boost to plants growing in low-nutrient soils, legume species that engage in this symbiosis are very often dominant competitors in early stages of ecological succession, which take place on young and often nitrogen-poor soils (Vitousek and Walker 1989). For millennia, humans have co-opted this aspect of habitat construction by cultivating legumes across vast areas of temperate and tropical agricultural and pasture lands, not only for the legumes' nitrogen-rich edible seeds and tissues but also, specifically, to enhance soil fertility (Oldroyd et al. 2011). The soil-enriching impact of leguminous plants is impressive even by comparison with modern-day industrial substitutes. The additional nitrogen provided to *Zea mays* by intercropping peanut, *Arachis hypogaea*, at a planting ratio of four peanut plants per corn plant adds as much nitrogen to the soil as does applying 96 kg of fertilizer per hectare (references in Zahran 1999). Indeed, rhizobia–legume symbioses alone are the primary source of biologically available nitrogen entering biotic systems via the soil (Tate 1995), fixing 200–300 kg of atmospheric nitrogen per hectare annually (Peoples et al. 1995).

A remarkable feature of these vitally important symbioses is that the bacterial partner is integrated within living, functional cells of the host plant's root (Sprent 2001), where it takes on the role of a nitrogen-fixing organelle (Oldroyd et al. 2011). To attain this intimate association, legumes and their rhizobial symbionts have coevolved to exchange highly specific chemical signals that provide mutual recognition (D. Werner and Müller 2003; see Sprent 2007 on macroevolutionary history). This biochemical conversation begins with root exudates that stimulate rhizobial chemotaxis toward the root surface (Marschner 1995). In response to colonization by appropriate rhizobial bacteria, the root secretes specific compounds (often consisting of flavonoids) which induce the expression of an extremely diverse group of nodulation genes in the bacteria; the products of these genes are lipochitin oligosaccharides known as *Nod factors* (D. Werner and Müller 2003). Even at extraordinarily low concentrations in the soil (10^{-8} to 10^{-11} M), these bacterial gene products induce rapid morphogenetic responses in the epidermis, cortex, and pericycle of the legume host plant. These responses initiate the formation of enlarged, hollow tumors or *nodules* on the host's roots to house the bacterial colonies (details and references in Oldroyd et al. 2011; Figure 5.7).

These induced root nodules solve a critical functional dilemma for the rhizobial symbionts. The enzyme that reduces atmospheric N_2 to ammonia, nitrogenase, is inactivated by oxygen even at low concentrations, but oxygen is necessary for the ATP-generating reactions (i.e., oxidative phosphorylation) that rhizobia employ to break the strong nitrogen–nitrogen bond. The internal environment of legume nodules provides precisely the conditions that maintain this delicate balance, by combining a high rate of cellular respiration (to maintain very low oxygen concentrations) with a specific carrier molecule that rapidly ferries oxygen to fixation sites (Sprent 2001). Through a complicated chain of developmental and physiological interactions, the metabolic integration of plant and bacterial cells is completed. It is only within the root nodule that the bacterial *nif* genes that initiate nitrogenase synthesis become active and those for assimilating ammonia suppressed, while the plant's *nodulin* genes initiate changes to the plant cell membranes that allow metabolites to be exchanged with its endosymbiont and produce additional enzymes for assimilation of the newly available ammonia (details and references in Oldroyd et al. 2011).

Numerous species of nonleguminous flowering plants, including many trees, also participate in highly productive symbiotic interactions with nitrogen-fixing root bacteria (Zahran 1999), interactions that shape soils and the biological communities those soils support. Like other organisms that markedly alter environmental conditions, these symbionts can be particularly disruptive when introduced into new areas. One such case has been studied in Hawai'i, where the plant partner, *Myrica faya*, is an invasive evergreen tree native to the Canaries and other Atlantic islands (Vitousek and Walker 1989). In their native range, *Myrica* trees host symbiotic nitrogen-fixing species of bacteria from the cosmopolitan genus *Frankia*. In their introduced range in Hawai'i, where they are also colonized by *Frankia*, *Myrica* trees substantially increase the total nitrogen content of soil in which they grow, in part through the deposition of nitrogen-rich leaf litter. This local soil enrichment enhances their own

(a)

(b)

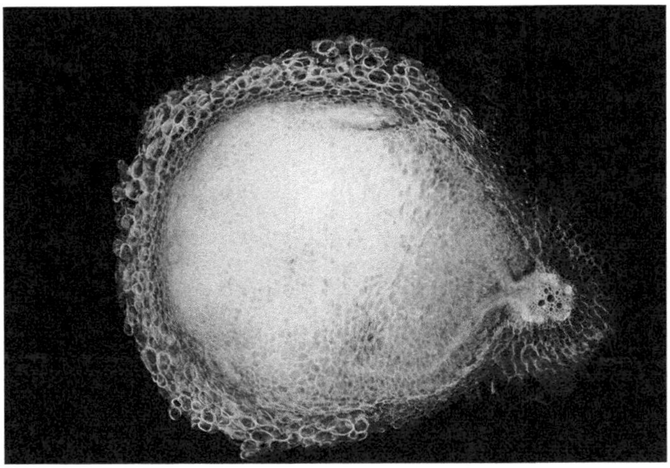

Figure 5.7 Root-colonizing *Rhizobium* bacteria and their legume host plants participate in symbioses that bring atmospheric nitrogen into biological systems. Bacterial signals induce the root to form hollow nodules. Within the nodules, dense bacterial colonies convert inert nitrogen to a form usable by the plant. (a) Rhizobium-induced nodules formed by roots of a fava bean plant, *Vicia faba*. (b) Micrograph of a cross-section through a single, fluorescently stained root nodule, showing the plant's (comparatively narrow) root at the right, the loosely packed plant cells that enclose the symbiont population, and the dense mass of nitrogen-fixing bacteria within. Photo credit for both images © Phil Gates/Biological Photo Service.

growth and reproductive output relative to native neighbors that do not carry out this important type of habitat construction (Vitousek and Walker 1989). In addition to facilitating their own invasive success, the enhancement of soil quality by introduced *Myrica* trees may eventually promote invasion of Hawai'i by nonnative species that would otherwise be excluded from its nutrient-poor, volcanic soils (Vitousek and Walker 1989).

5.4.3 Environmental impacts at the microorganismic scale

In addition to their broad-scale impacts on terrestrial, atmospheric, and marine systems, bacteria shape environmental conditions at their own extremely fine functional scale, for instance, by enriching oxygen content within the structural pores of flooded soils (Koza et al. 2011). Because this functional scale includes the many macroorganisms whose surfaces, tissues, and cells constitute bacterial habitats, bacterial habitat construction can include physical and chemical changes to those host organisms. As a result, it is not always possible to distinguish abiotic from biotic environmental effects at the bacterial scale.

From a microorganismic vantage point, one exceptionally important habitat is the *phyllosphere*, which consists of the surfaces of leaves and other above-ground parts of plants. Although this habitat exists at a relatively fine, within-plant scale, when summed across terrestrial plants, the total area of the phyllosphere is estimated to be from 600 million to over one billion square kilometers (C. Morris and Kinkel 2002; Vorholt 2012). This vast living habitat is colonized primarily by specialized communities of bacteria and by certain yeasts and fungi (Lindow and Brandl 2003). After many decades of comparative neglect, the ecology and microbial diversity of the phyllosphere have recently become topics of intensive investigation (Vorholt 2012 and references therein).

Bacterial inhabitants experience the phyllosphere as a heterogeneous place with a pronounced topography formed by the bulges, outgrowths, and indentations of epidermal cells (Figure 5.8). This microtopography causes water droplets to be unevenly distributed on the surface, while nutrients are found mainly in irregularly distributed "oases" of glucose, fructose, and other carbon-rich compounds that leak from plant tissues via minute injuries or glandular trichomes (Leveau and Lindow 2001).

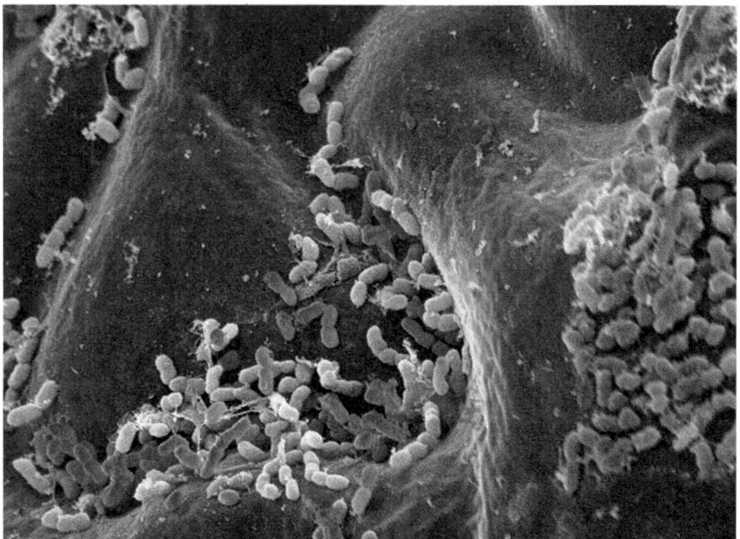

Figure 5.8 Leaf surfaces create a heterogeneous microtopography which structures the phyllosphere habitat. This unseen but globally vast habitat is occupied by specialized microbial communities. This scanning electron micrograph provides a close look at the surface of an *Arabidopsis thaliana* leaf, showing *Sphingomonas* spp. bacteria clustering at sites of local resource availability. Image courtesy of Gerd Innerebner and Roger Wepf, ETH Zurich, Switzerland.

Although phyllosphere bacteria aggregate at these local sites of resource availability, their activities also change both moisture and nutrient conditions at their own functional scale. The large, diverse aggregations of bacteria that form on leaf surfaces alter moisture conditions by producing extracellular polysaccharides, creating a consistently moist slime matrix that protects them against desiccation stress (Lindow and Brandl 2003). These authors speculate that bacterial cells in these aggregates participate in density-dependent "quorum-sensing" signaling to neighboring cells by releasing small molecules that induce collectively beneficial traits such as the synthesis of moisture-trapping substances. Phyllosphere bacteria also release biosurfactant compounds that increase the solubility and diffusion of nutrients and may, in addition, create a water film on the leaf's hydrophobic cuticle layer that helps to distribute the bacteria to high-nutrient patches (Vorholt 2012).

Bacteria may also alter chemical conditions in their phyllosphere environment by inducing the host plant to release more nutrients (Leveau and Lindow 2001). This has been studied in both pathogenic and nonpathogenic strains of *Pseudomonas syringae*, a bacterial species which inhabits diverse plant hosts. *Pseudomonas syringae* strains secrete a compound called *syringomycin* that is a potent surfactant (Hutchison et al. 1995). In the high quantities released by virulent strains of *Pseudomonas syringae*, this compound induces the formation of ion channels in the plant cell plasma membrane, leading to the release of cell metabolites and eventually to the lysis of host plant cells. Nonpathogenic strains of *Pseudomonas syringae* also possess the complex biosynthetic machinery to produce syringomycin (Hutchison et al. 1995). Such strains are thought to release smaller amounts of the compound, amounts that are sufficient for surfactant activity and just high enough to induce the low-level release of plant metabolites for consumption by surface bacteria. Using a different biochemical approach, many phyllosphere bacteria synthesize and release a compound that is identical to the plant growth hormone indole-3-acetic acid (IAA). When applied exogenously, this substance stimulates the release of saccharides from the walls of plant cells. Greenhouse tests show that IAA-deficient mutant bacteria are outcompeted by identical but IAA-producing strains, confirming the beneficial habitat-constructing effect of this bacterial product (Lindow and Brandl 2003 and references therein).

Some pathogenic bacteria alter the phyllosphere physically to allow their entry into plant tissues, often via the leaf's own stomates (Zeng et al. 2010). Although a leaf's stomates can close in response to molecular signatures indicating the presence of specific pathogens, many bacteria have evolved counterdefenses that keep stomates open, either via specific toxins or by mimicking the plant's own self-regulatory signals (Underwood et al. 2007; Hogenhout et al. 2009). The pathogenic bacterial species discussed above, *Pseudomonas syringae*, creates entry points into leaf tissue by releasing a phytotoxin called *coronatine* that reverses stomatal closure (Melotto et al. 2008 and references therein). Similarly, many fungal pathogens create entry points into plant tissues by releasing compounds that induce stomatal opening, secreting enzymes that actively degrade the leaf cuticle and epidermal cell walls, or exerting mechanical force (Underwood et al. 2007; Zeng et al. 2010).

Pathogenic bacteria also may manipulate a host plant's signaling pathways (often by producing analogs of the plant's own hormones) to derail its defense responses and create a more hospitable host environment (Melotto et al. 2008; Hogenhout et al. 2009; Schmid-Hempel 2009). A dramatic example of such biotic habitat construction is provided by pathogenic *Agrobacterium* spp., which commandeer the host plant's developmental signaling machinery to create neoplastic disease phenotypes that constitute favorable bacterial conditions. Most well-studied is the rhizosphere bacterium *Agrobacterium tumefaciens*, which has the broadest host range of any bacterial plant pathogen. As its specific name indicates, this species causes the tumorous distortion known as crown gall disease through the following sequence of events (see Pacurar et al. 2011 for details and references). First, free-living soil *Agrobacterium* cells sense sugars and phenolics released from minor wounds in plant roots and, using these compounds as guides, move chemotactically into the host's intercellular spaces. The bacteria then release DNA into host plant cells, which are rapidly dividing in the process of wound healing. The transferred DNA becomes integrated into the plant's genome, where it transforms host function to benefit the bacteria in two ways. One set of bacterial genes alters the cell's hormone production and responses so as to initiate cell proliferation and tumor growth. A second set of transferred genes alters the host cell's secondary metabolism to synthesize compounds used by the bacterial invader as carbon and nitrogen sources. The ability of *Agrobacterium* cells to effectively transfer DNA across kingdoms into plant cells is also used in producing transgenic crops and other commercially exploited plant populations, so these bacteria also contribute to large-scale habitat construction by humans.

CHAPTER 6

Community-level consequences of habitat construction and eco-devo responses

Organisms interact with their environments in two key ways. Through their various modes of life, organisms alter environmental conditions (external niche construction or *habitat construction*; see Chapter 2, Section 2.3.3, and Chapter 5). Conversely, external conditions influence phenotypic expression in organisms, shaping their individual growth, morphology, allocation, life history, physiology, and behavior (*eco-devo responses*;[1] see Chapter 1, Section 1.3, Chapter 2, Section 2.3.1, and Chapter 3). Both types of organism–environment interaction have community-level consequences: first, because changes to external environments affect all organisms that inhabit those environments; and, second, because individual phenotypic responses shape ecological interactions. Through these community-level effects, both habitat construction and eco-devo adjustments create ecological feedbacks on the originating organism. This chapter examines how (a) habitat-constructing effects and (b) individual eco-devo responses such as plasticity contribute to the emergent complexity of ecological communities as functional systems. (Although for convenience they are discussed separately, it is important to note that these two reciprocal processes are not independent; eco-devo responses, including plastic responses, themselves influence the impact of organisms on their environments, by mediating how individuals engage with and succeed in those environments [Chapter 4].) To begin, it is useful to set out a working view of ecological interactions in real communities.

6.1 A closer look at ecological interactions

Ecological communities are extraordinarily complex, since they typically consist of thousands of species that are each engaged in multiple direct and indirect interactions (Hacker and Gaines 1997; E. Werner and Peacor 2003; Holt 2006). A terrestrial forest community, for example, extends from variously commensal, competing, and facilitating soil bacteria, archaea, and fungi, to plants that respond to and shape this soil biota and each other, to pathogens, herbivores, defenders, pollinators, and dispersers that respond to and elicit plant defenses, attractants, and rewards and that compete with or facilitate each other, to further levels of predators and their parasites, pathogens, and symbionts. The positive versus negative directions of these interactions, and their intensity, all vary as a result of abiotic and biotic conditions, resulting in flexible webs of multidirectional causation. Moreover, these myriad ecological variables interact across temporal and spatial scales to produce hard-to-quantify

[1] As explained in Chapter 1, Section 1.3 (first paragraph), the inclusive terms *ecological development* or *eco-devo* are used (following S. Gilbert 2001) to denote all aspects of phenotypic expression that are influenced by environmental conditions, whether phenological, morphological, physiological, or behavioral. *Eco-devo responses* include both adaptive and inevitable phenotypic plasticity (changes in phenotypic expression across environments; see Chapter 3, Section 3.1.1), as well as polyphenism, canalization, or any environmentally contingent aspect of phenotypic expression.

Organism and Environment. First Edition. Sonia E. Sultan.
© Sonia E. Sultan. 2015. Published 2015 by Oxford University Press.

"multi-scaled cause-effect relationships" (Soranno et al. 2014, 67); the strength of any given interaction is conditioned by, and will indirectly influence, the rest of the community (Inouye and Stinchcombe 2001; E. Werner and Peacor 2003). This complex, multileveled interconnectedness is considered to be central to ecosystem-level functions, including carbon storage, nutrient and water cycling, and total biomass productivity (Hooper et al. 2005).

By necessity, ecologists have traditionally focused on simplified components of these systems, such as pairwise species interactions under constant conditions. More recently, the frame of such studies has opened to allow for more complex causation, revealing a fine-scale context dependence. For example, Kniskern and Rausher (2006) found that the nature and impact of a specific host–pathogen interaction (between the common morning glory, *Ipomoea purpurea*, and the fungus *Coleosporium ipomoeae*) varied dramatically in the field, depending on both abiotic and biotic factors. Fungal damage strongly reduced the size, longevity, and reproductive output of plants growing in poor-quality microsites (i.e., those where moisture and nutrient levels were reduced by the presence of competitors), but the same amount of tissue damage had no negative impact on plants in more favorable microsites, where the relationship was commensal (and possible even mutualistic) rather than parasitic. A negative effect on plants was also lacking in all microsites during growth seasons that were cut short by early frost. The impact of *Coleosporium ipomoea* on host *Ipomoea* plants was also influenced by the activities of other natural morning glory enemies in the community. Leaf damage from the fungus was reduced when plants experienced insect herbivory but increased when the plant had been previously infected by a different pathogen active earlier in the season, presumably because the first infection had depressed the host's systemic immune response (Kniskern and Rausher 2006).

The surprisingly variable relationship between this fungus–host plant pair illustrates two important points. First, *ecological interactions are substantially mediated by the impacts of co-occurring organisms on their external environments*—in this case, by the depletion of microsite resource levels by plant neighbors, and by herbivore or pathogen effects on leaf tissues. It is well known that niche-constructing impacts can be fundamental to ecological communities, as exemplified by "habitat-forming" species such as forest trees, marine kelp, terrestrial or seagrasses, and colonial corals (Bruno et al. 2003; see Chapter 5; also see Chapter 6, Section 6.3.1). However, less obvious organismic impacts on shared habitats can also play key community roles.

For instance, the saltmeadow rush, *Juncus gerardii*, has two key abiotic effects on its intertidal environment (Hacker and Gaines 1997). *Juncus* plants grow in dense stands that prevent evaporation from the soil surface (such evaporation would lead to high substrate salt concentrations). In addition, they deliver oxygen into the soil via specialized aerenchyma channels that run from the aerial parts to the roots. Without the substrate-altering effects of this species, several plant taxa would be excluded from intertidal habitat zones, as would the herbivores and higher trophic levels associated with those taxa; it seems the entire community rests on the habitat-constructing effects of *Juncus* plants (Hacker and Gaines 1997). Such effects of organisms on environmental stress levels and resource availability can also shift community composition through effects on competitive balance (Callaway et al. 2003). Biotic habitat-constructing impacts as well can shape communities: for instance, the parasitic plants known as dwarf mistletoes, *Arceuthobium* spp., induce the trees they infect to produce massively branched vegetative "witches brooms" that increase the diversity and abundance of insect and bird inhabitants (Bennetts et al. 1996 and references therein; see Figure 6.1). The plant-mediated presence and density of soil fungi (*mycorrhizae*) alters soil nutrient availability in ways that indirectly affect competitive interactions among cohabiting plants (Hodge and Fitter 2013). Because co-occurring species share local conditions and resources, even small-scale organismic effects on external environments ripple outward via direct and indirect interactions, increasing connectivity among members of ecological communities (Bogert et al. 2006).

The second fundamental point illustrated by the *Ipomoea* example is that *ecological interactions are mediated by traits of individual organisms* (see Bolnick et al. 2003). Previously, trophic and other functional

(a)

(b)

Figure 6.1 Trees infected by parasitic plants known as dwarf mistletoes, *Arceuthobium* spp., produce massively branched "witches' brooms," an eco-devo response that indirectly alters the associated community of insects and birds. (a) An infected black spruce, *Picea mariana*, in northern Minnesota exhibits the characteristic response to dwarf mistletoe infection. Joseph O'Brien, USDA Forest Service, Bugwood.org, Creative Commons Attribution 3.0 License; <http://www.forestryimages.org/browse/detail.cfm?imgnum=50330326.2>. (b) The highly reduced reproductive shoots of *Arceuthobium pusillum* are the only parts of the parasite that emerge from beneath the host's bark. Joseph O'Brien, USDA Forest Service, Bugwood.org, Creative Commons Attribution 3.0 License; <http://www.forestryimages.org/browse/detail.cfm?imgnum=5032074>.

interactions were conceived in terms of constitutive species-level traits (Messier et al. 2010), and their impact calculated as a function of density based on fixed per capita effects of one species on another (see the critique in E. Werner and Peacor 2003). It is now broadly recognized that per capita trophic, competitive, and other ecological effects are *not* constant. Rather, these effects vary as a result of the behavioral, morphological, physiological, and life-history adjustments made by interacting individuals in response to each other and to other aspects of their environments (Callaway et al. 2003; Luttbeg et al. 2003; Miner et al. 2005; O. Schmitz et al. 2008; see Chapters 3 and 4). In other words, community-structuring processes are mediated by continually generated, individual ecological developmental variation, including both adaptive and inevitable plastic eco-devo responses (see Chapter 3, Section 3.1.1).

To take a broad example, the effect of a predator population on its prey will be different if prey individuals alter their morphology, phenology, or behavior so as to minimize predation (e.g., Agrawal 1999; Raimondi et al. 2000 and references therein). By reducing the prey's vulnerability to the predator, the trait change will shift multiple ecological interactions in the community, such as those between the predator and other possible prey species, and those between the altered prey and its own competitors or prey organisms. These changed interactions can in turn alter patterns of overall resource use and other cumulative environmental impacts (E. Werner and Peacor 2003). Even trophic transfer efficiencies, classically viewed as fixed, can be variable, for instance, if predators cause increased stress that leads to reduced assimilation efficiency or lowers tissue quality for downstream consumers (O. Schmitz et al. 2008). Theoretical investigations show that the ecological consequences of individual trait changes, termed *trait-mediated interactions* (or *trait-mediated indirect interactions*), will influence community dynamics in both the short term and the long term (B. Bolker et al. 2003 and references therein).

Many trait-mediated community effects involve behavioral shifts in response to a negative ecological interaction. For instance, when fish-eating bass are present in ponds, potential bluegill prey shift away from more open water to littoral vegetation; this shift then releases those populations of zooplankton that inhabit pond centers from predation by the bluegill. Due to this effect on bluegill feeding behavior, the presence of bass indirectly causes two- to sevenfold increases in zooplankton abundance (A. Turner and Mittelbach 1990).

In other cases, community effects are mediated by morphological trait changes (although these cases are generally less recognized; Miner et al. 2005). An exceptionally well-studied example involves the predatory whelk *Acanthina angelica*, which through direct contact induces young juveniles of its primary prey species, the ordinarily conical acorn barnacle, *Chthamalus anisopoma*, to develop an altered, "bent" defensive morphology (Lively 1986; Mokady et al. 2000). Increased frequency of this altered morph suppresses recruitment of the mussel *Brachidontes semilaevis*, as the mussels have difficulty finding suitable attachment sites among bent barnacles and thus themselves become prey items for the thwarted whelks. Reduced mussel populations in turn benefit brown algae, *Ralfsia* spp., which compete with mussels for attachment space (Raimondi et al. 2000). A five-year study revealed that the precise timing of the whelk–barnacle interaction affected long-term community composition: when whelks encountered juvenile barnacles capable of developmentally shifting to the bent form, the presence of whelks had a *negative* indirect impact on the mussel population, and a *positive* indirect impact on the algal population. When whelks came into contact with adult, conic barnacles unable to change morphologically, their presence had exactly the opposite effects on mussel and algal populations (Raimondi et al. 2000; Figure 6.2). Consequently, depending on the ontogenetic timing of whelk–barnacle contact, the community takes shape either as one with few mussels and substantial algal cover, or one with plentiful mussels and few algae. In this case, induction of the altered morph was by physical contact, so the density of the predator along with timing contributed to the trait-mediated effect; in cases of defensive change induced by chemical cues, predator density may bear no relation to the prey population's response and its downstream impact on the community.

In addition to trait changes induced by the ecological interaction itself, phenotypic responses to

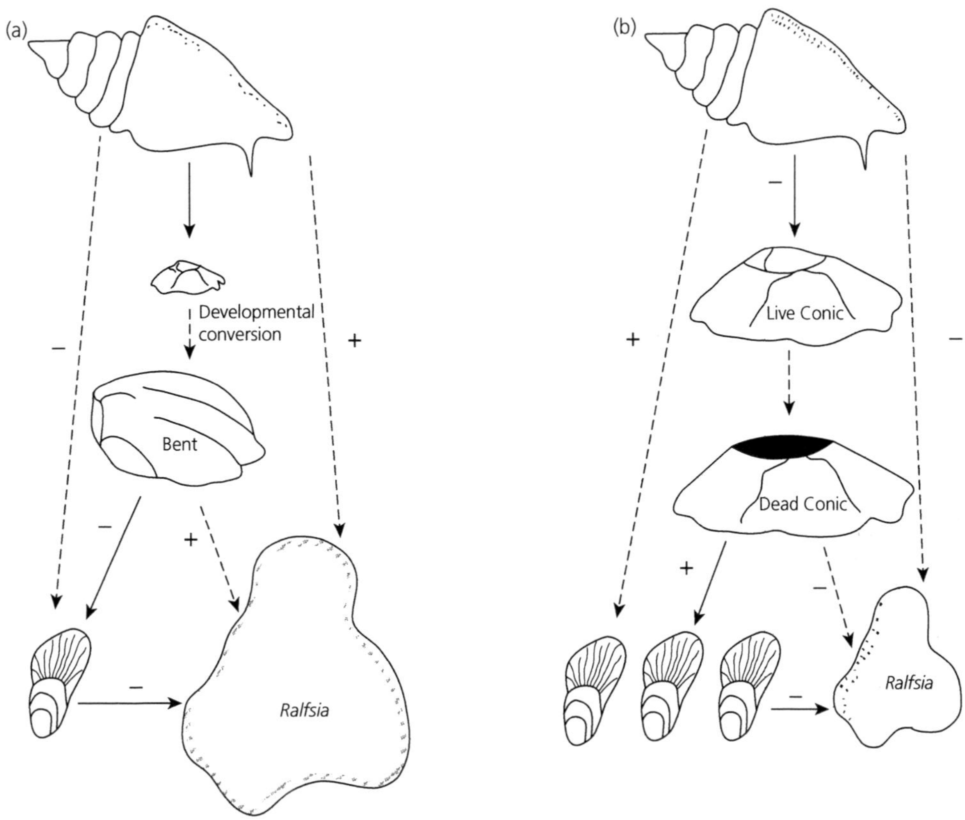

Figure 6.2 The predatory whelk *Acanthina angelica* induces plastic expression of a "bent" defensive morphology in juveniles of its prey, the ordinarily conical acorn barnacle, *Chthamalus anisopoma*. Because this plastic change alters species interactions, the presence of whelks leads to opposite effects on community composition depending on the timing with respect to barnacle development. (a) When whelks come into contact with juvenile barnacles, they induce the barnacles to develop into a "bent" shape and consequently indirectly reduce mussel populations and promote the growth of *Ralfsia* algae, which compete with mussels. (b) When whelks come into contact with adult barnacles, which are unable to change morphologically, their presence has exactly the opposite effect: by preying on the barnacles, whelks promote larger mussel populations, thus suppressing the algae. Redrawn from P. T. Raimondi, S. E. Forde, L. F. Delph, and C. M. Lively, 2000, Processes structuring communities: evidence for trait-mediated indirect effects through induced polymorphisms, *Oikos* 91: 353–361; copyright John Wiley and Sons.

abiotic conditions can influence the magnitude and direction of an interaction and, consequently, its community effects (Callaway et al. 2003). For example, oak seedlings, *Quercus douglasii*, respond developmentally to the vertical distribution of moisture in the soil by producing a shallow, highly branched root system when deep soil layers are dry, and a single, deep taproot when deep soil layers are moist (Callaway 1990). These alternative individual phenotypes have very different community impacts as the oaks grow: the diffuse, shallow root systems compete fiercely with herbaceous understory grasses and herbs, strongly reducing herbaceous plant biomass and shaping community composition by excluding weak competitors. By contrast, deep-rooting oaks do not compete with herbaceous plants for soil resources but instead *promote* growth of a diverse herbaceous understory via protective leaf litter (Callaway 1990; see Chapter 5, Section 5.3.1).

By virtue of individual plastic adjustments, environmental influences on development play a key role not only in shaping phenotypes but also in structuring biological communities (see

Section 6.2.3). Recognizing this level of variation expands our view of functional biodiversity to the individual and trait levels. Trait-mediated effects on ecological interactions cannot be assumed to be minor; indeed, they can be as great in magnitude as effects mediated strictly by density (references in Luttbeg et al. 2003). However, these are not exclusive alternatives; the community impact of competitive, trophic, or other interactions generally reflects *both* the density of interacting populations *and* trait adjustments by individuals (Miner et al. 2005).

A recently studied trophic cascade provides an example that also shows how ecological interactions can knit together aquatic and terrestrial ecosystems. Larval dragonflies are an important prey item for freshwater fish, while adult dragonflies are major predators of bees and other insects that pollinate terrestrial plants. As a result, the presence and abundance of fish in ponds indirectly promotes the reproductive success and resulting size of nearby plant populations by reducing the number of dragonflies that reach adulthood and consume local pollinators (T. Knight et al. 2005 and references therein). In addition to consumption per se, this trophic chain is mediated by behavioral trait changes in two of the participants: the indirect positive impact of fish on plant populations occurs not just because fish eat larval dragonflies but also because mother dragonflies avoid ovipositing near fish-containing ponds. Likewise, the indirect, negative effect of fish-free ponds on pollinator visits is due not only to greater pollinator predation by dragonflies but to preemptive avoidance by foraging insects of dragonfly-frequented locations (T. Knight et al. 2005).

Environmentally mediated trait changes that influence communities can be developmental as well as behavioral and, in some cases, can reflect anthropogenic disruptions to shared habitats. In one recently studied case, an abiotic change was found to alter development of a native prey species but not its invasive predator. Sanford et al. (2014) studied the response of Olympia oysters, *Ostrea lurida*, which are native to the west coast of North America, to acidified seawater produced by a doubling of dissolved carbon dioxide, as is predicted to occur by the late twenty-first century. Juvenile oysters that had been raised in carbon dioxide–enriched seawater grew 29%–40% smaller in size; this reduction in size increased their attractiveness as prey items to an introduced, invasive drilling snail, *Urosalpinx cinerea*. The smaller oysters experienced a 20%–48% increase in predation, evidently because they were easier for *Urosalpinx cinerea* individuals to drill and consume (possibly due to weaker shell structure as well as reduced size). Surprisingly, the snail's behavior and effectiveness as a predator were unchanged by rearing in acidified water. While the oysters' ability to build their calcium carbonate shells was compromised in acidified conditions, the radula with which snails drill through oyster shells is composed of chitin, an extremely tough polysaccharide that is likely unaffected by water acidity (Sanford et al. 2014). This example illustrates a more general concern about future ecosystems: because the phenotypic impacts of novel, human-mediated environments are species-specific, ecological interactions that structure communities are likely to be broadly disrupted in ways that are impossible to predict without knowing norms of reaction to those changed environments.

It is increasingly recognized that both organismic impacts on local environments and individual phenotypic responses to those environments influence the myriad interactions that structure biological communities. These individual effects and responses arise from norms of reaction for habitat-constructing and eco-devo traits that can vary at the levels of species, population, and genotype (E. Werner and Peacor 2003; Sultan and Stearns 2005; Sultan 2007; and references therein). As noted previously, these two aspects of organism–environment interaction are inseparable in a community context, because eco-devo responses to the environment alter the biotic and abiotic environment encountered by other community members. Encompassing this variability creates several daunting new challenges for community and ecosystem ecologists. Although developmental plasticity in plants is relatively well studied, these studies have rarely examined either direct or downstream community effects of such plastic changes (Callaway et al. 2003). In general, short- and long-term community effects of trait-mediated interactions depend on precise patterns of response (and on response trade-offs, which are rarely measured); such realistic levels of community complexity have been difficult to model in ways that lead to general results (B. Bolker et al. 2003). The effects of trait changes on resource distribution and abiotic stress factors are also relatively little studied to date. Integrating responses to biological and abiotic factors, and doing so across spatial and temporal scales, is both important and difficult (O. Schmitz et al. 2008).

This chapter next examines how functional diversity structures ecological communities and how both habitat construction and eco-devo responses contribute to that functional underpinning. Subsequent sections present selected case studies. Particular attention is paid to interactions that involve climate change, biological introductions, and other anthropogenic impacts on contemporary and future environments.

6.2 Functional diversity, facilitation, and ecological complementarity

6.2.1 The property of complementarity

At a time when both regional and global species losses are accelerating due to human activities (Hooper et al. 2012 and references therein), ecologists are urgently focusing on the role of biodiversity in maintaining the stability and productivity of ecosystems. The relation of biodiversity to these key ecosystem properties is understood to result from an emergent property of multispecies assemblages known as *complementarity*.

Complementarity arises from two major features of biological communities: *functional diversity* of species, as such diversity causes different types and spatiotemporal patterns of habitat use that combine to maximize resource-use efficiency, and *positive interactions* among species, as such interactions enhance the ability of species to utilize resources (Tilman et al. 2001; Cardinale et al. 2007; Fründ et al. 2013; and references therein). Due to these two types of functional synergism, species-rich mixtures generally produce greater total biomass (compared with the summed weighted averages of the component species when grown in monoculture), a phenomenon known as *overyielding* (Ashton et al. 2010 and references therein) that leads to high, stable productivity over time. The effect of complementarity on ecosystem productivity is well established. For example, a meta-analysis of 44 experiments over two decades confirmed that the positive effects of plant community species richness on biomass production arose from complementarity (Cardinale et al. 2007). Species diversity at higher trophic levels can also affect productivity via functional synergisms, as occurs in agroecological systems. A diverse group of insect predators may more effectively suppress herbivorous crop pests through complementary direct and indirect effects, resulting in increased crop yields (e.g., Cardinale et al. 2003).

Although such combinatorial effects have largely been attributed to functional differences among species (Ashton et al. 2010), positive species interactions such as *facilitation* also contribute importantly to the stability and high productivity of species-rich communities (Tilman et al. 2001; Callaway et al. 2002; Bruno et al. 2003). Facilitation occurs when the presence of a given species *facilitates* the occupancy of a shared habitat by one or more other species in the community (Bertness 1989 and references therein). For example, the presence of seaweeds reduces heat load and desiccation stress for marine intertidal invertebrates, enlarging suitable habitat space for these animals on rocky shores (Bertness et al. 1999). Similarly, the stabilization of New England cobble beaches by *Spartina* shore grasses makes the beaches suitable for occupation by a diverse community of other plants and their associated animals, taxa that would otherwise be excluded (Bruno and Kennedy 2000). A species' realized distribution encompasses not only habitat that *lacks negative interactors* that would exclude that species (such as powerful predators and competitors) but also habitat the species can occupy *due to facilitation by positive interactors*. One consequence of facilitation, then, is that a species' "realized niche" can actually be *larger* than its autecological "fundamental niche," by virtue of the beneficial impacts of co-occurring species (Bruno et al. 2003).

After several decades of focusing largely on antagonistic and competitive effects of species on each other—on "conflict and privation" as the major structuring principles for biological communities—facilitation is being increasingly integrated into an understanding of community dynamics (Bruno et al. 2003). As the above examples of seaweeds and

Spartina grasses make clear, facilitation includes many aspects of habitat construction: specifically, those cases where the presence and activities of certain organisms change shared environments in ways that improve those environments for co-occurring species. Specific community members may mediate physically stressful conditions, add to or redistribute resources, or provide nest or attachment sites (see discussion and examples in Chapter 5). Organisms also consume resources, occupy space, and otherwise alter their environments in ways that can negatively affect cohabitants. Not surprisingly, then, many species interactions have both positive and negative aspects that vary in relative intensity depending on circumstances (Bruno et al. 2003).

For example, a field experiment across a global sample of montane sites found that interactions among plant species were predominantly positive rather than competitive at high elevations: survival, biomass, and reproductive output of alpine plants were higher when nearby neighbors of other species were present (Figure 6.3; Callaway et al. 2002). At those sites, the main factors limiting growth were abiotic stresses such as low temperatures and desiccating conditions, both of which were mitigated by plant neighbors. However, the effect of plant neighbors on growth (although not survival) was negative at low elevation sites, where more favorable conditions allowed individuals to grow to the point where they encountered resource limits and hence competed with neighbors. Like other types of species interaction (see Section 6.1), the effect of facilitation on community composition will vary depending on abiotic conditions and other factors (Bertness et al. 1999).

6.2.2 Habitat construction contributions to complementarity

Along with their comparatively well-studied modes and patterns of habitat and resource utilization, species have differing impacts on their environments. This habitat-constructing dimension of functional diversity contributes to species complementarity and, consequently, to ecosystem productivity and stability. For instance, recent models indicate that such stability arises from a

Figure 6.3 At exposed montane sites such as this limestone cap (Cataract Mountain, Glacier National Park), closely spaced neighboring plants facilitate each others' survival and growth by mitigating the cold, dry conditions. Interactions among heterospecific individuals may be predominantly positive rather than competitive in such stressful habitats, while in more favorable conditions individuals may instead compete with close neighbors for resources. Photo courtesy of R. Callaway, Glacier National Park, Montana, USA. For the color image, see Plate 17.

specific type of community complexity: distinct, unequal inputs of energy and materials from organisms with contrasting niche-constructing roles (e.g., primary producers vs. microbial detritivores; Holt 2006). These distinct inputs are used by co-occurring organisms and integrated at higher levels of trophic and other interactions (Holt 2006). In the classic field experiment of Tilman et al. (2001, 2006), plant species diversity had progressively stronger positive effects on the total biomass of experimental plots over a decade. The species combinations in this study included taxa with differing niche-constructing impacts, such as nitrogen-fixing legumes and C_4 grasses that produce very dense, shallow root systems. This aspect of biodiversity may contribute to facilitation and hence complementarity; specific beneficial habitat-constructing effects such as soil nitrogen enrichment may be one reason that both species diversity per se and the particular composition of a multispecies mixture determine combinatorial stability and productivity.

The habitat-constructing dimension of species functional diversity likewise led to complementarity in an experimental microbial ecosystem. Wohl et al. (2004) created different "microcosm" communities consisting of between one and eight randomly chosen species of cellulose-degrading bacteria isolated from leaf litter and then measured the cellulose decomposition they accomplished in a closed environment (i.e., one with finite resources). Because cellulose is an abundant but inaccessible source of carbon in many ecosystems, breaking it down into chemically available forms is a key bacterial niche-construction role that substantially benefits cohabiting organisms as well. In these microcosms, greater bacterial species richness resulted in enhanced rates of total cellulose breakdown, because the bacteria created fine-scale differences in biochemical function that allowed more individuals to be maintained and permitted species to coexist over numerous bacterial generations (Wohl et al. 2004). Species diversity in this system promoted complementary modes of habitat-constructing function, allowing for a complex community structure as well as greater ecosystem productivity.

In such cases, functional complementarity for niche-constructing effects can contribute to habitat amelioration. A recent stream bioremediation experiment (Cardinale 2011) provides a practical example. A key management objective for aquatic systems is the reduction of excess nitrogen (added by agricultural fertilizers and fossil fuel combustion), since high nitrogen levels lead to the eutrophication of aquatic systems and coastal dead zones (V. Smith and Schindler 2009). The remediation study directly tested the impact of biofilm species diversity on the ability of stream communities to take up and store excess nitrogen. Researchers manipulated the number of diatom and green algal species growing in the biofilms of 150 stream mesocosms, recreating the types of habitat variation that influence the coexistence of algal species, such as heterogeneity of flow and periodic disturbance regimes. Both nitrogen uptake rates and algal community biomass increased linearly with species richness, evidently owing to functional complementarity (Cardinale 2011).

In view of these biodiversity insights, the species losses predicted over coming decades can be expected to reduce the efficiency with which many different types of communities capture and use resources (Cardinale et al. 2007). In a global synthesis of meta-analyses, the impact of moderate-to-high rates of species extinction (40%–60% species loss) on ecosystem function is estimated to be as large as that of major drivers of environmental change, such as nutrient pollution, ozone production, and climate change (Hooper et al. 2012). Even if global levels of species loss are less drastic than these estimates, moderate-to-high levels of extinction are likely to be reached locally in many regions of intense human activity (especially with respect to top predators; Estes et al. 2011). These anticipated biodiversity losses can be expected to lower ecosystem productivity and stability through lost habitat-constructing effects as well as reduced functional diversity.

6.2.3 Eco-devo contributions to complementarity

The functional and habitat-constructing diversity that underlies coexistence, complementarity, and other ecosystem properties reflects variation in phenology; location and timing of foraging and metabolic activities; morphology, size, and placement of body parts (e.g., of coexisting root systems); and resource processing (Hooper at al. 2005). These aspects of development, physiology, and behavior are generally quite plastic and may change repeatedly within an individual's life cycle (see Chapters 3 and 4). Along with variation due to genetic diversity, intraspecific variation due to plastic trait expression is now understood to contribute to the structure and function of ecological communities, and to resource complementarity in particular (Jung et al. 2010; Burns and Strauss 2012; and references therein). This recognition represents a departure from the long tradition in ecological theory and research of treating ecological roles in a community, including patterns of resource use leading to complementarity, as the result of constitutive traits that are fixed at the species level (Ashton et al. 2010). Trait variation expressed by individual organisms is increasingly emphasized as a factor that shapes ecological interactions in general (see Section 6.1). However, as yet, the impact of context-mediated trait variation on community dynamics and structure is little understood (Miner et al. 2005).

Individual eco-devo variation in trait expression for resource-gathering organs has been experimentally shown to increase community productivity while reducing the degree to which constitutive species-level traits determine community assembly (Burns and Strauss 2012). In an unusual field study, researchers in a lowland tropical forest examined variation for critical leaf functional traits across nested ecological scales from individual leaf, to sun and shade strata within a tree canopy, to an entire tree, to species, plot, and site (Messier et al. 2010). Since trait expression was influenced by variable microclimate, insolation, and soil conditions as well as individual genotype, within-species trait variation was found to be greater than trait variation among species. Intriguingly, local habitat plots with quite different species compositions had very similar leaf trait means and distributions; this observation suggested the existence of a nonrandom assemblage based on functional diversity that did not arise at the species level (Messier et al. 2010). A different aspect of functional complementarity was studied in experimental communities that included from one to five pollinating bee species (but a constant total number of individuals). When more than a single bee species was present, seed production in a nine-species plant community increased significantly; thus, pollinator services appeared to be functionally complementary (Fründ et al. 2013). As in the previous case, this community-level complementarity reflected individual eco-devo responses rather than fixed species-level traits: bees of each species changed their floral preferences depending on the presence of heterospecific bees, creating context-dependent pollinator complementarities (Fründ et al. 2013).

How exactly do such individual trait adjustments contribute to complementarity at the community level? One key process is *facultative character displacement*, in which trait expression in co-occurring individuals is more divergent than in individuals growing alone (D. Pfennig and Murphy 2002; Burns and Strauss 2012). Such flexible, eco-devo based displacements may be far more common in natural communities than evolved, constitutive trait divergence (D. Pfennig et al. 2006). In animals, such transient displacements can include biotically induced shifts in foraging behavior, as well as changes in morphological traits involved in resource use (D. Pfennig et al. 2006). For instance, co-occurring animals may avoid direct competition by switching dietary preferences for host plants or prey items, or as pollinators may switch to plant species with less bounteous rewards but fewer other visitors. Theoretical work indicates that such individual flexibility can better promote species coexistence across a broad range of conditions than fixed traits can (e.g., Kriven 2003; additional references in Miner et al. 2005). Plastic trait expression can thus be a major mechanism for resource partitioning, and consequently complementarity, by means of shifts in resource use by individual animals (D. Pfennig et al. 2006).

Plants as well may shift patterns of resource use in response to competitive conditions (Callaway et al. 2003). Indeed, differentiation in resource use, leading to coexistence and complementarity in plant communities, may be largely due to individual plastic responses that allow plants to partition resources despite their broadly overlapping resource requirements (Ashton et al. 2010). For example, physiological plasticity in the use of different forms of a common chemical resource can mediate competitive interactions, as demonstrated in a study of four alpine herb species (Ashton et al. 2010). When grown singly, these species had the same nutrient-use patterns for different chemical forms of nitrogen, preferentially taking up nitrate, and then ammonium, over glycine. In competition, however, individuals of the competitively dominant species switched to increase their uptake of ammonium, the most available form of soil nitrogen, while the competitively inferior species continued to take up nitrate. Interestingly, the identity as well as the presence of neighbors can induce such changes in resource use (references in Ashton et al. 2010). Eco-devo responses to microsite conditions that entail changes in morphological traits (e.g., shoot height, root allocation and form, and leaf shape and structure) can also lead to fine-scale resource-use differentiation and reduced competition among plant individuals (Callaway et al. 2003). Such trait-based differentiation was found to occur, for example, among individuals in meadow communities located along a soil flooding gradient (Jung et al. 2010).

Individual plasticity in the expression of defensive traits can also contribute to community properties in both animals and plants (O. Schmitz et al. 2004). Induced defenses can reduce the amplitude of population fluctuations due to predators or pathogens, hence increasing the stability of complex ecological systems (Miner et al. 2005). For instance, sharp fluctuations in the abundance of algae, herbivorous zooplankton, and carnivorous zooplankton were much dampened when the algae at the base of this trophic system expressed an inducible defense, as compared to a system based on primary producer algae that lacked the capacity for induced defense (Verschoor et al. 2004). The expression of induced defenses can also alter competitive interactions among individuals, especially when resource stresses exacerbate the costs of defense (Callaway et al. 2003 and references therein).

6.2.4 Integrating the effects of individual trait changes

The effects of individual eco-devo responses on ecological communities are even more complex in the many cases that involve reciprocal plasticity—that is, cases where "the response of one species to the other creates the environment to which the other species may then respond" (Agrawal 2001, 321). These reciprocal trait changes unfold in time as a Markovian process in which each individual's response is conditioned by the previous response of its interactor(s). For example, when confronted by gape-limited predators such as the aquatic protozoan *Lembadion bullinum*, the tiny ciliate *Euplotes octocarinatus* produces lateral "wings" that dramatically increase its effective size (Kuhlmann et al. 1999). In response, *Lembadion bullinum* enlarges its own cell size and hence the gape size of its feeding peristome, thus allowing it to feed more successfully on the enlarged prey (Kopp and Tollrian 2003a). The ability to adjust its cell and gape size can be considered an "inducible offense" that allows this generalist predator to feed efficiently on either large or small prey items, including prey species that, like *Euplotes*, express induced defensive morphologies (Kopp and Tollrian 2003b). In some cases, either species' response can initiate a reciprocal cycle. For instance, crabs that eat intact mussels (with their calcium-rich shells) grow larger and stronger claws than do crabs that eat mussels without shells, while mussels respond to attack cues from crabs and other predators by defensively increasing shell thickness (Leonard et al. 1999). The lag time, reversibility, and persistence (including across generations) of ecologically important phenotypic responses all can vary depending on the taxon and trait, making it very challenging to integrate these reciprocal effects (Agrawal 2001).

Along with changing ecological interactions such as competition and predation, plastic trait adjustments alter the environmental impacts of those interactions. Such eco-devo effects on habitat construction must be taken into account to fully capture the dynamics of ecological systems. For instance, the aquatic larvae of chironomid flies are major zoobenthic inhabitants of many freshwater systems. These larvae respond to the presence of predatory fish by retreating into refuge burrows instead of foraging at the sediment surface, where they would be at risk. By changing the site of their waste deposition, this behavioral response to predator presence results in a spatial redistribution of the organic nutrients chironomids provide to aquatic systems (Stief and Hölker 2006). As this simple case indicates, ecological context-dependent trait variation can influence the properties of ecosystems, such as primary productivity, distribution of organic matter, and nutrient cycling (O. Schmitz et al. 2008).

A more complicated, terrestrial example (discussed in O. Schmitz et al. 2008) further illustrates the way that eco-devo adjustments to ecological circumstances interact with habitat construction to shape communities. The herbivorous grasshopper *Melanoplus femurrubrum* prefers to feed on the grass *Poa pratensis*, but in the presence of the predatory hunting spider *Pisaurina mira* it will switch to the less nutritious dominant goldenrod, *Solidago rugosa*, a dense-growing clonal plant that offers better protective cover (O. Schmitz 2004; Figure 6.4) By inducing this foraging change in its plant-eating prey, the spider causes a major biophysical change in the community, because the spider's presence results in an increased abundance of grass and a reduced abundance of goldenrod. This diminishes the goldenrod's habitat-constructing effect of creating tall, dense stands that shade the soil. Consequently, the presence of this spider creates a more open, patchy habitat that promotes the growth of a variety of light-demanding herbs that (in the absence of the spider) would be excluded by goldenrod shade. Because the leaf litter produced by these high-light herbs decomposes more rapidly than does goldenrod leaf litter, the plant community that this spider promotes is characterized by increased rates of nitrogen cycling through soil as well as by greater herbaceous species diversity (O. Schmitz 2006). Interestingly, this predator's impact reflects the fact that it is a "sit and wait" hunter, such that the grasshoppers can survive simply by moving into goldenrod patches and avoiding confrontations. In contrast, the presence of a different predatory spider that hunts by active jumping results in grasshopper mortality, reducing the grasshoppers' impact as herbivores and indirectly increasing the density of both goldenrod and grasses (O. Schmitz and Suttle 2001 and references therein). As this example makes clear, behavioral and other eco-devo responses of individual organisms can play a surprisingly large role in shaping ecological communities, via direct effects on trophic and other interactions as well as by indirect effects mediated by habitat construction.

Figure 6.4 Plastic adjustments by individual organisms may play important roles in structuring ecological communities. In the presence of a predatory hunting spider, the herbivorous grasshopper *Melanoplus femurrubrum* will switch from its preferred host plant, a grass, to the less nutritious goldenrod *Solidago rugosa*, a densely growing plant in which it is more well hidden (as shown). This change in foraging behavior diminishes the abundance of goldenrod and consequently that plant's soil-shading effect; this promotes a different community of herbaceous plants that produce different leaf litter, resulting ultimately in changes to nutrient cycling. Photo courtesy of Colin Donihue. For the color image, see Plate 18.

6.3 Case studies: community-level effects of habitat construction

6.3.1 Reef-building corals, and habitat deconstruction due to their decline

A dramatic example of individual niche-constructing activity that gives rise to an entire ecological community is the classic case of reef-building corals. The main builders of undersea reefs are colonial cnidarians in the order Scleractinia, also known as "stony corals." These tiny, sessile animals secrete calcium carbonate to create huge three-dimensional structures that provide a porous, topographically complex habitat for astonishingly diverse marine communities. The coral polyps themselves comprise the habitat for symbiotic photosynthetic dinoflagellates called *zooxanthellae*, which inhabit cells of the animal's endoderm. This coral–phytoplankton symbiosis extends from shallow, well-illuminated waters to waters well over 100 m deep, where it is evidently limited more by cold temperatures than by low light (Muscatine et al. 1991; Moberg and Folke 1999 and references therein). Indeed, the system accommodates depth-related changes in irradiance through phenotypic adjustments at several levels: both the form of host coral colonies and the tissue morphology and behavior of individual coral animals change in apparently adaptive ways when light availability is reduced (Muscatine 1990 and references therein), while their zooxanthellae inhabitants alter pigment content and other biochemical features related to light harvesting (e.g., Chang et al. 1983).

These one-celled photosynthetic symbionts are major primary producers of tropical reef communities: zooxanthellae use the plentiful carbon dioxide dissolved in seawater, as well as that produced by host coral respiration, to fix an estimated 0.5–5.0 kg of carbon per square meter per year (Muscatine 1990 and references therein). Only a small fraction of the fixed carbon is used for growth and respiration of the symbionts themselves. Instead, up to 97% of this abundant organic carbon is translocated to the animal host in the form of sugars, lipids, and other compounds that "subsidize" host growth and respiration (Muscatine 1990; Moberg and Folke 1999). Ultimately, approximately half of the photosynthetically fixed carbon is either incorporated as carbonate into the skeletal matrix of the reef or released into the water (Muscatine 1990; Moberg and Folke 1999). This symbiotically derived carbon is thus key to the corals' two habitat-constructing impacts: calcification that builds the physical reef structure, and primary productivity for the reef community.

The resulting coral reefs give rise to highly productive and diverse ecosystems. For instance, although this habitat covers less than 0.5% of the ocean floor, almost one-third of all marine fish species are believed to occur on coral reefs (Moberg and Folke 1999; see also Messmer et al. 2011). Along with a characteristic rainbow of associated fish and other vertebrates such as sea turtles and marine mammals that visit to graze and hunt, reef communities include a rich and diverse invertebrate fauna of sponges, echinoderms, mollusks, crustaceans, and polychaete worms. Primary producers in coral reefs include diatoms and other planktonic microalgae (in addition to the zooxanthellae described above), as well as green and red macroalgae. Among the latter are calcium carbonate-producing *crustose coralline algae* that also contribute to physical reef formation.

Like other complex ecological communities, this system self-organizes based on dynamic interactions among primary producers, herbivores, predators, competitors, and facilitators. Corals maintain their dominance or codominance in the reef community by virtue of several such interactions. For instance, the intense competition for space (and hence light) between corals and macroalgae is mediated at several trophic levels (Mumby et al. 2006; Powell et al. 2014; and references therein). By preferentially grazing on macroalgae, herbivores such as sea urchins, order Echinoidea, can prevent reef-building corals (with their photosynthetic symbionts) from being overgrown and outcompeted. As a result, these algal grazers help to maintain a coral-based community, as do predators of coral-eating animals (although some direct feeding by predators on coral is needed as a distribution vector for zooxanthellae; Moberg and Folke 1999). Conversely, predators that reduce herbivore abundance (e.g., fish that eat sea urchins), or invertebrates that release algae from grazing by providing an alternative food source for generalist fish (e.g., sponges), may promote the dominance of algae over corals (González-Rivero et al. 2011). Marine reserves that aim to restore large predators must be managed so as to avoid overpredation on populations of grazers, if the coral reefs they encompass are to persist (Mumby et al. 2006). Outside of reserves, overfishing of herbivorous fish can have the same result, leading to local replacement of corals by algae (Anthony et al. 2011; Pratchett et al. 2011).

Along with grazing herbivores, the ecological dominance of corals is strongly influenced by sponges, phylum Porifera, which are ancient metazoan animals that comprise a major part of the reef fauna. Sponges compete with corals and other sessile animals for space and food resources, while themselves providing food for fish, turtles, and starfish (Powell et al. 2014). Sponges can be fierce competitors with corals, often displacing them through direct, antagonistic interactions. "Excavating" sponges such as *Cliona delitrix* (from the Latin *delitor*, meaning "obliterator") establish themselves in the reef by boring deeply into the calcium carbonate matrix and then killing adjacent coral tissues by releasing allelochemicals directly from sponge cells to living coral cells (Chaves-Fonnegra and Zea 2007; González-Rivero et al. 2011).

Beyond supporting this ecologically rich marine biome, coral reefs play a role in building two associated types of ecological community. Because reefs physically buffer the coastline from storms, waves, and ocean currents, they create lagoons and sedimentary areas that (over geological time) facilitate the development of mangrove forests and seagrass beds (Moberg and Folke 1999). Reefs are functionally interwoven with these two adjoining habitats; all three interact to bind sediment and to exchange nutrients. These ecosystem processes are influenced in complex ways by individual feeding and migrating behavior (e.g., Acosta and Butler 1997). For instance, grazing sea urchins and herbivorous fish that move from reefs to seagrasses affect community dynamics at the seagrass beds (in part by weeding out large algae) but may migrate back to deposit their nitrogenous wastes at the reef. Migrant reef animals may also use the grass beds as protected breeding, spawning, or nursery grounds. These links between coral reefs and associated marine communities ultimately lead to the export of nutrients, organic materials, and plankton to surrounding waters (Moberg and Folke 1999).

Because of the far-reaching impacts of coral reefs in supporting marine biodiversity and ecosystem function, the well-documented decline and loss of these ecosystems comprise a particularly troubling aspect of contemporary environmental change (references in T. Hughes et al. 2010; Powell et al. 2014). That decline, along with those of seagrass and mangrove systems, exemplifies how the loss of major habitat-constructing organisms (e.g., scleractinian corals, marine grasses, and mangrove trees) can result in habitat degradation and loss of structural complexity, both of which lead to reduced community biodiversity (Messmer et al. 2011; Pratchett et al. 2014).

A number of natural and anthropogenic factors, including disease outbreaks, hurricanes and other severe storms, coastal development (with resulting fragmentation, eutrophication, and sedimentation), marine pollution, and mining of reefs for building materials, can contribute to the decline and loss of colonial reef-forming corals (Powell et al. 2014). Global climate change is a particular concern, because high water temperatures disrupt the symbiotic relationship between corals and their photosynthetic symbionts. Periodic increases in seawater temperature (e.g., of 1.0° C above normal local maxima) cause corals to expel their zooxanthellae, a phenomenon known (for the loss of pigment color) as *coral bleaching*. Such "bleaching" has been increasing in frequency and geographic extent for the past three decades (Pratchett et al. 2011 and references therein; see Figure 6.5). The loss of the photosynthetic symbionts changes the corals' metabolic function and nutrition, leading to prolonged stress, reduced vigor, and eventually mortality; loss of the mutualism also changes the calcium flux in the reef system (Moberg and Folke 1999). Coral bleaching can also occur in response to heavy metal contamination or increased coastal runoff due to forest clear-cutting (Moberg and Folke 1999), a poignant reminder that environmental management decisions in inland cities or forests can disrupt cellular processes in marine organisms critical to entire marine communities.

Figure 6.5 Reef-building corals collectively produce structurally complex underwater habitats that support diverse communities. The coral polyps themselves provide habitat for unicellular photosynthetic symbionts, zooxanthellae, which are major primary producers. Increased seawater temperature, pollutants, and other stresses can cause corals to expel these pigmented symbionts, causing the coral to appear "bleached" and, eventually, to die. This photograph of a reef in Hawai'i shows its three-dimensional complexity; the dark-colored corals retain living symbionts, while the "bleached" coral (top right) has lost its zooxanthellae. Photograph by Raphael Ritson-William, courtesy of Ruth Gates. For the color image, see Plate 19.

Environmental changes can also destabilize the ecological interactions that allow reef-forming corals to remain dominant in reef communities and that consequently serve to help maintain the diversity of habitats corals create. As noted above, overfishing or loss of reef habitat can cause population declines in the herbivorous fishes that regulate macroalgal populations so as to insure suitable sites for the settlement of coral polyps; a reduced abundance of these fishes can accelerate the process of reef habitat degradation by rendering structural reef renewal more difficult (Pratchett et al. 2011, 2014). Increased ocean acidification due to higher atmospheric carbon dioxide levels is predicted to reduce rates of calcification in corals (Anthony et al. 2011), with consequences for corals' competitive success as well as their collective habitat-forming activities. Physical and biotic disruptions to reef ecology can interact to dramatically alter these diverse systems. One pristine, remote Caribbean reef was transformed over a period of just 40 months from a diverse community codominated by corals and macroalgae to a depauperate system dominated entirely by algae, because of a combination of three distinct events: mass coral bleaching (likely due to stresses such as increased temperature and pollution), physical damage to the reef structure by a hurricane, and a pathogen outbreak that decimated populations of an herbivorous (algae-eating) sea urchin (Ostrander et al. 2000).

Disruptions to the reef habitat are accompanied by a reduced abundance and diversity of coral reef-associated fish species and invertebrates, with specific changes contingent upon each species' "resilience" to coral loss (Messmer et al. 2011 and references therein). For instance, reef fish that specialize on particular coral species for food, habitat, or both are likely to be steeply reduced in number, while generalist fish may increasingly come to dominate the altered reef community (Pratchett et al. 2014). The general outcome of these environmental stresses and associated trophic and competitive disruptions is a profound change in habitat-constructing activities, as corals are replaced as dominant ecological actors by either sponges or macroalgae, both of which give rise to less structurally and functionally complex habitats and less biologically diverse communities (Gonzalez-Rivero et al. 2011; Pratchett et al. 2011; Powell et al. 2014; additional references in T. Hughes et al. 2010).

6.3.2 Community impacts of native and invasive plants via effects on soil

Plants change the soil environment in numerous ways, primarily via the activities of their roots (see Chapter 5, Section 5.3.2). The most obvious of these habitat-constructing effects are changes to soil resource levels. By taking up nutrients, water, and oxygen and by occupying space, a plant's root system generally reduces the availability of these essential, shared resources to other co-occurring plant individuals (Marschner 1995; Fitter and Hay 2002). Root systems can also *increase* the availability of soil resources to neighboring plants, for instance, by redistributing water or promoting populations of beneficial microorganisms (H. Schenk 2006; Hinsinger et al. 2009; Hodge 2010). Through associations with mycorrhizal fungi, for instance, roots can enhance the decomposition of dead plant and insect tissues in the soil so as to increase the availability of nitrogen and possibly other nutrients as well (Hodge 2004), potentially enriching the system for cohabiting species. In some instances (discussed below), such effects can facilitate invasion by nonnative taxa (Theoharides and Dukes 2007; Stanton-Geddes and Anderson 2011).

Plants that host symbiotic nitrogen-fixing bacteria (see Chapter 5, Section 5.4.2) can also alter soil nutrient conditions in ways that influence community composition and dynamics. For example, the native yellow bush lupine, *Lupinus arboreus*, facilitates the invasion of its California coastal habitats by nonnative annual herbs by enhancing the nitrogen content of the local sandy soils (Maron and Connors 1996). By means of bacterial symbionts together with rapid growth rates, these shrubby perennial legumes fix substantial quantities of atmospheric nitrogen, adding up to 185 kg per hectare to the soil every year via their leaf litter (Maron and Connors 1996). Individual *Lupinus* plants killed by occasional outbreaks of insect herbivores leave nitrogen-enriched microsites that are readily colonized by fast-growing, nonnative annual plants. The combination of soil enrichment by *Lupinus* and these plant introductions results in a productive but less diverse community of weedy annuals (Maron and Connors 1996). More broadly, when plants either increase or reduce soil resource levels, they influence the physiological and developmental responses of neighboring individuals, shaping plant competitive interactions and trophic cascades and ultimately influencing community productivity and composition (see McCully 1999; Wolfe and Klironomos 2005; H. Schenk 2006).

In addition to effects on shared abiotic resources, plant root systems also influence the soil environment through a diverse repertoire of chemical root exudates (see Chapter 5, Section 5.3.2). These copiously produced substances have a major impact on the composition, distribution, and abundance of soil microbial and invertebrate communities (Walker et al. 2003; Badri and Vivanco 2009; Philippot et al. 2013; and references therein). By shaping the soil biota, root exudates have important indirect effects on the diversity and productivity of the entire system: soil bacteria, fungi, and invertebrates are essential to nutrient cycles because of their positive effects on decomposition and mineralization; they create soil aeration and aggregation; and they play key ecological roles as nitrogen-fixing (bacterial) and mycorrhizal (fungal) plant mutualists (McCully 1999; Wolfe and Klironomos 2005; and references therein).

The microbe-mediated effects of plant root exudates on ecological communities can be surprisingly complex. For instance, beneficial mycorrhizal associations can affect competitive interactions among plants in three ways: by changing the availability of soil resources, by exacerbating size inequalities among competing conspecifics, and by interacting differently with alternative host plant species (H. Schenk 2006). Negative plant–fungal associations mediated by root chemical exudates can also have long-term effects on community dynamics. Over time, species-specific soil pathogens can accumulate near a host, as occurs in the case of pathogenic *Pythium* fungi associated with native black cherry trees, *Prunus serotina*, in North America. The buildup of this specialized pathogen in the soil eventually reaches a level that suppresses black cherry seedlings, promoting successional change (Reinhart et al. 2003). Interestingly, in Europe, where black cherry is nonnative, *Pythium* fungi are absent and, in the absence of this enemy, the tree is invasive (Reinhart et al. 2003). Species-specific associations between beneficial fungi and plant hosts can also influence a community's vulnerability to biological invasion (Bruno et al. 2003). If mutualistic mycorrhizae favor native species over introduced cohabitants, this can provide the native with a competitive advantage based on enhanced resource access (e.g., Bray et al. 2003). In other cases, native mycorrhizae can facilitate the entry of an alien species into a community (Richardson et al. 2000), especially if greater mycorrhizal affinity for an exotic species puts native competitors at a disadvantage (Callaway et al. 2004 and references therein).

A fascinating literature has recently emerged that investigates the particular habitat-constructing impacts of nonnative invasive plants on soil environments. These effects occur in part through novel interactions between introduced plant taxa and soil biota (reviewed by Wolfe and Klironomos 2005). For example, introduced plants can form new associations with native nitrogen-fixing bacteria to increase soil nitrogen content and thus can alter the native plant community, facilitate the entry of additional, nitrogen-demanding invasives, or both (Theoharides and Dukes 2007). Both pathogenic and beneficial soil microbes can influence the success of a given plant invasion and its ultimate community impact (Reinhart and Callaway 2006). In turn, nonnative species can alter the soil's chemistry, and hence its microbiota, in ways that will feed back to affect the macroorganismic community (Wolfe and Klironomos 2005). For instance, diffuse knapweed, *Centaurea diffusa*, which is a Eurasian *Asteraceae* species that is invasive in western North America, produces the antimicrobial allelochemical *8-hydroxyquinoline*. This root exudate changes the composition and density of the microbial community in North American soils so as to reduce growth in native plants near knapweed individuals (Callaway et al. 2004). Interestingly, this microbe-mediated competitive effect does not occur in Eurasian soils, where the (presumably coevolved) biota is 8-hydroxyquinoline resistant (Vivanco et al. 2004).

Roots of certain introduced plants release allelopathic exudates that directly suppress heterospecific (e.g., native) plant individuals so as to create a weakened competitive environment, thus rendering the plant community in the introduced range more invasible (Thelen et al. 2005 and references therein). A well-studied case is that of spotted knapweed, *Centaurea maculosa*, which is a formidable invasive in North America (and congener of the species discussed above; also see Chapter 5, Section 5.3.2). *Centaurea maculosa* roots release a phytotoxic compound called *(±)-catechin* (Bais et al. 2003), which has been experimentally shown to inhibit germination, root elongation, and plant growth in a number of native North American species (Ridenour and Callaway 2001; Thelen et al. 2005). This biochemical habitat-constructing activity has evidently contributed to the successful spread of *Centaurea maculosa* in North America: within approximately 120 years of its introduction, *Centaurea maculosa* has spread across nearly three million hectares, forming dense stands that almost completely exclude native species (Callaway et al. 1999). Perplexingly, the species' root-exudation behavior renders it invulnerable to typical "biocontrol" measures used against invasive plants: when *Centaurea maculosa* plants are attacked by either the root-boring specialist knapweed root moth *Agapeta zoegana* or the common cabbage looper, *Trichoplusia ni*, which is a voracious leaf-eating generalist, the herbivory stimulates the plants to produce far greater amounts of the phytotoxic (±)-catechin, resulting in even greater suppression of native plant growth (Callaway et al. 1999; Thelen et al. 2005).

An even more diabolical case of biotic habitat construction by an invasive plant is that of garlic mustard, *Alliaria petiolata*, a species native to Europe and which was also introduced to North America in the nineteenth century. Garlic mustard carries out a multipronged biochemical assault on communities in its novel, introduced range. The species has recently invaded forest understories in much of North America, suppressing populations of native plants through the production of allelopathic chemicals that directly inhibit seed germination (Prati and Bossdorf 2004; Figure 6.6). In addition to producing herbivore feeding deterrents such as cyanide, garlic mustard plants release *allyl isothiocyanate* (AITC), which is a potent toxin for soil fungi (Cantor et al. 2011). Specifically, AITC suppresses both spore germination and hyphal growth in the arbuscular mycorrhizal fungi that form beneficial associations with the roots of native plants (Cantor et al. 2011 and references therein).

Figure 6.6 The European herb garlic mustard, *Alliaria petiolata*, is a formidable invasive in North America by virtue of changes it makes to its competitive environment. *Alliaria petiolata* plants produce allelopathic compounds that inhibit germination of native plants. They also release a powerful fungal toxin, suppressing the growth-promoting mycorrhizal fungi that colonize the roots of native plants. Vegetative tissues produce toxins that deter herbivory. As a result of these biochemical modifications to the environment, *Alliaria petiolata* creates dense and extensive monospecific stands. Steven Katovich, USDA Forest Service, Bugwood.org, Creative Commons Attribution 3.0 License; <http://www.forestryimages.org/browse/detail.cfm?imgnum=5468250>.

Along with introduction history, dispersal traits, and habitat-constructing effects on competitive interactions such as those discussed above, the spread and ecosystem impact of invasive plants reflects individual ecological developmental properties (Baker 1965). These include the ability to efficiently transform limited resources into reproductive output, to colonize disturbed sites and exploit their often resource-rich conditions, and to express adaptive plasticity for growth and survival under novel stresses (Maron and Connors 1996; Sultan 2001, 2004; C. Richards et al. 2006; Theoharides and Dukes 2007; Matesanz et al. 2012). In native as well as nonnative taxa, individual phenotypic responses play a major role in shaping ecological communities, as seen in the case studies below.

6.4 Case studies: community-level effects of individual eco-devo responses

6.4.1 Plastic aspects of plant–pollinator mutualisms

Interactions between plants and their pollinators have long been recognized as essential for both partners. The great majority of terrestrial plant species (an estimated 75% of angiosperms) depend on animal vectors—primarily insects—to effect fertilization by carrying pollen from one conspecific individual to another; for their part, animal pollinators depend on food and chemical rewards such as nectar, excess pollen, and other floral products which they obtain from the plants they visit (reviewed by R. Mitchell et al. 2009). Following a long-standing emphasis on specialized pairwise relationships, the recognition that these important mutualisms can be of a generalized nature as well (Fenster et al. 2004 and references therein; see historical overview by Waser 2006) has underscored their complex interactions with community structure (Burkle and Alarcón 2011).

Both the features through which plants attract and reward their animal pollinators, such as floral morphologies, markings, and biochemical products, and the structural, biophysical, and behavioral traits of the animals themselves as floral foragers and pollen vectors are generally viewed as constitutive, species-level characteristics that result from a coevolutionary history (R. Mitchell et al. 2009; Schiestl and Johnson 2013). Certainly, this viewpoint has yielded fascinating insights into macroevolutionary innovations and patterns. However, at an ecological scale, these traits can also be surprisingly labile, although studies of floral plasticity in particular are few. For instance, in the hummingbird-pollinated California chaparral shrub *Epilobium canum*, rates of nectar production were found to vary substantially depending on light and moisture levels, in genotype-specific ways (Boose 1997). It is not known how this among-plant variation for carbohydrate-rich nectar rewards affects the pollination behavior of hummingbirds, which have high energetic requirements. Even floral scent (which is generally assumed to be constitutive) can be influenced by environmental factors such as air temperature, light availability, humidity, and soil moisture availability (Majetic et al. 2009). Individual plants of the insect-pollinated perennial herb *Hesperis matronalis*, which is in the mustard family, changed both the composition of their floral scent (i.e., the proportions of different volatiles, e.g., aromatics and terpenoids) and the rate of scent emission, depending on nutrient and moisture availability in the field (Majetic et al. 2009). In this study, environmental effects contributed to differences in pollinator attraction traits among field populations in different locations; further studies could elucidate the plant-reproductive and potential community consequences of these differences.

Flowering phenology (the seasonal timing of flower production) is an environmentally labile trait that not only influences but indeed makes possible plant–pollinator interactions in real communities. Although species' phenologies have evolved to coincide with the activity of their major pollinators, the developmental transition to flowering is often hastened at warmer temperatures. Moreover, individuals of different species advance flowering time to different extents in response to a given temperature increase (Wolkovich et al. 2012). Consequently, increases in temperature under global climate change may profoundly disrupt plant–pollinator mutualisms at both the species and community levels (Parmesan 2006 and references therein; see Both et al. 2009 on phenological mismatches more generally). Potential disruptions of this sort were investigated in a central North American plant–insect community for which detailed phenological data were available. By the late twenty-first century, mean annual temperatures in this region are expected to rise by 3.5°C to 5°C and, depending on species, individual plants are predicted to flower from one to three weeks earlier (Memmott et al. 2007). Using a simulation model based on known plant–pollinator relationships in this system, these researchers predicted substantial, direct, immediate impacts on the community as a whole: results showed that many pollinator species would face disrupted food supplies and hence have lower fecundity and reduced population growth rates, and that plant outcrossing success and hence reproductive output would likewise decline (Memmott et al. 2007).

Longer-term ecological and evolutionary effects of species-specific shifts in floral phenology due to climate change are less clear. Competitive interactions will clearly be destabilized, at all trophic levels (i.e., among plant taxa, pollinators, and those species that feed on or parasitize pollinators or their predators). Pollinators may change their behavior to forage more broadly, a change which would lead to selection for less specialized morphologies and possibly create more plant hybrids; plant taxa may evolve toward self-fertilizing or apomictic breeding systems or go locally extinct. Indeed, temperature-induced phenological plasticity may lead to the "extinction of interactions" between plants and their pollinators (Memmott et al. 2007, 715; see Parmesan 2006; Van der Putten et al. 2010; and see Ovaskainen et al. 2013 on climate change effects on biotic interactions across trophic levels).

In addition to temperature, environmental variables such as light and moisture availability, attack by pathogens or herbivores, soil chemistry, and even soil microbiome can influence floral phenology (M. Wagner et al. 2014 and references therein). The foraging activities of animal pollinators are also influenced by a broad range of abiotic and biotic factors (Polatto et al. 2014 and references therein). Like plant phenological plasticity, behavioral plasticity of pollinators can also contribute to community-level impacts. For example, the introduction of nonnative plants to an existing natural community can create new patterns of foraging behavior in generalist pollinators, thus inducing changes in competition, facilitation, and ecological network properties (Memmott and Waser 2002; Albrecht et al. 2014).

Pollinator flight and foraging behavior in both natural and crop systems is under intense investigation, as populations of major pollinators such as bees are declining worldwide (evidently due to a synergistic combination of anthropogenic stresses including agricultural chemicals, habitat loss, and novel pathogens spread by introduced insects [Potts et al. 2010]). In view of the likely regional extinctions of some taxa, a pressing question is the extent to which the loss of particular pollinator species will cause plant communities to decline (Biesmeijer et al. 2006). Some simulation models (references in Brosi and Briggs 2013) have found that, even when several pollinator species are lost, plant communities are predicted to remain stable because of existing redundancies in pollinators' plant preferences. However, these models perhaps unrealistically have assumed that pollinator behavior would remain constant in the event of changes to a community's set of floral visitors. In contrast to this assumption, a manipulative field study found that loss of even a single insect species led to major behavioral shifts by remaining pollinators (Brosi and Briggs 2013). Even more worrying, these changed floral visiting behaviors reduced effective pollen transfer among conspecific plants and hence their reproductive output. In this innovative study, the most abundant species of bumblebee, genus *Bombus*, was temporarily removed from each of 20 subalpine meadows in the Rocky Mountains of western North America, and then the floral foraging behavior and pollen loads of the remaining several *Bombus* species in each site were examined. (Note that there was no decline in the total number of bees present, since the number of individual pollinators in each site was augmented by nearby bees.) In response to the removal of the most abundant pollinator species, the remaining bees immediately changed their behavior: they reduced their fidelity to particular plant species and dramatically increased their foraging visits to plants of other species, even within a single foraging flight. As a result, their pollen loads were more mixed: in the single wildflower species tracked, the proportion of conspecific pollen that had been deposited diminished, and the flowers later produced 32% fewer seeds than when the full complement of local bumblebee fauna was present (Brosi and Briggs 2013).

Both plants and their pollinators respond to abiotic and biotic circumstances by adjusting aspects of their phenotypes, including (but not limited to) phenology, biochemistry, and behavior. As a result of these (not necessarily adaptive) eco-devo adjustments, plant–pollinator interactions vary in ways that can cause community-wide changes via effects on plant fitness and trophic cascades. The examples presented above were chosen to focus on anthropogenic environmental changes—higher temperatures, species introductions, and biodiversity losses. Because of this, they also serve to illustrate a more general and indeed crucial point about the individual plastic adjustments organisms may express specifically to novel conditions. Phenotypic traits, including those that underlie ecological interactions, may be expressed in entirely unpredictable ways in the new abiotic and biotic circumstances that are rapidly arising worldwide (Sultan 2007). The immediate and long-term community consequences of anthropogenic changes are profoundly uncertain in part because of these unknown individual responses (and unknown patterns of genotypic and population variability in response; see Chapter 7, Section 7.2.2).

6.4.2 Community impacts of induced plant defenses

Many of the biochemical and morphological defenses of plants (and animals as well) are produced only in the event of attack (see Chapter 3, Section 3.4.3). Because such *induced plant defenses* can alter herbivore nutritional status, reproductive output, presence, or all three, these individual plastic responses can influence biotic interactions throughout the community—for instance, among the herbivores' competitors, predators, and mutualists. Induced defenses thus exemplify individual eco-devo responses that are at the same time changes to the biotic environment encountered by other community members. Attack- or threat-induced defenses can have an even longer-term impact on herbivore populations and their communities when defenses persist across generations (e.g., Agrawal et al. 1999; Holeski 2007; further references in Holeski et al. 2012; see Mondor et al. 2005 for an insect example).

Induced defensive responses, such as the production of repellent or toxic chemicals or structures, typically alter the biotic environment by directly reducing the activity, growth, or abundance of herbivores (or a combination thereof; Karban and Baldwin 1997). In certain cases, these responses instead initiate reciprocal plasticity in the form of a counterdefense by the herbivore, such as, for example, the activation of tissue detoxification mechanisms that overcome plant chemical defenses (references in Kopp and Tollrian 2003a). Such counterdefenses can permit herbivores to inflict substantial damage on plants, altering competitive and facilitative interactions at all trophic levels. Such one-on-one "arms races" have been particularly well studied in several crop plant–pest interactions. Many plants in the economically important and biochemically rich potato and tobacco family Solanaceae express complex, specific biochemical responses to herbivore attack, including the production of compounds that make leaf tissue indigestible (e.g., Steinbrenner et al. 2011). For instance, when leaves of the potato plant *Solanum tuberosum* are damaged by a specialist herbivore, the leaf-chewing larvae of the Colorado potato beetle, *Leptinotarsa decemlineata*, plant genes responsible for the production of chemicals that inhibit the enzymes used by the beetles to digest leaf proteins are upregulated (via a methyl jasmonate signaling pathway; references in Bolter and Jongsma 1995). The beetle larvae meet this targeted inhibition of their digestive activity with their own precise chemical response: they begin synthesizing *different* proteinases, ones that are not sensitive to the induced chemical inhibitors produced in the leaves. As a result of this counterdefense, larval development and growth are not affected by the plant's induced defensive response (Bolter and Jongsma 1995).

Such precise reciprocal plasticity might be expected to occur only in highly specialized, coevolved plant–herbivore relationships. However, a surprisingly similar biosynthetic switch is expressed by larvae of the beet armyworm, *Spodoptera exigua*, in response to induced proteinase inhibitors in tobacco leaves, *Nicotiana tabacum* (Jongsma et al. 1995). The beet armyworm, originally from Asia, has been introduced worldwide and is now a major agricultural pest on crops as phylogenetically diverse as asparagus, corn, soybean, and tobacco. Possibly this generalist herbivore has a large biochemical repertoire of alternative, inhibitor-resistant proteinases; alternatively, it may be able to synthesize certain enzymes that are insensitive to numerous inhibitors (Jongsma et al. 1995). Through effects on each other's growth, biochemistry, behavior, and abundance, pairwise plant defenses and herbivore counterdefenses can influence other ecological interactions in a community, such as competitive interactions among plants or herbivores.

In response to herbivore attack, a plant may initiate the production of toxins, compounds that reduce the nutritional quality or digestibility of plant tissues, structures that render plant tissues difficult to process or metabolize, or some combination of these (Karban and Baldwin 1997; see Chapter 3, Sections 3.4.3.1 and 3.4.3.2). In addition to their immediate impact on the attackers, induced defensive responses alter plant tissues as potential food for later herbivores on those plants (e.g., insects that might actively feed later in a growth season; Van Zandt and Agrawal 2004). Many plants express highly specific defenses in response to particular herbivores, so these tissue effects can depend on precisely which herbivore inflicted the initial attack. As a result, early episodes of herbivory can create species-specific changes to the amount and quality of available plant tissues and thus alter the feeding behavior, population dynamics, and competitive interactions among later herbivores and their associated communities (Van Zandt and Agrawal 2004). For example, early season herbivory by different insect species on common milkweed, *Asclepias syriaca*, had different effects on subsequent host plant use by another herbivore: monarch butterfly larvae, *Danaus plexippus*, avoided plants that had been previously attacked by the weevil *Rhyssomatus lineaticollis*; however, they fed intensively on plants that had been damaged by the leaf beetle *Labidomera clivicollis*. These secondary feeding changes in turn influenced the structure, composition, and species richness of the insect community (Van Zandt and Agrawal 2004).

Plants of many taxa respond to herbivore attack by releasing "bouquets" of volatile organic chemicals. These airborne bouquets mediate a number of ecological interactions in natural communities (see Chapter 3, Section 3.4.3.2). Herbivore-specific changes in the composition and abundance of these volatile compounds can be induced by insect oviposition or by feeding damage to leaf, stem, seed, and root tissues (e.g., via signaling pathways differentially regulated by salivary components used in chewing versus those used in sucking; de Vos and Jander 2009, 2010; A. Kessler and Heil 2011). In addition to having a direct defensive function as herbivore repellents, these highly specific induced emissions often constitute signals to enlist enemies of the herbivores (reviewed by A. Kessler and Baldwin 2001; Baldwin 2010). In such cases, plant volatile emissions have evolved specifically as a means of interacting with members of the broader community to create *indirect* defenses.

One of the first cases of indirect defense to be investigated was the response of lima bean plants, *Phaseolus lunatus*, infested by the voracious, generalist two-spotted spider mite, *Tetranychus urticae.* This tiny arthropod feeds by puncturing leaf epidermal cells and sucking out the contents. Although the wounds are individually quite small, spider mite reproduction on host plants is rapid and prolific; if populations grow unchecked, the mites can collectively inflict substantial damage. Bean plants attacked by these phytophagous mites release a combination of volatile compounds that attract the predatory mites *Phytoseiulus persimilis*. The recruited predators serve as plant "bodyguards" by feeding on spider mites and thus reduce the size and hence the destructive impact of the spider mite population (Dicke and Sabelis 1987).

Induced volatile signals provide location cues for a broad range of herbivore enemies involved in indirect plant defenses, including nematodes, predacious mites and insects, parasitoid wasps and other hymenopterans, and even birds (A. Kessler and Heil 2011). Individual predators and parasitoids can learn to associate particular volatile compounds with the presence of their herbivore prey, leading to reciprocally plastic behavioral responses that influence trophic webs (Agrawal 2001). An example is the mutualism between Neotropical *Cecropia obtusifolia* trees and *Azteca* ant species. Resident ants actively patrol the trees (in return for nectar and food bodies), attacking insect and vertebrate herbivores as well as encroaching vines (references in Agrawal 1998a, b). In this well-studied system, plant chemicals induced by herbivore damage lead to behavioral changes in the ants. When *Cecropia* leaves were experimentally damaged, the trees emitted volatile signals that (together with other damage cues) caused a fivefold increase in the number and patrolling activity of ants, within just eight to twelve minutes (Agrawal 1998b). In other ant–plant mutualisms, damaged plants respond to leaf herbivory by increasing the volume, sugar concentration, or amino acid content of nectar in extrafloral nectaries, thus enhancing rewards and inducing rapid ant recruitment (Agrawal and Rutter 1998). (A similar indirect defense response is expressed by larvae of the honeydew-producing acmon blue butterfly, *Plebejus acmon*: individual larvae respond to simulated attack by increasing glandular "nectar" production, thus causing Argentine ants to double their protective tending behavior [Agrawal and Fordyce 2000].)

In a number of crop–pest systems, indirect defensive signals such as volatile attractants have been precisely characterized. When individual plants in the mustard and cabbage family (Brassicaceae) are infested with any one of several (specialist or generalist) aphid species, the plants release glucosinolate-derived volatiles that attract the generalist aphid parasitoid *Diaeretiella rapae* (as shown through both behavioral studies and gas chromatography–electroantennography studies; de Vos and Jander 2010). Several crops in other plant families, such as potato and bean, express a different kind of volatile response to aphid attack. In these cases, plants that are infected by pea aphids, *Acyrthosiphon pisum*, release *E*-β-*farnesene*, the major component of the aphid's own alarm pheromone (references in Petrescu et al. 2001). This biochemical trickery may provide defense either directly by repelling aphids, or indirectly, by habituating them to the alarm cue so that they become more vulnerable to their own predators (Petrescu et al. 2001). Induced volatile emissions can provide information to plant neighbors as well as to animal members of the community. For example, young cotton plants, *Gossypium hirsutum*, infested with

Tetranychus urticae spider mites produce volatile compounds that serve as a warning signal to their nearby conspecific neighbors (Bruin et al. 1992). The uninfested plants experience a reduced rate of mite oviposition, presumably owing to two responses to the chemical warning: the production of a direct defense, and the release of preemptive volatile emissions of their own that attract predatory mites (Bruin et al. 1992).

The community contexts of tri-trophic defense systems have been little studied as yet, although initial findings point to some fascinating interactions. For instance, certain insect herbivores can disable plant defense responses to subsequent attacks. One recent study (Tooker et al. 2008) investigated a suite of insects that exploit the same host plant species, the widespread native goldenrod, *Solidago altissima*. The researchers were particularly interested in the effect of the gall-inducing tephritid fly *Eurosta solidaginis* on plant defenses, because these parasitic insects strongly affect other aspects of host plant phenotype. (Like other gall-forming insects, *Eurosta solidaginis* co-opts the goldenrod plant's shoot development to induce a tumor-like, food-rich, protective gall as home to the insect's larval brood.) As expected, leaf chewing by the generalist caterpillar *Heliothis virescens* elicited a strong up-regulation of volatile production by goldenrod plants, presumably recruiting natural caterpillar enemies such as parasitoids (Tooker et al. 2008). However, infestation by the gall-inducing fly failed to elicit this response. Even more surprising, when plants were first infested by the fly, subsequent feeding by the caterpillar failed to induce the usual production of volatiles, leaving the plants undefended against both attackers. In other words, the gall-inducing fly appeared to systemically suppress the goldenrod plant's indirect defense system (Tooker et al. 2008). By disabling the plant's volatile-production response to attack, *Eurosta solidaginis* could effectively reduce the mortality of leaf-chewing insects and thus shape the community of goldenrod competitors and herbivores. The fly could also prevent an attacked plant from inducing preemptive defense in neighboring plants or clonal ramets (Stireman and Cipollini 2008).

The tri-trophic defense system between plants, herbivorous spider mites, and predatory mites (see above) can also be altered when plants are concurrently attacked by a second herbivore. Plants infested by both spider mites and the whitefly *Bemisia tabaci* were found to be less attractive to predatory mites, because whiteflies evidently suppress the usual spider mite-induced up-regulation of JA that initiates the gene signaling pathway for the biosynthesis of predator attractants (P.-J. Zhang et al. 2009). In other cases, plants attacked simultaneously by different insect herbivores can produce a different volatile blend in response to each one that remains effective (references in de Vos and Jander 2010).

Like plants, fungi can respond to predator and parasite attacks through biochemical plasticity. When attacked by invertebrate grazers such as nematodes, collembolans, and insects, many fungi produce harmful or feeding-deterrent compounds, although the transduction of attack cues is not yet understood (Kempken and Rohlfs 2010). The ecological consequences of these animal-induced fungal responses too are only beginning to be studied. In one manipulative experiment involving the soil mold *Aspergillus nidulans*, a mutant strain in which the production of induced defensive chemicals had been disabled was found to be preferentially eaten by springtails and showed reduced fitness (Rohlfs et al. 2007). A second type of inducible fungal response, the production of cytoplasmic lectins, is toxic to both predators and parasites (Bleuler-Martínez et al. 2011). (Fungi can also react to animal competitors by releasing deterrent compounds; e.g., fungi may exude secondary chemicals into the feeding substrate of *Drosophila* larvae that are competing for the same decaying plant resources [Kempken and Rohlfs 2010].) Induced defenses are a well-recognized aspect of plant eco-devo response; it is clear that fungi likewise exhibit adaptive plasticity in this regard and that this type of fungal plasticity may influence community interactions, both in the rhizosphere and above ground.

CHAPTER 7

Natural selection in the light of the organism–environment relationship

A central part of Darwin's conceptual legacy is the powerful metaphor of "natural selection," an external force that—by analogy with the deliberate selection practiced by plant and animal breeders—chooses favorable traits over time (Darwin 1859; Ewart 1899). "Selection" worked brilliantly as a heuristic device, conveying the cumulative sorting process carried out by "nature" based on reproductive differences among individuals in their environments. During the New Synthesis, this sorting process was readily transferred to the level of Mendelian allelic variants, impervious to environment, that were understood as the heritably transmitted sources of the individual phenotypic variation underlying reproductive differences (see Section 7.1.1). But our growing understanding of the environmental context dependence of gene expression (and consequently of genetic variance), of the incorporation of environmental effects into individual inheritance (through epigenetic and cytoplasmic mechanisms), and of the niche-constructing effects of organisms on their selective environments makes clear that the environment's role in evolution is far more than a simple selective filter.

The preceding chapters have argued that environments shape the phenotypes of organisms and that, conversely, phenotypes shape the environments that organisms (and cohabiting neighbors) encounter. These dual processes are mediated by individual ecological developmental responses (i.e., environmentally contingent aspects of phenotypic expression), characterized as *norms of reaction*. Eco-devo responses, together with their niche-constructing impacts, create a nexus of reciprocal interactions within and across generations that constitute the organism–environment relationship. In a broad sense, this active, dynamic relationship itself evolves; it is both the consequence of natural selection and its cause. By incorporating environmental inputs and impacts, this perspective departs from a New Synthesis view of phenotypic information as fully contained within the nuclear genome, to be subsequently sorted by external environmental forces.

How can we integrate these reciprocal causal influences into our view of the evolutionary process, to study the "descent with modification" of the organism–environment relationship? It is too soon to know how much change to prevailing concepts, methods, and models this shift in perspective will require. But the explosion of new data regarding the complex, environmentally mediated regulation of phenotypic expression, the multiple types of heritable information, and the organismic feedback on selective environments makes clear that "the stuff of evolution" includes a much richer set of mechanisms and interactions than change in DNA sequence alone. Perhaps most compellingly, this moment in history demands a more integrated framework for asking whether organisms will functionally adapt to an unprecedented barrage of impending environmental changes. To address the crucial question of future adaptation, the scope of inquiry must be expanded to include a broader set of organism–environment causes and effects: ongoing changes to environmental cues and developmental inputs; the expression of phenotypes and of genetic variance in future conditions; complex realignments in

Organism and Environment. First Edition. Sonia E. Sultan.
© Sonia E. Sultan. 2015. Published 2015 by Oxford University Press.

ecological interactions; and altered effects of future organisms on their environments and communities.

How can contemporary insights about ecological development, epigenetics, and niche construction be integrated into evolutionary studies? Although exciting progress is well under way on each of these issues (as exemplified by many of the references here cited), integration is far from complete. This chapter offers a brief overview of some key factors to be considered. Section 7.1 begins with new research insights into the nature of variation, heredity, and fitness, insights that cannot be reconciled with a strictly allelic currency for adaptive evolution. Section 7.2 briefly summarizes a norm of reaction perspective that explicitly incorporates the widespread context dependence of gene expression and, consequently, of genetic variance. In place of a "special case" approach to the evolution of plastic norms of reaction, a unified eco-devo perspective can bring this context dependence to the general study of adaptive evolution and diversification. The chapter goes on to explore the consequences for selective evolution of epigenetic factors, of niche-constructing effects, and of reciprocal organism–environment influences.

7.1 Variation, heredity, and fitness

7.1.1 The allelic currency of evolutionary change

The twentieth-century's New Synthesis framework for adaptive evolution rests upon the marriage of two key tenets (Mayr and Provine 1980). First is Darwin's contribution: the principle that local conditions shape populations by allowing individuals with favorable traits to predominate numerically over time, generation by generation, in a cumulative process of "natural selection" (Darwin 1859). The second is Mendelian genetics: the notion that trait variation is produced by particulate genes (i.e., allelic DNA sequence variants) with fixed phenotypic effects, genes that are transmitted unchanged from one generation to the next. Together, these conceptual elements led to what R. A. Fisher (1930) definitively termed "the genetical theory of natural selection"—a deterministic, quantitative approach that defined adaptive evolution as change in allele frequencies. Apart from minor environmentally based "noise" around these signals (and the potential interference with selective frequency change occasioned by random drift, gene flow, and genetic architecture), causality in this selective process was clear: the genotype produced the phenotype, and the environment enforced selection owing to the relative success of those genetically based phenotypic variants. In this view the organism itself, its development, physiology, and behavior, and even its ecological impact were understood as the by-products of this largely deterministic process (e.g., Dawkins 1982; see historical discussion and references in Provine 1971; Hamburger 1980; Nijhout 1990; Sultan 1992; Amundsen 2001; Sarkar 2006).

This powerful conceptual unification worked because inherited allelic differences, understood both as the *cause* of phenotypic variation and as the faithfully transmitted *record* of relative success across generations, could now be construed as a single unifying evolutionary currency. In this twentieth-century view (based on transmission genetics and the discovery of DNA's informational code), unitary genetic variants entirely determined fitness-relevant phenotypic variation, and these genetic variants alone were heritable (Keller 2000; Neumann-Held 2006). As long as phenotypic variation was understood to result from genetic differences that were stably transmitted across generations, evolution could be defined as change in allele frequencies (e.g., Dobzhansky 1937; Futuyma 2009), including adaptive evolution due to (presumably genetically based) fitness differences (Haig 2007).

In this view, despite the quite evident environmental effects on realized fitness differences, only those aspects of adaptive variation that are consistently associated with DNA sequence variants can be considered relevant to evolutionary change (e.g., Orr 2009). This principle remains in place despite impressive progress in analytical approaches to the study of phenotypic selection within generations (references in Scheiner et al. 2000). As a result, the primary focus of evolutionary research remains allelic diversity, rather than phenotypes in environmental context, and even causally sophisticated studies are framed by the restrictive question, *what is the genetic basis of adaptive variation?* (e.g., Anderson et al. 2013; Leinonen et al. 2013). A second research legacy is the problematic distinction between "genuine" adaptation based on allelic change, and immediate adaptation due to plastic phenotypic expression of individuals (see Section 7.2.4). While alleles are certainly essential factors in both development and evolution, a research tradition that has focused on studying a subset of mutations with very large phenotypic effects, in controlled environments, and in lines selected to epistatically magnify rather than buffer these effects may have distorted their precise role (Lewontin 2000). Recent insights at all levels of biological organization have challenged the *uniquely* allelic basis of both phenotypic variation and heredity that has characterized the New Synthesis approach (for extended discussion and references, see Müller and Newman 2003; Odling-Smee et al. 2003; West-Eberhard 2003; Jablonka and Lamb 2005; S. Newman and Müller 2010; Pigliucci and Müller 2010; Hallgrímsson and Hall 2011).

An explosion of discoveries about gene regulation (including the roles of internal and external conditions, epigenetic modifications, noncoding RNAs, etc.) have made it clear that these complex environmental and molecular effects on gene expression are critical in determining phenotypic outcomes and hence individual fitness variation (see Chapter 1, Sections 1.1 and 1.2; also see Section 7.1.2). Molecular regulatory changes to gene effects include epigenetic modifications (see Chapter 1, Section 1.2.1), epistatic interactions among genetic elements (e.g., Plucain et al. 2014; see Lewontin 2000; Wolf et al. 2000; Wade et al. 2001; Phillips 2008), and stochastic variation (Z. Wang and Zhang 2011). Even more startling is the ubiquity of environment-specific gene expression (Turelli and Barton 2004; Paaby and Rockman 2014). In evolutionary terms, this context dependence means that genetic variance available to selection also depends on the environment (Husby et al. 2011; see Section 7.2.2). Microarray studies of both naturally evolved and model systems show that, depending on the species and environmental factor, 5%–50% of genome expression varies as a function of different environmental conditions (Snell-Rood et al. 2010). Likewise, pleiotropic effects of specific alleles on phenotypes also can also exhibit environment dependence: in the well-studied three-spined stickleback fish, for instance, the negative pleiotropic effect of "high-armor" alleles of the *Eda* locus on adult growth rate occurs in fresh water but not in salt water (Barrett et al. 2009).

The context dependence of gene expression is now beyond dispute; prominent New Synthesis evolutionists (Wray et al. 2014, 163) agree that "technological advances in the past decade have revealed an incredible degree of plasticity in gene expression in response to diverse environmental conditions." Evolutionary theoreticians are increasingly incorporating this context dependence into predictive models (e.g., Chevin et al. 2010; Gomez-Mestre and Jovani 2013). But does the growing recognition that gene expression is context dependent fundamentally undermine the unified allelic currency of evolution proposed by the New Synthesis? There is no consensus as yet. Thus, for example, Wray et al. (2014) remain committed to the primacy of DNA alleles as the source of adaptive variation (see also Scott-Phillips et al. 2013), while others argue that this new information fundamentally alters the role of genes in development and, consequently, in selective evolution (West-Eberhard 2003; Jablonka and Lamb 2005; Depew and Weber 2011; Laland et al. 2014; Noble et al. 2014).

The current view of heredity has also been profoundly transformed by recent findings. In addition to Mendelian variants (in DNA sequence and copy number), a number of developmental factors can be transmitted across generations (Danchin et al. 2011; Bonduriansky 2012; see Chapter 1, Section 1.2.1; also see Chapter 4, Section 4.4, and references therein). These include (a) molecular epigenetic modifications and small RNAs; (b) maternally transmitted elements, including hormones, nutritional states, proteins, RNAs, immune factors, and endosymbionts; and (c) directly inherited aspects of the environment ("ecological inheritance" sensu Odling-Smee et al. 2003), such as locality, habitat, climate, and often microsite; host plant, nest, or burrow; and parental care (Day and Bonduriansky 2011; Fitzpatrick 2014; Noble et al. 2014; and references therein). What is most destabilizing to a neo-Darwinian framework is the way that epigenetic, parental, and ecological information streams incorporate environmental inputs to the process of heredity (indeed, these phenomena are sometimes termed "Lamarckian," in the sense of discredited or invalid). If environments shape internally transmitted developmental factors in specific, repeatable ways, then adaptive evolution cannot be explained as a statistical sorting of randomly produced variants. These multiple modes of inheritance differ in fidelity and persistence (see Section 7.3). Thus, DNA sequence variants can persist unchanged across many thousands of generations; epigenetic modifications may last from several to hundreds of generations; maternal cytoplasmic factors persist across only a single generation; and directly inherited environmental factors can persist or, in some cases, accumulate over any possible number of generations.

To take a simple hypothetical example, imagine two individuals in a plant population: one growing in an open, sunlit microsite, and one in a shaded spot. What developmental information will their respective progeny inherit? Obviously, they will inherit DNA sequence differences due to parental genotypes and recombination. These genetic differences will bear a signature of previous evolutionary and reproductive history within their respective lineages but will carry no record of the maternal (and possibly paternal) microsite difference. However, these inherited genomes may be different epigenetically as a result of contrasting parental environments, depending on exactly how such epigenetic changes are induced and (maternally and paternally) transmitted. In addition, maternal cytoplasmic factors will be transmitted that vary strongly with microsite, including nutrient and hormone composition and quantities. Finally, although the sunny versus shaded parental environment is not transmitted internally per se, offspring inherit many developmentally relevant aspects of habitat, and possibly of microsite as well, by virtue of dispersal and germination characteristics that are themselves influenced by parental environment (see Chapter 4, Section 4.1).

One reason these nonallelic aspects of inheritance were so long overlooked is that they are easily confounded with genetic (i.e., sequence-based) factors, either because of similar patterns of inheritance or because of complex interactions between them (Johannes et al. 2009; Danchin et al. 2011). For instance, common-garden studies comparing individuals collected from different field families or populations routinely subsume possible parental environmental effects (due to microsite or local habitat) within presumed genetic differences. To avoid confounding such parental environmental effects with genetic differences, some researchers impose on field-collected experimental material an intervening generation under uniform conditions (e.g., Sultan et al. 2009). Yet, if these parental effects are not transmitted cytoplasmically but instead mediated by epigenetic switches, they may persist for several generations, resulting in an indefinite, longer-term confounding of genetic and environmentally induced epigenetic variation. This type of confounded causation may pose a problem in the search for inherited factors in human diseases that exhibit a familial component. Based on a working assumption that those inherited factors were DNA sequence variants (acting singly and epistatically), the primary research approach has been genome-wide association studies (GWAS) designed to identify specific alleles associated with phenotypic outcomes (Mattick 2012). Yet despite a considerable research investment, GWAS studies have identified only a small proportion of even the estimated *heritable* component of diseases and other complex human traits (Slatkin 2009). Depending on their patterns of induction and loss (and resulting familial distributions), epigenetic modifications may explain a good deal of the missing causality of human diseases (Slatkin 2009; Gluckman et al. 2011).

Distinguishing genetic from other inherited factors is made more difficult because the different types of inherited factors may interact. Not only do environmentally induced epigenetic and cytoplasmic (e.g., egg or seed provisioning) factors influence how genetic information is expressed (Day and Bonduriansky 2011), both cytoplasmic and epigenetic effects of parental environment on offspring phenotype generally vary among genotypes (i.e., $G \times E$ interactions for transgenerational plasticity; Schmitt et al. 1992; Sultan 1996; Herman et al. 2012). Complex *genotype* × *environment* × *epigenotype* interactions have been identified in several studies of maternal environmental effects in mammals: the familial component of obesity and type II diabetes is evidently mediated largely by epigenetic changes induced by maternal nutrition (Gluckman, Lillycrop, et al. 2007), and the epigenetic consequences of uterine conditions vary among human genotypes (Teh et al. 2014; see Chapter 1, Section 1.2.1).

Because genetic and nongenetic inherited factors have different modes of transmission across generations, they can be viewed as multiple, interacting, short- and long-term information streams (Bonduriansky 2012). Although these different types of information contribute jointly to development within any single generation, they differ in their capacity to record evolutionary history. Only genomic DNA sequence, with its exceptional fidelity of transmission, provides a robust, cumulative record of the evolutionary and reproductive past within any lineage. Other types of inherited information (e.g., epigenetic and cytoplasmic factors induced by specific stresses) generally bear a shorter-term signature of previous environmental conditions. Accordingly, measuring changes in the frequency of genetic alleles provides a uniquely stable way to *track the consequences* of past evolution, including the genetic impact of previous selection. However, because selection acts on *phenotypes*, allelic frequency data provide only partial information regarding the phenotypic variation that *caused* that prior selection and that will provide the material for future selection. Because it conflates the utility of allele frequency changes as a cumulative historical record, with a causal role for alleles as determinants of phenotypic variation, the single allelic currency omits from consideration essential aspects of evolutionary causation.

7.1.2 An eco-devo perspective on fitness

A central tenet of the New Synthesis framework was a rigid causal wall between the organism's internal genetic information and its environment. Because the environment does not directly influence the genetic information transmitted from one generation to the next, environmental effects on the phenotype have been seen as irrelevant to the cumulative process of evolutionary change (e.g., Futuyma 2009; Orr 2009). Yet, because selection acts on realized phenotypic differences in every generation, the environment is of central relevance to the evolutionary process (Wade and Kalisz 1990). The insights summarized above regarding variation and heredity (Section 7.1.1) suggest two important amendments to the conventional dismissal of environmental influences from evolutionary consideration. Together, these two points explain why studying allelic variation alone cannot predict evolutionary outcomes, even in terms of allelic variation. In other words, these amendments pertain even when the goal is to track evolution strictly as the change in the genetic composition of populations.

First, environmentally mediated aspects of inheritance can influence evolutionary outcomes, not by changing the DNA but by modifying phenotypic expression in individuals and hence affecting their relative fitness and the resulting intensity and direction of selection on the alleles they carry (Day and Bonduriansky 2011; Bonduriansky et al. 2012; Anderson et al. 2014; e.g., Stanton and Galen 1997). Several models demonstrate that nongenetic inherited factors can influence selective outcomes through effects on individual fitness (e.g., Kirkpatrick and Lande 1989; Donohue 2005; Geoghegan and Spencer 2012, 2013a, b). These selective effects do not require stable inheritance across multiple generations because they arise from effects of fitness differences expressed within a generation. However, integrating these diverse aspects of inheritance into evolutionary theory is a considerable challenge, because—unlike DNA sequence—they do not comprise a single continuous trajectory over time.

Second, as with nongenetic inherited factors, *immediate* environmental factors influencing phenotypic expression also shape evolutionary outcomes, by influencing the absolute and relative fitness of individuals within each generation (Odling-Smee et al. 2003; Laland et al. 2008; Kingsolver et al. 2012). As a result of these fitness effects, eco-devo responses to the environment affect selection (S. Wright 1931; Sultan 1987; Barton and Turelli 1989), even though those responses per se are not transmitted to the next generation. As noted above, these inherited and immediate effects on phenotypes require that we decouple allelic change ("Darwinian evolution," sensu Orr 2009) as a *record* of microevolutionary history, from the much more complex, environmentally mediated *causes* of fitness variation that contributed to that history (see Frank 2013 on separating causes of selection from fidelity of transmission).

In discussing how new information may expand our understanding, evolutionary biologists have largely aimed to harmonize recent insights with the New Synthesis model of changes in gene frequency over time. For instance, persistent epigenetic changes can be viewed as "epimutations" that may act very similarly to conventional gene sequence mutations (Haig 2007; Slatkin 2009). With respect to developmental plasticity, great interest has been focused on the possibility that phenotypic states initially produced as environmental responses may become genetically "assimilated" or constitutive and hence gain evolutionary validity as gene-based traits (West-Eberhard 2003; Moczek et al. 2011; see Section 7.2.3). While these questions are undoubtedly interesting, fully incorporating development into evolution by including immediate and inherited environmental influences on phenotypic expression *as such* presents a greater challenge. By means of these individual eco-devo effects, the environment as a whole plays a causal role in producing the fitness variation on which the environment as the selective agent then acts. Fitness thus becomes environmentally dependent in two distinct ways: the environment shapes the phenotype, and the environment determines which phenotype works best (e.g., M. Wagner et al. 2014). In some cases, different factors may effect these two, often successive roles of the environment (Scheiner 1993); often environmental effects on the phenotype are positively correlated to performance in that environment. By virtue of niche construction, the environmentally influenced phenotype also *changes* the environment as a selective milieu (see Section 7.4), reflecting the reciprocal causality that characterizes the organism–environment relationship.

Eco-devo effects on phenotypic expression can alter individual fitness in two distinct ways. First, an individual's environmental circumstances can enhance or diminish its fitness relative to others in a population through *inevitable effects* of resource or stress levels on growth and physiology (Mitchell-Olds and Shaw 1987; further references in Chapter 2, Section 2.3.4). These effects reflect evolved norms of reaction and hence genetic and phylogenetic as well as developmental constraints. As a result, inevitable environmental effects will vary depending on taxon, population, and genotype, as well as on the type of environmental factor, its level, and its timing (e.g., Chakir et al. 2002). Traits related to fitness typically show pronounced environmentally based differences (Haldane 1946; Barton and Turelli 1989). Favorable conditions during development often lead to a suite of positively correlated fitness traits, including total growth and offspring number and quality, known as the "silver-spoon effect" (Grafen 1988). Since resources are often irregularly distributed both spatially and temporally in natural populations, inevitable positive and negative phenotypic effects can create a mosaic of fitness differences that outweigh the typically more subtle average differences among genotypes (Stratton 1994; Stanton and Galen 1997; Sultan 2003a; Emery et al. 2011; and references therein) and consequently may act to maintain genetic variation in a form of evolutionary stasis (Gillespie and Turelli 1989; Nager et al. 2000).

Second, individual organisms also express environment- and trait-specific plastic adjustments based on *cue and response systems* that mediate environmental conditions in ways that benefit the organism (see Chapters 3 and 4). Such functionally adaptive eco-devo responses may also act to buffer the impact of selection. In an innovative model, Chevin et al. (2010) tested the combined effects of evolutionary potential (genetic variance) and adaptive plasticity on a population's predicted persistence in the face of sustained environmental change. When individuals expressed plastic responses that approached the optimal norms of reaction for an environmental gradient, the strength of directional selection was reduced (see Section 7.2.3). Although the distinction can be difficult to implement empirically, functional and ecophysiological information can help distinguish adaptive plastic responses from inevitable ones (Sultan 1995; Kingsolver and Huey 1998; see Nijhout 2003 on the joint roles of selection and constraint in norm of reaction evolution).

Notwithstanding Darwin's elegantly simple concept of "success in leaving progeny" (1859, 62), evolutionary fitness is extremely difficult to measure because it depends on the production of offspring that themselves survive and reproduce, making fitness dependent on phenotypes expressed over several generations (Slobodkin 1968; De Jong 1994; Reznick and Travis 1996; Kingsolver et al. 2012). Because fitness arises from the organism–environment interaction in each generation (Wade and Kalisz 1990), one generation's reproductive output often correlates surprisingly loosely with the subsequent generation's success (Thoday 1953; Biere 1995). Moreover, fitness emerges from the entire phenotype (Anderson et al. 2014), so it is problematic to identify a particular trait as a "target" of selection (Kingsolver et al. 2012).

The challenge of identifying the selective "target" is made all the more difficult by the fact that trait contributions to fitness are themselves environmentally contingent. This contingency arises from three distinct mechanisms. First, the *expression* of many traits is directly influenced by the individual's environmental conditions (see Chapter 1, Section 1.2.2; also see Chapters 3 and 4). When both the trait and fitness are influenced by the environment, measures of selection are distorted (Rausher 1992; Travis 1994; Scheiner et al. 2002; Stinchcombe et al. 2002; see Chapter 2, Section 2.4). Second, the precise *impact* of particular traits and trait states on fitness varies from one environment to another (Travis 1994; Chevin et al. 2013; and references therein). For instance, larger seeds are advantageous in nutrient-poor or intensely competitive seedling microsites, where the ability to produce extensive resource-collecting tissues right away may be key to survival and establishment (Leishman et al. 2000; Larios et al. 2014). But in other situations—where seed predators or pathogens are abundant, for instance—seed size may less important than shape or seed coat thickness (Moles et al. 2000; Schutte et al. 2014). Similarly, selection on defensive traits will vary depending on resource availability, competition, and other conditions that affect the fitness consequences of predation or pathogens (Huhta et al. 2000; Kniskern and Rausher 2006).

Third, selection on a trait can vary among environments as result of plastic expression of other, functionally interacting traits (see Section 7.2.1 on the separate issue of pleiotropic constraints). In animals, for instance, environmentally induced morphological and anatomical effects such as altered feeding apparatus or defensive structures can change the adaptive impact of related physiological traits or behaviors (D. Pfennig et al. 2010; e.g., Moczek and Emlen 2000). In fact, the fitness impact of a given trait can vary among environmental states even if expression of that trait remains constant and even if the same emergent phenotype is favored in the different environments, owing to functional interactions with plastic traits. For example, T. Griffith and Sultan (2006) studied selection on seedling height at contrasting field densities in annual *Polygonum* plants. Height emerges jointly from two plant developmental traits: the number of stem nodes and the elongation of internode stem segments. Although seedling height was positively related to fitness at both low and high density, selection on a constantly expressed trait, node number, was significant only at low density, because height at high density was more strongly influenced by plastic internode elongation responses (T. Griffith and Sultan 2006). In general, functional relationships among traits can be expected to vary across environments (Chevin et al. 2013), complicating causal approaches to detecting selection, such as path analysis (Kingsolver and Schemske 1991).

7.2 Norms of reaction and natural selection

7.2.1 Evolution of reaction norms

The expression of genes, and of genomes, is influenced by environmental conditions in all biological systems (Paaby and Rockman 2014; see Chapter 1, Section 1.2.2). As discussed in Chapter 1, the *norm of reaction* refers to a genotype's repertoire of environmentally contingent phenotypes—its pattern of eco-devo responses to varying conditions (see Chapter 1, Section 1.3). Like other properties of the genome, norms of reaction evolve (Schmalhausen 1949; Stearns 1989; Gavrilets and Scheiner 1993; Schlichting and Pigliucci 1998; Lande 2009; see Leimar et al. 2006 for a simulation approach that combines environmental and genetic inputs). As products of evolution, norms of reaction are shaped by phylogenetic history, genetic constraints, selection, and drift (Chapter 1, Section 1.3.1). Consequently, norms of reaction differ at the level of populations, species, and clades, as well as among individual genotypes (e.g., Gupta and Lewontin 1982; Bell and Sultan 1999; Sultan 2001; Heschel et al. 2004; Hodge 2004; L. Chapman et al. 2008; Ledón-Rettig and Pfennig 2011; Matesanz et al. 2012).

The genomic norm of reaction is further conditioned in the short term by inherited epigenetic and cytoplasmic factors that influence the phenotypic response of an individual to its immediate environment (Chapter 1, Section 1.3.3). These factors reflect the environment experienced by one or more previous generations, along with possibly stochastic molecular switches and resettings (see Section 7.1.1). A particular phenotypic outcome can thus be interpreted as a highly complex interaction between genotype, epigenotype, maternal environment, and current environment (e.g., Teh et al. 2014). The *G × E (genotype-by-environment) interaction* terminology that describes phenotypic causation has not yet been altered to accommodate this added complexity. To the extent that epigenotype is the result of environmental induction at some point, the term *G × E interaction* could be maintained, with the *E* term including recent/maternal environmental history along with the individual's immediate conditions. This usage allows the evolution of genotypic norms of reaction to be addressed in conventional terms as change in allelic components (for clarity, this approach is taken here).[1]

The evolution of highly plastic versus relatively constant norms of reaction for fitness-related traits has been the subject of considerable theoretical and empirical study (e.g., see reviews and discussion of alternative modeling approaches by Scheiner 1993; de Jong 1995; Via et al. 1995; Nussey et al. 2007; Chevin et al. 2010, 2013). Models agree broadly that plastic norms of reaction are favored by fine-scale spatial and temporal environmental variation, depending on the organism's ability to respond accurately to its environment (i.e., the existence of reliable environmental cues, and the organism's accurate perception and timely transduction of those cues to appropriate phenotypic expression; Moran 1992; Tufto 2000; Sultan and Spencer 2002; Berrigan and Scheiner 2004; Baythavong 2011; Scheiner and Holt 2012; and numerous references therein). Despite the central role of environmental variability in these selective dynamics, the scale and distribution of such variability in natural systems is seldom known (Stinchcombe et al. 2010).

The possible "costs" of adaptive plastic responses could also influence selective outcomes, but "despite a determined search for costs of plasticity" (Scheiner and Holt 2012, 752; e.g., Dorn et al. 2000), very little empirical evidence exists for such costs. Theoretically, only substantial costs of plasticity that are expressed in every environment appreciably alter selective outcomes favoring plasticity (Sultan and Spencer 2002). Such costs would result from distinctively expensive developmental machinery (DeWitt et al. 1998; Auld et al. 2010). In an eco-devo view, development in general arises from complex regulatory interactions that include environmental information; although response patterns in some taxa and traits are more plastic while others are less so, there is no reason to assume that more plastic norms of reaction reflect qualitatively different mechanisms (Sultan and Stearns 2005 and references therein). This unified view of regulatory machinery is supported by a meta-analysis conducted on 27 studies; the analysis found that (generally very weak) fitness costs of *canalized* response were as frequent as those associated with greater plasticity (Van Buskirk and Steiner 2009). Nevertheless, expressing a particular phenotypic response may entail energetic or materials costs, for instance, the production of thicker shells by snails in response to predation (Trussell 1996). Such environment-specific developmental costs will depend on the underlying mechanisms of phenotypic expression (Snell-Rood et al. 2010); many trait adjustments may bear no added costs (e.g., the spatial deployment of plant roots in resource-rich soil zones). In other cases, there may be weak costs

[1] As thus understood, norms of reaction may be empirically characterized by maintaining uniform environmental conditions in one or more previous generations (i.e., a common garden approach to previous environments). Altered expression of genotypic norms of reaction due to environmentally induced inherited effects can then be studied by controlled manipulation of conditions during one or more preceding generations (e.g., Sultan et al. 2009; Herman et al. 2012). Alternatively, the *genotype* component of the interaction term could be taken to refer inclusively to the individual's entire inherited response system (including epigenetic and maternal effects), since it is this system that interacts with the immediate environment. In practice, this approach has been taken whenever experimental material in norm of reaction studies has undergone uncontrolled variation for maternal and earlier environments. To the extent that epigenetic modifications are stochastic and spontaneous, they have been broadly confounded with genotype in norm of reaction studies.

associated with repeated phenotypic changes (Van Buskirk and Steiner 2009). It may be most realistic to assume that, in naturally evolved systems (i.e., those subject to selection in real environments), any costs of plastic norms of reaction or their expression are outweighed by the benefits of environmental adjustment (Chevin and Lande 2009).

Because the genetic architectures underlying norms of reaction are not yet known (Gutteling et al. 2007; Anderson et al. 2014), key questions about the evolution of plasticity remain open (Via and Lande 1985; Gomulkiewicz and Kirkpatrick 1992; Scheiner 1993; Chevin et al. 2010; Engen et al. 2011; Scheiner and Holt 2012; Scheiner et al. 2012; Z. Wang et al. 2013). A central question concerns the degree of pleiotropy versus modularity in regulatory systems. Pleiotropic gene effects could limit the expression of environmentally specific phenotypes (Via and Lande 1985; Papakostas et al. 2014). Several aspects of eco-devo response may be mediated by common pathways, as with the JA pathway in plants and the vertebrate endocrine system (see Chapter 3, Sections 3.4.1 and 3.4.2). Shared regulatory pathways may act as developmental constraints on the evolution and expression of new norms of reaction (Moczek et al. 2011). However, if different environmental signals initiate alternative, modular regulatory pathways, pleiotropic constraints would be reduced, and alternative phenotypes could be independently refined by selection (Snell-Rood et al. 2010). Genetic architecture is also important to the evolution of epigenetic change and stability systems as regulators of environmental response (Herman et al. 2014). Further information about both the genetic architecture of environmental response and the role of epigenetic modifications will allow for more realistic models for evolution of reaction norms (Draghi and Whitlock 2012).

One evolutionarily important property of reaction norms in general is their regulatory complexity as environmental cue and response systems (see Chapter 3). In part, this complexity reflects the diverse signals organisms in nature process in modulating their phenotypic expression (see Chapter 3, Sections 3.2 and 3.3). To take one well-investigated example, individual *Arabidopsis thaliana* plants adjust flowering time (and attendant life-history traits) on the basis of environmental factors known to include photoperiod, cold temperature, fluctuating temperature, water availability, atmospheric carbon dioxide concentration, soil nutrient levels, leaf herbivory, and seed predation (Brock et al. 2009; Wilczek et al. 2009; Anderson et al. 2014; and references therein). Even when the initiating cue for a particular eco-devo response is relatively simple, the underlying regulatory factor interactions provide multiple possible evolutionary change points, allowing for selective "fine-tuning" of the pathways for environmental sampling and perception, signal transduction, and expression of specific phenotypes (Refardt and Rainey 2010; Moczek et al. 2011).

For example, Refardt and Rainey (2010) demonstrated in phage that both sensitivity to an environmental signal and the threshold for response (prophage induction) could be readily "tuned" by experimental selection. Evidence for the evolutionary lability of reaction norms can also be found in natural systems (see references in Ghalambor et al. 2007). Based on local competitive conditions, new populations of *Onthophagus taurus* scarab beetles have rapidly evolved adaptive changes in the resource-based norm of reaction for producing the horns used in aggressive male–male encounters (Moczek and Nijhout 2003; see Chapter 3, Figure 3.1). In a resurrection study across 11 generations, replicate New England populations of the introduced plant *Polygonum cespitosum* were found to have evolved changed norms of reaction in response to full sun, including increased reproductive output and greater photosynthetic and allocational plasticity (Sultan et al. 2013). As human activities rapidly alter environmental cues, conditions, and distributions, the evolvability of eco-devo response pathways may be a crucial element in the future ability of organisms to produce adaptively matched phenotypes (J. Hutchings 2004; Gienapp et al. 2008; Kelley et al. 2012; Merilä and Hendry 2014; see the epilogue to this book, Section E.5).

7.2.2 The environmental context dependence of genetic variance

Selective change in reaction norms requires that genotypes differ in their eco-devo response patterns for fitness-related traits, a mode of variation characterized as *genotype × environment interaction variance* or *G × E variation* (see Chapter 1, Section 1.3.2; note that selective change could in theory occur if genotypic norms were strictly parallel across environments, but such variation is rare in natural systems; see below). Different genotypes may be more or less sensitive to a particular environmental signal, for instance, or express different degrees of phenotypic adjustment in response. Just as the norm of reaction is a different way to understand the genotype—as a set of responses rather than a single determinate outcome—norm of reaction diversity offers a more nuanced perspective on genetic variation and its selective consequences.

Genotypic differences in environmental response patterns are "a general feature" of continuously varying (e.g., growth and fitness) traits (Barton and Turelli 1989; see also Kruuk et al. 2008; Des Marais et al. 2013). However, as with any aspect of genetic variation, the amount and type of *G × E* variation available to fuel selective change in the form of adaptive norms of reaction to changed conditions will differ among taxa and populations (e.g., Matesanz et al. 2014). This potential is of particular concern with respect to global climate change and other novel environmental challenges that may lie beyond the tolerance range accommodated by existing norms of reaction—that is, beyond the ability of existing cue and response systems to produce functionally appropriate phenotypes that would allow populations to persist (Ghalambor et al. 2007; Visser 2008). As the onset of spring becomes earlier in temperate climates, a crucial question is the availability of *G × E* variation for norms of reaction that comprise earlier responses to seasonal cues. The presence of this kind of life-history variation was confirmed in a Dutch population of great tits, *Parus major*, in which female birds expressed heritable variation for reproductive timing in response to a warm previous spring (Nussey et al. 2005). As spring warming changes further, continuing selection on such diverse norms of reaction might allow this population to adaptively hasten reproduction to keep pace with its earlier-hatching caterpillar food supply (Nussey et al. 2005).

The term *G × E variation* can be defined as nonparallel norms of reaction across a given environmental range. Since genotypic norms are rarely parallel in natural systems (Haldane 1946; Gillespie and Turelli 1989), the *amount* of genetic variation, the *rank order* of relative fitness of genotypes, or both generally differ from one environmental state to another (see Chapter 1, Figure 1.6c, d). As a result of this environmental context dependence, the strength and direction of selection depend not only on the set of genotypes present but also on the distribution of environmental states (Chapter 1, Section 1.3.2 and references). In quantitative genetic terms, trait variances and covariances (e.g., among fitness components) are strongly environmentally dependent rather than constant (Via and Lande 1987; Barton and Turelli 1989; Falconer and Mackay 1996; Weinig and Schmitt 2004; Brock et al. 2009; Emery et al. 2011; and references therein). Consequently, as explained below, genetic variation is available to selection only in certain environments (Wilson et al. 2006; Snell-Rood et al. 2010).

One aspect of *G × E* variation (nonparallel norms) is that the *differences among a given set of genotypes can vary in magnitude*: genotypes can express similar phenotypes in certain biotic or abiotic environments yet express ones that differ substantially in others (Conner and Hartl 2004; Kruuk et al. 2008; and references therein). For example, genetic variation for water-depth niche and niche breadth in a population of vernal pool plants was detected only in the absence of interspecific competitors (Emery and Ackerly 2014). It is largely due to environmental effects on genetic variance that the potential for selective change in response to future conditions is so uncertain. A new environment can either trigger a selective event by releasing genetic variation or buffer selection, depending on whether norms differ or converge in that environment (Sultan 2007). A study of the leopard gecko, *Eublepharis macularius*, which has temperature-based environmental sex determination, revealed some good news with respect to anticipated climate warming: substantial *G × E* variation was found for the proportion of males produced at several realistic incubation temperatures. This result suggests that the potential exists in this species for environmental sex determination thresholds to continue to selectively evolve (Janes and Wayne 2006). At present, however, norms of reaction to predicted future conditions, and consequently the availability of $G \times E$ variance, are generally unknown.

As noted in Section 7.1.2, the fitness benefit of a given phenotype can also vary among environments; this benefit (i.e., the strength of selection) may interact with environmentally specific genetic variance to determine selective response (Falconer and Mackay 1996; e.g., Chevin et al. 2010). In a wild population of great tits, for example, earlier reproduction gave a stronger selective advantage in warm temperatures, under which conditions greater genetic variance for reproductive timing was also expressed (Husby et al. 2011). In this case, the environmentally dependent change in genetic variance would enhance a response to selection for changed reproductive norms of reaction (Husby et al. 2011). In contrast, the response to selection on offspring body weight in wild Soay sheep was reduced because maternal genotypes expressed *less* genetic variance in the poor-quality environment, in which size more strongly affected fitness (Wilson et al. 2006). Novel conditions may also elicit the expression of maladaptive or developmentally disrupted phenotypes, for instance, in response to a newly extreme level of a physical stress or a novel toxin (Ghalambor et al. 2007). The potential evolution of adaptive norms of reaction to future environments will be limited if those environments tend to elicit eco-devo responses that are convergent, dysfunctional, or both (Merilä et al. 2001; Sultan 2007; see below regarding "cryptic" variation that is expressed only in novel environments).

The second aspect of nonparallel norms of reaction is that *genotypic differences can vary in rank order from one environmental state to another* (e.g., see Chapter 1, Figure. 1.6d): "genotype A may be superior to genotype B in environment X, but inferior in environment Y" (Falconer 1989, 135). In this case, even if genotypic differences are expressed in all environments, selection can be constrained (Via and Lande 1985). Specifically, if the relative fitnesses of genotypes switch from one temporal state or microsite to another, and those states occur within a population or its dispersal range, $G \times E$ variation can maintain genetic diversity (Gillespie and Turelli 1989; see also Star et al. 2007a, b; shown empirically by Sultan and Bazzaz 1993b; Stratton 1994). When norms differ but do not cross, such "fine-grained" environmental variation can promote the evolution of generalist (e.g., highly plastic) norms of reaction that express relatively high fitness across many environmental states (Moran 1992; Sultan and Spencer 2002; Baythavong 2011; and references therein; these selective outcomes will depend quantitatively on the relative frequency of alternative environmental states and the size of genotypic fitness differentials).

Norms of reaction for plastic eco-devo responses can have particular implications for selection on underlying genetic factors. When trait states are expressed conditionally rather than constantly, alleles that contribute to those states are exposed to selection only in the inducing environments (Kawecki 1994; Kawecki et al. 1997; Van Dyken and Wade 2010; Ledón-Rettig et al. 2014). In temporally fluctuating environments or those with fine-scale spatial heterogeneity, each individual is likely to experience all environmental states, so even environment-specific genes will be exposed to selection in every individual. However, genes associated with plastic responses that are expressed only by individuals in certain environments will experience weakened or "relaxed" selection, as compared with alleles that are expressed in all individuals (D. Pfennig et al. 2010; Snell-Rood et al. 2010).

Relaxed selection due to environment-specific gene expression can substantially shape selective dynamics in populations that occupy heterogeneous environments—that is, natural populations (Van Dyken and Wade 2010). In particular, conditional expression can lead to the accumulation of *cryptic genetic variation*—variation that is expressed only in rarely occurring environmental states and that is therefore largely "hidden" from selection (Ledón-Rettig et al. 2014; Paaby and Rockman 2014). In this way, deleterious mutations can accumulate that affect phenotypes expressed only in unusual conditions, limiting the evolution of adaptive plasticity for rarely encountered environments (Fry 1996; Van Dyken and Wade 2010; T. Griffith and Sultan 2012). By permitting greater standing variation, relaxed selection on environment-specific alleles can also *promote* the evolution of adaptive response to a novel environment if genetic differences that were cryptic in previous conditions are favorable in the new ones

(Snell-Rood et al. 2010; Paaby and Rockman 2014; e.g., Husby et al. 2011). This scenario is intriguing because, like $G \times E$ interaction–driven changes in genotypic variance, the initiating evolutionary event can be an environmental change that "releases" existing variation to selection, allowing rapid adaptive evolution of reaction norms in the absence of new mutations (Lande 2009). However, it is not yet known whether cryptic genetic variation is prevalent or rare in natural populations, or, more importantly, whether it is likely to contribute to adaptive responses or is generally deleterious (Ledón-Rettig et al. 2014).

7.2.3 The dual impacts of eco-devo responses on selective diversification

Functionally appropriate eco-devo responses can affect selective diversification in two distinct and apparently opposing ways. It is important to note that, in both cases, such responses (whether arising from shared genetic or extragenetic regulatory factors) allow many individuals—or even an entire population—to adjust at once to phenotypically match environmental conditions. Such widespread responses can allow populations to withstand environmental demands without undergoing selective bottlenecks and hence maintain genetic variation (Gomez-Mestre and Jovani 2013). Furthermore, when new adaptive phenotypes are produced in this way (e.g., in response to an environmental change or dispersal to a different habitat), the novel phenotype will not be lost, as would likely be the case for a phenotype dependent solely on a beneficial mutation that is initially at low frequency (D. Pfennig et al. 2010).

To the extent that eco-devo adjustments allow individuals to adaptively accommodate unfavorable environments (Chapters 3 and 4), these responses can buffer local selection so as to maintain genetic variation and inhibit divergence (S. Wright 1931; Sultan 1987; Chevin et al. 2010; Gomez-Mestre and Jovani 2013). Using a developmental gene network model, Draghi and Whitlock (2012) confirmed that, as a result of selective buffering, greater standing genetic variation was maintained in populations consisting of phenotypically flexible genotypes. Indirect empirical evidence that plasticity can obviate selective divergence was provided by a study in which adaptation to thermal regimes was examined in European graylings, *Thymallus thymallus*. Differences in gene expression in response to temperature were greater within individual fish, and in the same presumably adaptive direction, than the average evolved differences between populations of different thermal origins (Papakostas et al. 2014). By allowing individual organisms to tolerate various stresses, adaptive eco-devo response patterns may play a crucial role in buffering populations from extinction as natural habitats are increasingly altered and degraded (Chevin et al. 2010, 2013). However, this benefit will depend on whether existing norms of reaction offer sufficient phenotypic breadth and whether environmental cues become too disrupted to allow for accurate perception and response (Ghalambor et al. 2007; Sultan 2007). If future conditions cause cues to become decoupled from appropriate responses, plastic eco-devo response systems could even hasten a population's extinction (Reed et al. 2010).

In addition to their role in *buffering* selective diversification, plastic norms of reaction can *facilitate* such diversification by giving rise to novel phenotypes that instigate new adaptive possibilities (the "Baldwin effect," as discussed by West-Eberhard 2003, 2005b; Badyaev 2009; see also Laland et al. 2008; Lande 2009; D. Pfennig et al. 2010; Thibert-Plante and Hendry 2011). As noted in Section 7.2.1, the regulatory complexity of environmental cue and response systems provides for reusing and re-combining pathways to produce new outcomes in a new site or if local conditions change (Moczek et al. 2011). Even stochastic epigenetic modifications to norm of reaction expression permit "phenotypic exploration" that may prove beneficial in some novel context (Jablonka and Lamb 2005).

Following the expression of a new phenotype (possibly by many individuals at once), selection will likely modify the genetic components of eco-devo response systems to refine and integrate cue and expression pathways to produce coordinated phenotypes (*genetic accommodation*; Nijhout 2003; West-Eberhard 2005a; Bateson and Gluckman 2011). Comparative studies can highlight trait interaction changes that arise through plastically induced phenotypic novelty. For instance, in species of spadefoot toads that do not normally consume shrimp, a shrimp diet induces a shortened gut morphology. In related species, this environmentally induced change is accompanied by a suite of functionally integrated changes to behavior, physiology, and other aspects of morphology; phylogenetic analysis indicates that these modifications evolved subsequent to the initial gut plasticity response (Ledón-Rettig et al. 2008). Another, more controversial, possibility is that response pathways may evolve such that a phenotype initially produced as one of several plastic alternatives becomes constitutive (*genetic assimilation*); this outcome is theoretically predicted if the plastic norm bears a cost or if the new environment does not vary (Lande 2009; Chevin et al. 2013).

In either case, ancestral plastic norms of reaction may produce morphologies that provide the basis for subsequent evolutionary divergence at the population or species levels, possibly contributing to adaptive radiation (West-Eberhard 2005a; D. Pfennig et al. 2010; Muschick et al. 2011). Supporting evidence comes from cases where alternative plastic morphs produced by individuals of an ancestral taxon mirror phenotypic differences among individuals of derived populations or species (Ledón-Rettig et al. 2008). This process could explain the repeated evolution within lineages of certain phenotypes that characterize distinct taxa (D. Pfennig et al. 2010). An example of this pattern of diversification is provided by the Midas cichlid, genus *Amphilophus*, which has radiated in parallel in several geographically separate lakes, starting with populations from the Great Lakes of Nicaragua (Muschick et al. 2011). In these fish, foods of different hardness induce the plastic expression of lower pharyngeal jaw bones, a type of anatomical diversity that distinguishes Midas species and indeed cichlids more generally. This plasticity may possibly have provided the alternative phenotypes that fueled the adaptive, resource-related radiation of the group (Muschick et al. 2011).

Environmentally induced resource-use phenotypes could cause ecological isolation that in itself could lead to rapid, sympatric speciation (D. Pfennig et al. 2010). This process has been investigated in North American *Rhagoletis* flies. In this system, individuals that encountered apple rather than hawthorn trees after emerging from their pupae were able to feed and oviposit on this novel host by virtue of behavioral and physiological plasticity. This change effected reproductive isolation that has led to sympatric evolutionary divergence for host plant preference "races" (Filchak et al. 2000; Linn et al. 2012; and references therein). Similarly, an environmentally induced shift in reproductive timing, for instance, in a marginal or high-altitude plant population, would effectively cause reproductive isolation from source populations that could facilitate local adaptive change and speciation (Levin 2009).

Interestingly, these two apparently opposing evolutionary consequences of adaptively plastic norms of reaction can work together. Individuals with such norms can spread to new locations and habitat types and can withstand in situ changes. They may consequently encounter novel conditions in which subsequent selection can occur. As noted above with respect to cryptic variation, standing genetic variation maintained by individual plasticity can also provide the basis for selective differentiation when a new environment is encountered in which variation is expressed (Badyaev 2005; Ghalambor et al. 2007). In such cases, individual adaptive plasticity can be positively correlated with genetic change under subsequent selection. For example, in a large experimental population of marine microbes, *Ostreococcus* spp., metabolic plasticity combined with fluctuations in carbon dioxide levels allowed a diverse group of genotypes to persist; the presence of these highly plastic genotypes fueled a strong subsequent response to selection at constant high carbon dioxide (Schaum and Collins 2014).

7.2.4 Plastic adaptation versus genetic adaptation?

The relationship between adaptive genetic differences and the adaptive diversity arising from eco-devo responses is complex and poorly understood (see Chevin et al. 2013 for an excellent review). The literature reveals a good deal of confusion on this point, reflecting numerous examples of presumably "evolved" population differences later determined to result from individual plasticity (Merilä and Hendry 2014). This distinction has been considered critical for evaluating the evolutionary potential of a species to adapt to future environmental challenges. For instance, a careful review of long-term studies in vertebrates showed that most of the many reported adjustments to climate change probably were plastic responses and not, as reported, "adaptations" based on genetic change (Gienapp et al. 2008).

Although it is possible to distinguish these causes of variation at the population level, it is problematic to separate out adaptive variation arising from the norm of reaction from variation due to selection on genetic variants per se (Forsman 2014). Surely, an organism's evolutionary and selective history cannot be causally *separate* from its immediate phenotypic expression in response to its environment. After all, it is the evolved, genomic norm of reaction that determines that response, through its interaction with environmental and epigenetic factors. The distinction between evolved "adaptations" and individual phenotypic responses—a distinction which echoes an untenable view of genes versus environment as alternative mechanisms of phenotypic determination—has led to a research program to "disentangle" these two modes of adaptation in order to understand future adaptive potential (Anderson et al. 2014, 6). However, since it is always true that evolution of any trait state is the evolution of how that trait is *expressed* in a given set of conditions, it is not clear that distinguishing gene-based adaptation as somehow more real or lasting will best inform us about tolerance of future conditions.

In addressing this key question, a unifying approach could instead be used, in which the evolutionary unit and the source of adaptive diversity are the same—the eco-devo response system with its particular allelic components. If evolutionary change is studied as change in the genomic regulatory system (which is further conditioned within each generation by transitory elements), adaptation can be understood as the outcome of selection on this system, as expressed in a given context. Rather than deconstructing the phenotype into "adaptations" versus "plasticity," it is possible to reframe the question of future adaptation in norm of reaction terms: what is the adaptive repertoire expressed by existing genotypes in response to a given environmental range (e.g., Anderson et al. 2014), what is the diversity among populations or taxa for these adaptive repertoires, and what is the genetic variance for them, as expressed in predicted future conditions?

7.3 Epigenetic factors and selective dynamics

A profoundly important update to the allelic model of variation and inheritance is the suite of recently uncovered molecular epigenetic mechanisms that influence gene expression: DNA methylation marks, histone acetylation and other chromatin changes, and small RNAs involved in gene silencing (see discussion and references in Chapter 1, Section 1.2.1). Following a flood of exciting and often astonishing findings, these epigenetic factors are now understood to be major, nonallelic sources of phenotypic variation in natural (including human) populations. As such, they will necessarily influence the impact of selection on the genetic composition of populations.

In contrast to rare and initially low-frequency genetic mutations, epigenetic factors are labile and can rapidly generate substantial phenotypic variation. Although changes in epigenetic state can occur stochastically (e.g., Becker et al. 2011; R. Schmitz et al. 2011), an even more fundamental contrast to DNA sequence variants is that these changes are often influenced by cellular and external environmental conditions. In many known instances, these changes are induced by specific signals such as temperature, oxidative stress, pH, salinity, diet, or social behaviors (E. Richards 2006; Jablonka and Raz 2009; Bollati and Baccarelli 2010; Verhoeven et al. 2010; Bateson and Gluckman 2011; Paszkowski and Grossniklaus 2011; Gill et al. 2012; and references therein). Such specific induction cues suggest that epigenetic modifications may serve as (presumably cost-free) mediators of both adaptive and maladaptive eco-devo responses, including rapid, population-wide phenotypic adjustments (Herman et al. 2014; see Chapter 3 and Section 7.2.3). Defining the role of epigenetic factors in adaptive evolution requires answers to several as yet unresolved empirical questions: the degree to which epigenetic changes are environmentally rather than stochastically induced and reset; the fitness impact of their phenotypic effects; their stability across generations; and their interactions with genetic variants (Kalisz and Puruggunan 2004; Geoghegan and Spencer 2013a, b).

Transient epigenetic modifications have evolved as nodes in developmental pathways that shape individual phenotypes. As elements of eco-devo regulatory systems, these molecular switches can either direct cell differentiation along canalized pathways or mediate diverse plastic outcomes. In such cases of induced, functionally adaptive phenotypic effects, epigenetic mechanisms play the same evolutionary role as other eco-devo response mechanisms, generating within single genotypes an adaptive diversity that can obviate or contribute to selective diversification (see Section 7.2.3). A remarkable instance of epigenetically mediated individual plasticity in a plant–fungal interaction was studied by Herrera and colleagues (2012). In natural populations of the flower-living yeast *Metschnikowia reukaufii*, the floral nectars offered by nearly 100 potential host plants provide an extremely large range of differing sugar proportions and concentrations; yet, the yeast is able to occupy nearly half of these host taxa. When nectar sugar composition was experimentally manipulated, the different nectars induced specific methylation changes to the yeast that appropriately altered sugar metabolism. This epigenetically mediated functional diversity allowed each yeast genotype to successfully grow in a broader range of nectar environments, generating physiological plasticity that resulted in greater ecological breadth at the population level (Herrera et al. 2012).

If transmitted heritably, such environmentally induced modifications can shape descendent phenotypes on the basis of conditions perceived by a parental or ancestral individual and thus are sometimes considered "Lamarckian" (see historical discussion in Jablonka and Lamb 2005). Whether induced or stochastic, to the extent that epigenetic changes are heritable, they may play a unique evolutionary role (see C. Richards, Verhoeven, et al. 2012). Heritable epigenetic modifications can be viewed as *epialleles* because they result in diverse phenotypes, even in the absence of DNA sequence variation (Rapp and Wendel 2005; Dolinoy et al. 2006; Heard and Martienssen 2014). Yet unlike allelic variants, epigenetic changes are set and subsequently reset, so that they fluctuate across some number of generations rather than provide a stable substrate for extremely long-term, cumulative selective change (Becker et al. 2011).

In many cases, epigenetic factors can be stably inherited, via either maternal or paternal contributions, across several to potentially hundreds of generations (Henderson and Jacobsen 2007; Johannes et al. 2009; Soubry et al. 2014; and references therein). Transgenerational epigenetic inheritance has been documented in laboratory lineages of *Escherichia coli*, yeast, *Arabidopsis*, maize, *Caenorhabditis elegans*, *Drosophila*, mouse, and rat and in a range of other bacteria, protists, fungi, animals, and plants (extensively reviewed by Jablonka and Raz 2009). Epigenetic factors may be an especially prevalent source of inherited variation in biological systems such as plants and fungi, where molecular state changes in somatic cells can be transmitted to gametes when vegetative tissues transition to reproduction (E. Richards 1997), as well as in some asexual systems that "bypass" meiosis (Verhoeven and Preite 2014). Conversely, "model" taxa with relatively early germline segregation (such as *Caenorhabditis elegans*, *Drosophila* spp., and *Mus musculus*) may have unusually low levels of epigenetic inheritance (Jablonka and Raz 2009).

Nongenetic, inherited effects of maternal environment on offspring phenotypes can contribute to inconstant and complex evolutionary trajectories, for instance, by causing time lags in the response to selection, "evolutionary momentum" when a response continues after selection has ceased, or both (Kirkpatrick and Lande 1989). Epigenetic factors that are likewise transmitted across a single generation can be expected to play a similar role, while persistently stable epialleles may be subject to the same selective dynamics as gene sequence variants for hundreds of generations (e.g., Cortijo et al. 2014; also see Haig 2007; Shea et al. 2011). In many cases, however, epigenetic variants appear to convey phenotypic information across an intermediate span of long-term stability: greater than that conveyed directly by mothers to their offspring via cytoplasmic elements but shorter than the thousands of generations through which allelic variants are faithfully transmitted (Jablonka and Raz 2009; Herman et al. 2014). For example, methylation-state variation that influenced flowering time and plant height in *Arabidopsis* was stably inherited across 8 generations (Johannes et al. 2009), while stable inheritance of methylation epialleles was tracked for 30 generations in another *Arabidopsis* study (R. Schmitz et al. 2011). (As is usually the case, however, these values reflect the duration of the experiments rather than a biological limit to epigenetic state persistence, which is seldom determined [Gill et al. 2012; Heard and Martienssen 2014].)

Do inherited epigenetic factors influence phenotypic outcomes enough to affect fitness and, consequently, selection? A number of suggestive cases indicate so, although, as yet, studies of natural populations are rare (Scoville et al. 2011; C. Richards, Verhoeven, et al. 2012). To date, most information concerns the phenotypic impact of inherited DNA methylation modifications. In mammals, methylation changes evidently mediate the adverse eco-devo effects of several environmental stresses (Jablonka and Raz 2009; T. Franklin and Mansuy 2010; see Chapter 1, Section 1.2.1 for references on human health impacts). Disrupted early postnatal care in mice causes DNA methylation profile changes that are transmitted to the subsequent generation, where they are associated with altered gene expression in the brain (even in progeny who did not themselves experience stressful rearing conditions; T. Franklin et al. 2010). Exposing rats to *vinclozolin*, a common fungicide used in agricultural fields, vineyards, and golf courses, causes changes in the brain transcriptome that can be transmitted (via sperm epigenomic transmission) through to the third generation and which alter the descendants' responses to current stresses (Crews et al. 2012). Beneficial inherited effects of diet can also be mediated by methylation changes, as shown in an experiment using the *viable yellow agouti* strain of mice (Dolinoy 2008). A maternal diet rich in the soy plant estrogen *genistein* caused increased methylation of the agouti gene early in offspring development, resulting in decreased expression at maturity of both agouti coat color and associated obesity (Dolinoy et al. 2006).

Far more information is available about the effects of inherited DNA methylation changes on plant phenotypes. Although the inducing cues are seldom known, a number of fitness-related plant traits can be affected by methylation epialleles (possibly acting in combination); such traits include developmental and reproductive timing, floral morphology, seed and vegetative pigmentation, height, root length, and pathogen resistance (Kalisz and Purugganan 2004; Johannes et al. 2008; C. Richards, Verhoeven, et al. 2012; Cortijo et al. 2014). Methylation-state differences among nearly isogenic lines of *Arabidopsis thaliana* produced heritable variation for root allocational plasticity and other developmental responses to drought and nutrient levels that are known to influence fitness (Y.-Y. Zhang et al. 2013). The more general fitness impact of inbreeding may even be due to methylation changes in some cases, rather than to the accumulation of recessive deleterious Mendelian alleles (see Charlesworth and Willis 2009 for an example of the latter mechanism). In the primarily outcrossing plant *Scabiosa columbaria*, self-fertilization caused a 10% increase in DNA methylation, as compared with outcrossed plants; this change was associated with the characteristic "inbreeding depression" effects of reduced biomass and physiological performance (Vergeer et al. 2012). Through such effects of inbreeding, epigenetic systems could play a role in the evolution of dispersal behavior and mating systems (Vergeer et al. 2012).

Environmentally cued, heritable methylation-state changes may also mediate specific adaptive responses. This has been found in several cases where (actual or simulated) herbivory-induced methylation changes initiated defensive responses in offspring or primed them to more quickly respond to direct attack (Holeski et al. 2012). In *Mimulus guttatus* plants, simulated herbivore damage to leaves caused changes in the activity of specific transcription factors associated with the production of leaf trichomes, epidermal outgrowths that release a sticky (and possibly toxic) repellent fluid (Scoville et al. 2011). Evidently due to inherited changes in DNA methylation or other epigenetic states that influence transcriptional regulation (and possibly cytoplasmically transmitted hormones as well), the progeny of damaged plants showed altered expression of over

900 genes primarily associated with stress response and cell wall development, as well as greater leaf trichome production, even though these plants had themselves experienced no leaf damage (Colicchio et al. 2015). Grazing on *Viola cazorlensis* plants in the field also induced methylation-mediated defensive responses that enhanced herbivore resistance (Herrera and Bazaga 2011).

Although heritable epigenetic variation is only beginning to be integrated into evolutionary theory (Jablonka and Raz 2009; Geoghegan and Spencer 2013a), certain selective implications are becoming apparent. This type of variation is generated at rates that are far higher than rates of genetic mutation, possibly by orders of magnitude (Klironomos et al. 2013; e.g., Becker et al. 2011; R. Schmitz et al. 2011). At the same time, epigenetic changes may often affect phenotypes more subtly than major sequence mutations, modulating gene expression in ways that may be more likely to generate useful variants (Kalisz and Purugganan 2004). Because epigenetic state changes can quickly produce widespread phenotypic variation in the absence of genetic diversity, they can potentially create fitness-related variation in clonal or highly selfing populations or in those that have undergone a genetic "bottleneck" (Kalisz and Purugganan 2004; Verhoeven and Preite 2014).

Whether it arises from heterogeneous epigenetic states or from direct eco-devo effects of varying microenvironments, fitness variation that is unconnected to allelic differences renders selection on those differences less efficient (Herman et al. 2014). In a mathematical model in which (rapidly generated) epigenetic and genetic variation were both acted on by selection, adapted phenotypes arose well before they could occur purely by selective change in the population's genetic composition (Klironomos et al. 2013). In this model, fitness increases over time were "decoupled" from gene frequency changes, obscuring "the signature of selection in the genetic record" (Klironomos et al. 2013, 571).

Because epigenetic state changes are generated at such high rates relative to sequence mutation, and because they are at some point reset, the selective dynamics of these systems are labile and complex. These dynamics have been explored in a series of innovative "population-epigenetics" models (Geoghegan and Spencer 2012, 2013a, b). These models test the selective consequences of environmentally induced, heritable epigenetic variation that contributes to adaptive phenotypes but varies in fidelity across generations, based on a "reset coefficient." Results show that, because epigenetic variation is repeatedly "regenerated," these systems never reach selective fixation, despite consistent fitness differences. For a trait that is controlled entirely by epigenetic factors, such dynamics result in epiallelic polymorphisms and higher phenotypic variance at equilibrium.

In the more realistic case where both epigenetic and genetic variation are involved, selection leads to multiple locally stable and unstable equilibria, a result that may help to explain the otherwise perplexingly high levels of phenotypic and genetic variation observed in natural populations (Geoghegan and Spencer 2012). When reversible epigenetic variation is induced by two fairly common alternative environments, the combination of continued environmental induction and intermittent resetting will prevent selective fixation and instead maintain epiallelic and hence phenotypic variation (Geoghegan and Spencer 2013a, b). Critical parameters in this case are the frequencies of the inducing environments, and the resetting interval; it seems intuitively correct that, in the real world, these key factors will indeed regulate selective outcomes on epiallelic variation. These results also support the view that epigenetic systems may maintain phenotypic variation in natural populations even in the face of strong selection on allelic variation (Herman et al. 2014). Because resetting prevents selective fixation of particular epialleles, if epigenetic states are sensitive to numerous environmental signals rather than precisely cued or if they change stochastically, natural populations may be expected to maintain very high levels of epiallelic and consequently phenotypic variation.

To what degree is the heritable variation that underlies phenotypic diversity in natural populations epiallelic rather than allelic? Techniques now exist to identify such variation, although they have yet to be applied to natural populations. *Whole-genome bisulfite sequencing* provides single-base resolution of DNA cytosine methylation states, which are analogous to single-nucleotide polymorphisms (Niederhuth and Schmitz 2014). Genetically uniform but epigenetically diverse experimental material can be created in the form of *epigenetic recombinant inbred lines* (*epiRILs*), for instance, by crossing parent individuals that have contrasting DNA methylation profiles but very few DNA sequence differences (e.g., Johannes et al. 2009; Y.-Y. Zhang et al. 2013). These epiRILs can be used to identify epigenetic quantitative trait loci, which are genomic regions of methylation change associated with heritable variation for complex traits such as growth and phenology (Cortijo et al. 2014).

Even more challenging to study, epigenetic changes and DNA sequence variants function as interacting rather than parallel inheritance systems (Bateson and Gluckman 2011). Epigenetic factors affect the fitness consequences of the alleles with which they are associated, creating highly variable epistatic scenarios and potentially introducing uncertainty to adaptive matching achieved via plastic response (Scheiner and Holt 2012). Initial studies indicate that the location, induction dynamics, and persistence of methylation marks and other epigenetic changes can vary among genotypes, creating the possibility for complex genotype × environment × epigenotype interaction effects on phenotypic expression (e.g., Herrera et al. 2014; see Chapter 1, Section 1.2.1., for discussion and further references) and selection on correlated genetic variants (Jablonka and Raz 2009).

For instance, epigenetic factors play a role in the highly complex environmental regulation of flowering time in *Arabidopsis thaliana* (via silencing effects on the floral repressor *FLC*) but it is not yet known whether epigenetic effects are due to sequence polymorphisms that are intrinsic to the *FLC* gene itself and which alter its accumulation of epigenetic marks (C. Richards, Verhoeven, et al. 2012). In natural populations of another plant, *Helleborus foetidus*, genotypes differed (largely due to four particular loci) in the fidelity of transmitting DNA methylation states to male gametes (Herrera et al. 2014). Such variation confirms the fascinating possibility that systems of epigenetic induction and resetting, and resulting patterns of transgenerational persistence, can evolve under selection as an aspect of eco-devo regulatory systems (discussion and references in Herman et al. 2014). The evolution of epigenetic regulatory systems will be subject to genetic constraints such as low genetic variation at regulatory sites (Furrow and Feldman 2014). A simpler type of genetic–epigenetic interaction can occur if the same effect on gene expression is implemented by either an allelic variant or an epigenetic state change, since selection on either of these variants would be affected by the frequency and dynamics of the other (Klironomos et al. 2013). Epigenetic factors can also influence the amount and type of genetic variance that is expressed in particular environments, for instance, by changing the binding of transcription factors that are active only in certain conditions (Bateson and Gluckman 2011; see Section 7.2.2).

7.4 Niche construction and eco-evolutionary feedbacks

7.4.1 Niche construction and feedback dynamics

A recent revelation in biology is that selective evolution can take place in just a few generations, instead of the long spans once assumed to be necessary (Kinnison and Hendry 2001; S. P. Carroll et al. 2007; e.g., Cody and Overton 1996; Sultan et al. 2013; Agrawal et al. 2013). An essential corollary is that selective change and ecological interactions can occur on the same timescale (Pelletier et al. 2009; Schoener 2011). This congruence means that ecological and evolutionary processes can affect each other. Even more fundamentally, it means that they can do so in a cyclical pattern of causation, a two-way "feedback loop" between organisms and their environments. When an organism modifies its environment in some way, it alters the selection pressures it experiences; this altered selective milieu can lead to evolutionary change in the organism, which will in turn change its subsequent impact on its environment, this change then leading to further evolutionary response (Post and Palkovacs 2009; Schoener 2011; Reznick 2013; Travis et al. 2014). This reciprocal dynamic can apply whether the organism measurably changes its external conditions, as these authors discuss (see Chapter 2, Section 2.3.3., and Chapter 5), or (by means of eco-devo adjustments) the environment it experiences (Chapter 2, Section 2.3.4, and Chapter 4).

These "eco-evolutionary" feedback cycles can be very rapid or can unfold over centuries or millennia due to gradual, cumulative external impacts. In either case, they raise two key insights. First, "organisms may evolve in a way that is different from what one would predict without taking the organism's impact on its environment into account" (D. N. Reznick 2013, S3; for a comprehensive treatment of this thesis see the influential book by Odling-Smee et al. 2003). Second, the whole system of organism–environment interactions can be said to be evolving, rather than the organism alone. Excitement about these reciprocal dynamics has motivated research efforts to integrate ecology and evolution in new and important ways (Whitham et al. 2006; Wade 2007; Laland et al. 2008; Pelletier et al. 2009; Schoener 2011; Travis et al. 2014).

Because the phenotypes, activities, and byproducts of organisms necessarily change their environments (whether external or experienced; see Chapter 2, Section 2.3.2), niche construction in this broad sense can be seen as "a universal fact" (Schoener 2011; also see Odling-Smee et al. 2003; Post and Palkovacs 2009; Scott-Phillips et al. 2013; Chapter 2, Section 2.3.2, and Chapters 4, 5, and 6). Theoretical and empirical studies are just beginning to examine specifically how these changes may feed back to shape selective outcomes.[2] For instance, an organism's niche-constructing effects on its external environment may improve the conditions it encounters, such that selection will favor traits (and associated alleles) that promote fitness in the improved environment (see Holt and Gaines 1992; Kawecki et al. 1997); these new traits may then generate different environmental impacts (Donohue 2003). What distinguishes this process from an environmental effect that is simply an "extended phenotype" resulting from prior selection on the organism—a one-way causal arrow from evolved genotype to phenotypic effect on the environment (Dawkins 2004)—is that the organism's environmental impact *feeds back* to shape selection on the organism itself, granting the organism a role in its own evolution. How important is this role? The answer is not yet known, since few studies to date have tested for (and therefore can have found) these effects in natural populations (Travis et al. 2013). Reciprocal effects of an organism's ecological impact on its own selective change may be clear in an experimental test (Reznick 2013; e.g., Yoshida et al. 2003), but much further work is needed to determine how strong or widespread these feedbacks may be in complex natural settings (Post and Palkovacs 2009; Schoener 2011; Travis et al. 2014).

7.4.2 A few examples: wood cockroaches, alewife fishes, farmers, and guppies

Wood cockroaches, *Cryptocercus punctulatus*, nest in decaying tree trunks, where they enjoy a plentiful supply of their food source (wood, digested with the help of mutualistic gut symbionts). The drawback to this resource-rich, well-protected habitat with its favorable humidity and temperature levels is that these conditions also support a very dense and complex microbial community that includes parasitic fungi (Rosengaus et al. 1998 and references therein). The cockroaches alter this biotic environment by using their fecal matter and saliva to cement soil and wood fragments into a galleried nest structure that they continually cover with fresh feces. This constant, collective niche construction reduces the growth of fungal parasites in the insect colony because these feces contain (as yet unidentified) compounds that effectively suppress the germination of fungal spores (Rosengaus et al. 2013). Selection in this altered environment has evidently promoted not only cockroach fecal deposition behavior but also related behavioral and social traits, such as intranidal (within-nest) defecation and proctodeal (mouth-to-anus) feeding between individuals (perhaps even more unattractive to contemplate than the fecal decorating). With its diminished fungal threat, the "constructed" environment has presumably also obviated selection that might have favored the evolution of other cockroach traits to suppress fungal populations or to resist infection.

[2] Oddly, two parallel literatures have emerged that address this process as either "niche construction" (Odling-Smee et al. 2003, 2013) or "eco-evolutionary feedbacks" (Reznick 2013; Travis et al. 2014).

This suite of defecation behaviors is most effective against fungi when many individual cockroaches pool their feces—that is, in a colony setting. Following this insight, Rosengaus et al. (2013) proposed that this collective niche-constructing activity may have contributed to the evolution of coloniality in termites, which are derived from *Cryptocercus* (Cameron et al. 2012 and references therein). A collective benefit could arguably explain the otherwise puzzling evolution of eusociality in termites, which as diploid insects lack the kin-selective fitness benefits of the haplo–diploid genetic system found in bees, wasps, and other social hymenopterans. Sociality in termites might have evolved from a *Cryptocercus*-like ancestor from which they inherited this niche-constructing system (Cameron et al. 2012). In support of this hypothesis, phylogenetically basal termites such as *Zootermopsis* also nest in decaying wood, and they too cover the nest chambers and galleries with fungistatic fecal materials (Rosengaus et al. 1998). Note that, by suppressing fungal populations, *Cryptocercus* colonies also alter the tree-trunk microbial community's composition and competitive dynamics, which in turn will affect the web of ecological interactions and selection pressures in this forest "habitat within a habitat."

The wood cockroach exemplifies four important points about selective feedbacks. First, niche-constructing effects that increase fitness can directly promote selection for niche-constructing traits (as is inferred in this case). Second, such effects will feed back selectively on the organism's other traits as well, on the basis of the fitness value of those traits in the altered environment. Third, there may be macroevolutionary consequences to a selective feedback from niche construction. And fourth, one taxon's environmental impact may also influence the evolution of other members of the community.

Although few cases have been investigated to date, the first three points are confirmed by existing data. Post and Palkovacs (2009) demonstrated a direct selective feedback on several traits in a vertebrate, an observation with implications for evolutionary diversification. They studied the environmental impact of anadromous (seasonally migratory) populations of plankton-eating alewife fish, *Alosa pseudoharengus*, compared with populations that were trapped by dams. Alewife populations that had ceased to migrate exerted a constant feeding pressure that caused local zooplankton communities to become dominated by small-bodied taxa. This alewife-mediated change in the biotic environment in turn led to selection for alewife traits that improved their feeding efficiency on smaller prey items, such as reduced gape size, narrowly spaced gill rakers, and a behavioral preference for seeking small zooplankton (Palkovacs and Post 2008). However, this evolutionary change in foraging traits did not occur in populations that migrated for part of each year—populations which had a less pronounced niche-constructing impact on prey community composition. Consequently, alewife populations that exerted different effects on the biotic environment evolved different foraging structures and behaviors. If populations differ in environmental impacts (whether those differences arise from alternative alleles for niche-constructing traits or from local site effects such as dams, as in this case), the resulting differences in the strength, direction, or both of selective feedback might contribute to adaptive radiation (Post and Palkovacs 2009).

These cases underscore the point that an organism's environmental impacts alter selective conditions, even if those impacts are byproducts of its activities rather than beneficial results of adaptive behaviors. A case in human evolution makes this point even more emphatically. Early in the history of agriculture, West African peoples who cleared land to farm lowland crops of moist soils, such as yams, inadvertently created favorable breeding grounds for mosquitoes. Because these insects are the host and vector for the human malaria pathogen *Plasmodium falciparum*, farming led to higher rates of malarial infection (Laland and O'Brien 2012). Indeed, malaria has been a major disease of humans only since agricultural practices began, within the past 10,000 years (Tishkoff et al. 2001). This culturally mediated impact on the biotic environment evidently resulted in selective change within human populations inhabiting malaria-endemic regions. These populations exhibit higher frequencies of the *HbS* or "sickle cell" allele, as well as other blood cell variants that contribute to malaria resistance (Kwiatkowski 2005).

The reciprocal effects of environmental modifications on an organism's selective trajectory have been comprehensively studied in the Trinidadian guppy, *Poecilia reticulata*, and its stream habitat (Bassar et al. 2010; Reznick 2013; Travis et al. 2014; and references therein). These researchers have identified "diverse and easily quantified" effects of guppies on their ecosystems as potential selective feedbacks and then tested for selective response. Evidence to date confirms that both direct and indirect environmental impacts of guppy populations have contributed to the evolutionary outcomes of populations that establish in new streams. Interestingly, this research program was motivated by the unexpected finding that lack of predation as a selective pressure could not explain the life-history differences that repeatedly evolved in predator-free guppy populations. Instead, the key selection pressures arose from the community-dependent impacts of guppies themselves on their ecosystems, including changes in algal and invertebrate abundance due to greater population density in predator-free streams. This fruitful research program exemplifies the fact that, until potential organism-generated selective feedbacks are identified and tested, their importance in a given system cannot be known (D. N. Reznick, personal communication).

7.4.3 Theoretical insights into selective feedbacks

A reciprocal eco-evolutionary dynamic was first investigated in terms of the relatively simple feedback from an organism's increased population density and associated environmental impacts to a selective response based on relative fitness in the more crowded conditions. Such *density-dependent selection* has long been studied as a special case of evolutionary change, defined as a process in which population growth generates selection for different traits than those favored at low density (reviewed by Travis et al. 2013). Note that the key to this selective dynamic is not density per se but rather the fact that the organism alters its environment in a way that feeds back to influence its own selective response (Kokko and López-Sepulcre 2007). Accordingly, Travis et al. (2013, S17) propose that the evolutionary impact of density could actually represent "any alteration in the environment produced by a focal organism, including effects on ecosystem variables." In view of this recognition, they further suggest expanding density-dependent selection into a generalized concept of "ecological state-dependent fitness" (Travis et al. 2013, S17). Clearly, this conceptual step makes organism-generated selective feedbacks a general feature of evolution, since all organisms alter their environments. However, Travis and colleagues (2013, S18) argue that these feedbacks should not be considered "inevitable," because (a) the effects of organisms on their environments may be minor, and (b) genetic variation expressed under altered conditions may be insufficient for a selective response. (Note that the latter argument fails to include cases where an environmental effect eliminates the need for such a response, for instance, when the organism alters its environment so as to reduce stresses or to induce adaptive phenotypic adjustments.)

Niche-constructing impacts on an organism's selective milieu can thus be considered as a general case of density-dependent fitness. Such impacts are also likely to generate *frequency-dependent* fitness effects, because the selective feedback will reflect the intensity of the impact, and hence the *frequency* of the niche-constructing activity (W. F. Morris, personal communication). In the wood cockroaches discussed in Section 7.4.2, for example, positive frequency dependence could intensify selection in favor of fecal deposition traits, as the collective environmental impact becomes increasingly effective against pathogens. Conversely, in the case of a detrimental environmental effect, negative frequency dependence could prevent selective fixation of a niche-constructing trait. Although, like density dependence, frequency-dependent selection has often been considered theoretically as a special case, it is more realistic to treat relative fitnesses as generally frequency dependent (Maynard Smith 1989).

Beyond these general principles, models designed specifically to incorporate environmental modifications by organisms into selective trajectories are relatively new (Kylafis and Loreau 2008; note that models testing selective consequences of adaptive plasticity may provide useful approaches to feedbacks from experiential niche construction). Laland et al. (1999) and Odling-Smee et al. (2003) developed simple, two-locus population genetic models to study selective feedbacks due to niche construction. These models explore the feedback dynamics between allele frequency at a locus that influences the capacity for niche construction (and consequently the effect on some aspect of the selective environment), and selection at a second, "recipient" locus, at which relative genotypic fitnesses are influenced by the environmental state influenced by the first locus. The later (2003) models realistically allow the selective state (i.e., resource availability) to be set in part by the organism's niche-constructing activities and in part by independent ecological processes such as depletion and renewal.

These initial models show that, even at low frequencies, changes made by organisms to their environments can dramatically alter selection pressures on those organisms so as to generate unusual evolutionary dynamics, including inertia and accelerations (Geoghegan and Spencer 2012; similar dynamics have also been shown in models of environmental carryover from past generations due to maternal effects, e.g., Kirkpatrick and Lande 1989; see Laland and Sterelny 2006). These irregular selective dynamics can be understood intuitively. A time lag could occur between frequency change at a niche-constructing locus and the selective impact on a recipient trait, if the cumulative niche-constructing impact must reach a certain threshold before it is sufficient to influence selection (Odling-Smee et al. 2003). For instance, it may have taken many generations before the soil-processing activities of earthworms mitigated soil conditions sufficiently to affect selection on the worms' epidermal structure (Odling-Smee et al. 2003; see Chapter 5, Section 5.2). Similarly, a positive feedback might lead to a (frequency-dependent) selective acceleration.

Niche construction can also lead to counterintuitive selective outcomes, including fixation of otherwise disfavored alleles, and unexpected promotion or prevention of stable polymorphisms (Laland 1999; Odling-Smee et al. 2003). These unexpected consequences can occur if niche-constructing effects change the direction of selection on a trait to be the opposite of selection under unmediated environmental conditions (an effect termed "counter-selection" by Odling-Smee et al. 2003; for a theoretical test, see Kylafis and Loreau 2008). For example, plant thermoregulatory adjustments that maintained moderate leaf temperatures in desert climates (see Chapter 4, Section 4.2) might cause selection to favor alleles that would be deleterious to plants that experienced high, "unconstructed" heat loads. Depending on how the organism's environmental mediation affected its fitness, such "counter-selective" effects of niche construction either may overcome external selection pressures and cause otherwise deleterious alleles to be fixed or may lead to stable polymorphisms. A functional interaction between niche-constructing traits and "recipient" traits (the benefit of which depends on organism-altered conditions) can also drive evolutionary fixation of alleles that contribute to niche-constructing activities (Silver and DiPaolo 2006; Laland and Boogert 2008). These initial results show clearly that, in theoretical terms, organisms' effects on their selective environments can meaningfully and, indeed, dramatically influence their own evolutionary dynamics. Further theoretical investigations will likely generate specific, testable hypotheses for empirical studies in this largely unexplored area.

7.4.4 Niche construction as diffuse coevolution

As ecologists well understand, species evolve "within matrices of hundreds of species that coexist in variable environments" (Whitham et al. 2006, 510): evolution is always coevolution. When an organism changes some aspect of the external environment, it shapes not only its own selective conditions and response but also those of cohabiting species that either interact with the altered aspects of the abiotic or biotic environment or interact with the initiating organism itself (Odling-Smee et al. 2013). Consequently, niche construction is likely to effect both diffuse and direct coevolution, creating a "network of feedbacks" with potentially very high dimensionality and complex possible outcomes (Travis et al. 2013).

For example, leaf litter is a major way that trees alter the environment encountered by co-occurring species, especially those that interact strongly with the soil surface (see Chapter 5, Section 5.3.1). Stinchcombe and Schmitt (2006) tested its potential coevolutionary impact by manipulating leaf litter from oak trees, *Quercus* spp., to create bare-soil versus litter-covered sites for seedling emergence in a temperate woodland. They then examined selection for seedling traits in a naturally cohabiting species, the understory herb *Impatiens capensis*. The presence of oak litter altered the strength and direction of selection on ecologically important *Impatiens* traits such as germination timing and early shoot growth. When leaf litter was present, rapid stem elongation to emerge from beneath its cover was critical to seedling success. As a result, the frequency of the oak litter environment altered selection for elongation norms of reaction in *Impatiens*, demonstrating that heterogeneity in one species' niche-constructing effects can influence the evolution of plasticity in a co-occurring species (Stinchcombe and Schmitt 2006). As these authors note, the selective effect of niche construction by oak trees on *Impatiens* may generate further coevolutionary effects on species that interact with *Impatiens*, such as herbivores or competing plant taxa. Beyond such direct species interactions, changes by one species to a shared resource can indirectly affect selection on the many other species that utilize that resource. In this case, for instance, oak litter–induced changes to soil temperature and moisture, mineralization rates, and seedling growth will also alter selection on soil invertebrates, fungi, and bacteria. Although some models of coevolution implicitly include environmental effects of interacting species on each other, indirect effects via shared components of the environment are yet to be incorporated into coevolutionary models (reviewed by Odling-Smee et al. 2003).

Indirect, multispecies effects extend the possible evolutionary role of niche construction "beyond . . . pairwise coevolutionary processes to a broader context that includes complex community and ecosystem processes" (Travis et al. 2013, S18; see Whitham et al. 2006). These higher-order coevolutionary dynamics will also reflect the fact that, in contrast to the many traits studied in pairwise and often antagonistic coevolutionary scenarios, many organismic effects on the environment—such as depositing leaf litter, for instance—can be seen as "diffuse . . . costless . . . byproduct mutualisms" (Leigh 2010, 2520). Although niche-constructing impacts will inevitably have diffuse, and possibly surprising, coevolutionary effects that undoubtedly merit investigation, many aspects of these effects are likely to be extremely weak. Because evidence for diffuse coevolution is difficult to obtain, it is not known to what degree coevolution in general may be diffuse rather than pairwise (Wade 2007).

The evolutionary impact of an organism's niche-constructing activities on other species in its ecosystem will depend in part on whether these activities affect only the organism's own environmental and selective experience (i.e., through thermoregulatory behaviors or allocational shifts) or create an external impact on the ecosystem. In the former case (experiential niche construction; see Chapter 2, Section 2.3.4, and Chapter 4), coevolutionary effects on other species in the community could be mediated either indirectly, via the niche-constructing organism's altered resource use and waste production, or directly, due to behavioral or other phenotypic changes that alter its ecological interactions. External, habitat-constructing effects (see Chapter 2, Section 2.3.3, and Chapter 5) are likely to create stronger evolutionary feedbacks on both conspecific and heterospecific cohabitants (see Chapter 6, Section 6.3). Niche-constructing impacts can also initiate new ecological associations that may lead to pairwise coevolution (shown experimentally by Hom and Murray 2014). Further empirical studies are needed to determine the systems and ecological circumstances in which such organismic effects on the environment comprise significant coevolutionary forces.

7.5 Selection in context: reciprocal organism–environment effects

7.5.1 Integrating reciprocal organism–environment effects

Phenotypes bring about adaptation in two ways: by presenting a functional and life-history match to the environment and by influencing the environment the phenotype must match. Selective effects can result from adaptive or maladaptive phenotypic responses per se or from changes to the organism's (experienced or external) environment. As noted at the start of this chapter, because both eco-devo outcomes and the resulting niche-constructing activities influence selection, they are *causes* of further evolution as well as *results* of norm of reaction evolution (see Lewontin 2000; Odling-Smee et al. 2003; Sultan 2003a; Laland et al. 2008; Post and Palkovacs 2009). Moreover, these two processes are not independent: by virtue of eco-devo response pathways, current and (through epigenetic and maternal effects) prior environments participate in the realized phenotypes that, in turn, determine organismic impacts *on* the environment.

An updated framework for understanding adaptation must include the two-way, dynamic cycle of causation through which organisms and their environments shape each other and, at the population level, evolutionary trajectories (defined as change in the genetic components of norms of reaction; see Sections 7.1 and 7.2). Studies of "eco-evolutionary dynamics" have focused on the selective feedback created by the effects of organisms on their external environments (see Section 7.4), yet the reciprocal effects of these altered environments on the phenotypes exposed to selection have yet to be fully incorporated into dynamic models. Although this rich territory is largely unexplored at present, this section points to four evolutionary implications of reciprocal organism–environment effects, closing with some caveats regarding adaptive interpretation. (Unless otherwise specified, these points can pertain to changes made by organisms to their experienced as well as their external environments [see Chapter 2, Sections 2.3.2–2.3.4].)

First, *when organisms modify their external environments, this modification may influence subsequent selection by changing environmentally contingent aspects of their phenotypic expression* (Donohue 2005). This eco-devo feedback can occur within a generation, but it may also persist across generations if the environmental modification induces epigenetic or cytoplasmically transmitted effects or if it is itself long-lasting ("ecological inheritance" sensu Odling-Smee et al. 2003). Habitat-constructing effects combined with eco-devo responses to the resulting environment can cause either positive or negative selective feedbacks on environmentally influenced traits (Donohue 2005; see Section 7.4.3 on "counter-selection"). Changed trait expression can also influence selection in the altered environment on related traits (whether or not their expression changes), and on other organisms in the system.

Second, *niche-constructing activities can influence selection through effects on the expression of* $G \times E$ *variance* (Donohue 2005; Laland and Sterelny 2006). Recall that, owing to $G \times E$ variance, which is ubiquitous in natural systems, both the magnitude and rank order of expressed genotypic differences are environmentally dependent (see Section 7.2.2). As a result, by changing the environment(s) it encounters, a population may indirectly alter the intensity as well as the direction of selection to which it is subject. Donohue and colleagues (2005) tested the effect on $G \times E$ variance of environmental change due to seed germination timing, a form of habitat choice that determines the seasonal environment a plant encounters as it matures. The researchers examined how this type of niche construction, generated by plastic life-history responses to environmental conditions, affected the genetic variance that was expressed by a set of *Arabidopsis* genotypes. As a result of certain germination "choices," genetic variation for life-history traits was unavailable to selection because the *Arabidopsis* genotypes converged on similar life-history responses to the ensuing conditions (Donohue et al. 2005). This result illustrates how niche construction that results from plasticity, together with the plastic developmental responses that are elicited in the "constructed" conditions, may influence the further evolution of life-history or other eco-devo traits (Donohue 2005).

Third, niche construction affects the expression of genetic diversity in another way: *genetic variation for niche-constructing traits can result in a covariance between genotypes and environments*. In other words, "genotypes with different niche-constructing traits will systematically *experience* different environments" because they *create* different environments (Saltz and Nuzhdin 2014, 9). For instance, plant genotypes that copiously produce root exudates will inhabit a rhizosphere that has more abundant microbial populations than that inhabited by genotypes with less copious exudate production. When genetic variation is present for niche-constructing traits, this type of systematic covariance can decrease a trait's statistical heritability in a population by enhancing the environmental heterogeneity the population encounters (Saltz and Nuzhdin 2014; see also Kingsolver et al. 2012 on the effects of nonrandom environmental distribution on heritability).

Conversely, genotype–environment associations due to niche construction can increase apparent heritability through positive feedback effects.

The latter aspect of covariance, termed a "multiplier effect," has been examined with respect to the heritability of IQ and some human behaviors (Dickens and Flynn 2001). According to this hypothesis, an individual with a slight, genetically based behavioral predisposition might seek out an environment that reinforces that behavior. For instance, an individual with slightly more sensitive aural perception might spend more time listening to music, an enriched acoustical-information environment that could lead to a major difference in musical ability or related aspects of brain function. By magnifying initially subtle phenotypic differences, such positive genotype–environment covariance could make selection on a relatively minor genetic component of a behavioral difference more efficient (Saltz and Nuzhdin 2014). Statistically, such positive environmental feedbacks can inflate estimates of trait heritability (or, in the case of aversive effects, deflate them).

This insight may explain why high heritability estimates may be found for extremely complex behavioral traits for which direct genetic effects are likely to be small compared with the impact of individual experience (Dickens and Flynn 2001 and references therein). Epigenetic or maternally transmitted initial variants might also create a similar kind of environmental covariance. In a fascinating lab experiment, isogenic laboratory mice raised in an unusually complex cage environment showed individual behavioral differences that diverged increasingly with age and were accompanied by differences in brain (hippocampal) development (Freund et al. 2013; discussed by K. Lynch and Kemp 2013). Evidently, individual microhabitat choices very early in life (possibly influenced either by random, maternally transmitted stress or nutrient effects or by epigenetic factors) provided diverse, self-reinforcing environmental experiences that led to associated, later microhabitat and behavioral choices. Note further that, if niche-constructing variation among genotypes or epigenotypes causes environmental differences that affect the expression of other traits, the resulting covariance can influence the evolution of those traits via epistatic and pleiotropic feedbacks (Saltz and Nuzhdin 2014).

Fourth, *when niche-constructing effects improve the organism's environment, selection is expected to promote the evolution of specialized norms of reaction to this "constructed" environment.* When an organism's impact on its environment is beneficial and leads to higher fitness, selection will favor alleles that contribute to the niche-constructing traits or activities (Donohue 2005). In turn, this process may shift the distribution of environments the population encounters. In particular, if niche-constructing effects reduce the range or frequency of environmental states encountered, selection is expected to effect evolutionary specialization to the predominant, "constructed" state or states (Holt and Gaines 1992; Whitlock 1996; Kawecki et al. 1997). The persistence of generalist norms of reaction despite this predicted selective specialization might indicate a lack of functional trade-offs (e.g., due to antagonistic pleiotropy or mutation accumulation) in other, less frequently encountered environmental states (T. Griffith and Sultan 2012).

7.5.2 Adaptive interpretation

Recognizing the effects of environments on organisms, and vice versa, enlarges our perspective for both identifying selective pressures and interpreting adaptive phenotypes. Selection pressures may need to be differently defined and measured in the context of organismic effects that may exacerbate, modulate, or transform them (see Chapter 2, Section 2.3.3, and Chapters 4–6). Even when a critical environmental pressure is known, the key "targets of selection" may not be obvious. Despite many hundreds of selection studies, knowing the actual, ecological trait variants that generate selection "remains a key challenge" (Kingsolver et al. 2012, 1114). In part, this problem is due to the difficulties of distinguishing direct environmental impacts on fitness from those mediated by genetically influenced differences in trait expression (Kingsolver et al. 2012).

A niche-construction perspective may allow for more precise adaptive interpretation, by making it possible to distinguish between traits that comprise functionally adaptive responses to a given environmental challenge and those that instead remediate that challenge. For instance, in a greenhouse study of annual *Polygonum* plants from variable habitats, individuals that maintained fitness under severe drought stress and full sun did not produce small leaves, the predicted adaptation to reduce water lost from the transpiring leaf surface. These drought-stressed plants did, however, allocate proportionally more biomass to root tissue, thus increasing the availability of soil moisture (Heschel et al. 2004). The lack of an expected association between a certain (plastic or constitutive) trait state and fitness in a given environment (e.g., in a phenotypic selection analysis) may not mean that either the environmental factor or the trait in question is unimportant to fitness. Rather, this result may indicate that another trait (in the above case, plasticity in a different organ) is mediating the selective environment (Laland and Sterelny 2006).

Because related species, populations, and even genotypes may be differently constrained in these alternative aspects of phenotypic variation, adaptive comparisons based on single traits may be misleading. The ecophysiological plant trait *water use efficiency* (WUE; units of carbon fixed by photosynthesis per unit of water lost in the process) is a case in point. Low WUE may result from a plant's inability to adaptively minimize water loss by appropriately adjusting stomatal aperture or conductance rates. However, low WUE may also reflect an adaptive ability to maintain water *supply* (and hence obviate the need to conserve water) by means of various other traits, such as the size, spatial deployment, morphology, anatomy, or uptake potential of root systems, or some combination thereof (Sultan et al. 1998). Hence, higher WUE can be a sign of unsuccessful mediation of the moisture environment, rather than a functional adaptation. This poses a conundrum to crop breeders: although increased WUE would seem to be a clear adaptation to plants in water-limited conditions, such efficiency is notoriously as likely to be negatively associated with crop reproductive output as it is to be positively correlated (Blum 2005). In such cases, greater emphasis on niche-constructing traits may allow for a more comprehensive assessment of the adaptive aspects of phenotypes.

7.5.3 Studying adaptation and selection in context

After decades of studies focused on environmental pressures and their target traits, the bases of adaptation and selection in the real world remain only partly understood. Even in the case of traits that are recognized to be ecologically important and for which genetic factors and expression pathways are well established, the functional and fitness consequences of actual trait variation are remarkably little known in the context of natural environments (Anderson et al. 2014). An organism–environment research approach offers a way to more precisely characterize the interface of phenotype and environment that constitutes adaptation.

As Darwin recognized, selection arises from the organism–environment relationship. The dynamically co-constructing nature of that relationship has only recently become clear. New mechanisms for information transfer and unexpected feedbacks continue to be uncovered. In light of this rich causal interplay, our understanding of adaptation and selection must aim to encompass environmental effects on phenotypic expression and hence fitness, environmental mediation by organisms, and the potentially synergistic effects of these reciprocal processes. Without doubt, it is a substantial challenge to develop ways of studying adaptation and selection in this complex context, with respect to both theoretical and empirical research. But, more importantly, it is a challenge that can engage and energize researchers studying developmental pathways, epigenetic mechanisms, evolution, and community and ecosystem ecology, generating new understandings, connections, and questions.

Epilogue: Research directions and real-world challenges

... what biology needs to do to fulfill its program of understanding and manipulation is to take seriously what we already know to be true.

Richard Lewontin, *The Triple Helix* (2000)*

To know the nature, prevalence, and importance of environmental influences on organisms, and of organismic effects on environments, will require research protocols that explicitly test for these influences and effects. It is only through such tests that biologists can know how much our approaches, and our understanding, must shift in moving forward. Below, short sections summarize how eco-devo and niche construction perspectives, respectively, can lead to new research hypotheses and insights. Some medical and environmental applications are then briefly discussed, to demonstrate how this expansion of research directions may provide practical knowledge. The final section examines how an organism–environment focus can be brought to bear on the urgent question of future persistence of biological systems.

E.1 Characterizing phenotypic expression pathways and outcomes in environmental context

An eco-devo approach is centered on norm of reaction experiments, in which genotypes of interest are raised in (or exposed to) a set of environmental treatments that vary one or more ecologically relevant factors (see Chapter 1, Section 1.3.1 and Miner et al. 2005 on experimental design). Such experiments document how specific environments condition gene expression (e.g., transcriptome data based on microarray or RNA-seq techniques) or phenotypic traits in a given biological system. To better understand both development and variation in the real world, many more such data are needed—especially with respect to naturally varying factors (S. Gilbert 2001; Snell-Rood et al. 2010; Scheiner and Holt 2012). Designing and interpreting norm of reaction experiments calls for field data about the range, distribution, and covariance patterns of environmental factors that either cue or directly influence response pathways. However, simply identifying relevant factors can be difficult, as they may be far from obvious. Biotic, symbiotic, and social factors may play surprisingly important developmental roles, as in the cases of soil microbial community effects on plant phenology (M. Wagner et al. 2014), and gut microbiome inputs to mammalian brain development (Cryan and Dinan 2012). Once cue and response systems are identified, comparative mechanistic work on perception and signaling pathways can reveal precise components of adaptive diversity. Similarly, comparative norm of reaction studies can illuminate adaptive population

* Reprinted by permission of the publisher from THE TRIPLE HELIX by Richard Lewontin, p. 113, Cambridge, Mass.: Harvard University Press, Copyright © 1998 by Gius, Laterza & Figli Spa, Copyright © 2000 by the President and Fellows of Harvard College.

© Sonia E. Sultan. 2015. Published 2015 by Oxford University Press.

or species differences in terms of functional or life-history response patterns across a given environmental range.

An eco-devo approach must take into account the potential role of previous environments as well as current conditions. Since developmental information may be inherited in the form of induced maternal cytoplasmic factors, epigenetic factors, or both, it is necessary to test the effects of alternative parental or ancestral environments on offspring gene expression (e.g., Scoville et al. 2011). These nongenetic mechanisms of inheritance and their phenotypic consequences (often termed *transgenerational plasticity*) are only beginning to be understood. Much further work is needed to evaluate the generality and importance of these sources of variation and to understand their transmission dynamics. Heritable epigenetic mechanisms are a particularly compelling area for study: information is needed about induction versus stochasticity, maternal and paternal transmission, persistence across generations, and resetting conditions for various modifications. Studies linking naturally occurring cues to specific epigenetic changes and associated phenotypic adjustments are especially critical in order to evaluate the potential role of epigenetics in generating variation in natural systems (Geoghegan and Spencer 2013a; Herman et al. 2014). In addition, experiments designed to compare epigenetic dynamics of different genotypes are needed to address the important but largely unexamined question of genetic–epigenetic interactions.

Inherited environmental effects may also significantly change an individual's response to its immediate circumstances (*genotype × environment × parent environment interaction*; see Chapter 7, Section 7.2.1). A public-health study of Jamaican children provides a significant example: depending on maternal nutrition and its effects on birth weight, children who experienced famine tended to develop either the metabolically thrifty "marasmus" response or the maladaptive and often fatal "kwashiorkor" syndrome (Forrester et al. 2012). This type of causal interplay can be investigated in experimental organisms by testing the responses of genotypic replicates to factorial combinations of parental (or earlier) and offspring environments.

Along with the precise inclusion of environmental variables in experimental design, an eco-devo approach may lead to different criteria for choosing "model" organisms. Lab organisms have generally been chosen for minimal environmental responsiveness (J. Zhang and Lechowicz 1994; Lewontin 2000). Naturally evolved genomes, including those of taxa with broad environmental tolerances (e.g., Salinas and Munch 2012), may provide more representative developmental systems. With respect to theory, useful insights may result from incorporating environmentally sensitive phenotypic expression to population genetic models (e.g., Chevin et al. 2010; Gomez-Mestre and Jovani 2013; see discussion and references in Snell-Rood et al. 2010 and Chevin et al. 2013).

E.2 An example: how an eco-devo approach can generate adaptive hypotheses and insights

The author's own research program exemplifies how eco-devo studies of individual response patterns can produce a detailed picture of adaptive diversity, in this case, within a closely related group of annual plants (*Polygonum* spp.). Instead of tracking allelic or quantitative trait locus variation associated with specific trait states, this work documented genotypic norms of reaction in response to controlled ranges (based on field environmental measurements) of factors critical to plant growth and function. A suite of experiments revealed highly plastic repertoires of trait expression, including increased biomass allocation to organs when they were collecting a limiting resource (Sultan 2003b and references therein). In comparative norm of reaction experiments, individuals of four different species expressed characteristic patterns of functional and fitness plasticity; these patterns were consistent with broader versus narrower habitat distributions in the field, including differences in the ability to perceive or respond rapidly (or both) to a given change (Sultan et al. 1998; Bell and Sultan 1999; Sultan 2001; T. Griffith and Sultan 2005). Combined with analyses of fitness variance in the field, functional trade-offs, and costs of plasticity, these comparative data provided insight to the evolution of generalist versus specialist taxa in this monophyletic group (T. Griffith and Sultan 2012).

Individual plants also expressed species-specific differences in transgenerational environmental response. In *Polygonum persicaria*, an ecological generalist found in moist, dry, and variable habitats, parental drought stress resulted in enhanced root development and greater survival of offspring grown in dry soil (see Chapter 4, Section 4.4). These inherited, nongenetic adaptations persisted in the grandchild generation; this observation indicates an epigenetic rather than a cytoplasmic mode of transmission (the possible role of DNA methylation is currently being tested).

An eco-devo approach has also provided insights to plant invasiveness. *Polygonum cespitosum*, a member of the same species group, has recently become invasive in northeastern North America. Eco-devo resurrection experiments, combined with field studies, showed that a recent ecological expansion into open habitats was accompanied by rapid norm of reaction evolution for improved performance in full sun (Sultan et al. 2013; Matesanz et al. 2015). Large-scale norm of reaction experiments documented among-population differences in available $G \times E$ variation that could fuel future plasticity evolution, thus revealing a potentially critical aspect of invasive potential (Matesanz et al. 2014 and references therein). These rich eco-devo data sets can generate further investigations into various dimensions of the system—for example, the distribution of environmental states in natural habitats and their autocorrelation across generations; proximate transduction and transmission mechanisms for environmental effects; impacts of individual eco-devo responses on community interactions (e.g., with respect to emergence timing, tissue quality for herbivores, shading, and soil impacts); and selective response based jointly on $G \times E$ variation and environmental heterogeneity.

E.3 Incorporating niche construction into ecological and evolutionary research

Understanding the effects of organisms on their habitats, beyond the dramatic cases of "engineers" such as reef corals or beavers, is a new focus for ecological research (Post and Palkovacs 2009; Estes et al. 2011; Hodge and Fitter 2013). This more inclusive strategy is likely to be fruitful, because the species' impacts that are most critical to biological systems may not be the most evident ones (see Chapter 5, Section 5.1.4). Manipulative and observational field studies can determine how populations of particular organisms alter biotic and abiotic conditions, how the presence of particular taxa indirectly affects habitat diversity and community interactions, and the degree to which environmental impacts persist, at various scales, for subsequent generations of cohabiting taxa. Such studies can identify organisms of particular importance to community structure (such as substrate-stabilizing diatoms or grasses) and, in addition, can indicate population size thresholds that are necessary to conserve these effects (see Chapter 5, Section 5.1.3).

To understand species' distributions and adaptive diversity, it is also important to study how individual organisms change the environments they *experience* (see Chapter 4). As described above, norm of reaction experiments can document specific trait adjustments that mediate environmental stresses. To study how organisms alter their experience of natural habitats, environmental data can be collected so as to capture both the external conditions as such and the organism's mediated version of those conditions. For instance, imagine a study comparing how different plant taxa perform in contrasting light habitats. Along with characterizing ambient light conditions in each site by estimating mean incident photon flux density per square meter, a researcher could also examine leaf area data to calculate photon flux per unit of plant biomass or total photons captured, to compare the *access* to light achieved by individuals of each taxon (see Chapter 2, Section 2.3.4). Similarly, plant or animal body temperature could be measured in the presence or absence of thermoregulatory adjustments. Since many organisms alter their environmental experience by either creating or seeking out certain microsite conditions, field studies could compare the environmental states at those microsites to overall site conditions (measured at a random or spatially regular sample of microsites) (see Chapter 2, Section 2.3.3). Such sampling strategies could provide insight into subtle, individual-level changes to biotic and abiotic conditions, leading to more precise and meaningful measures of environmental tolerance and ecological distribution.

A niche-construction approach directs attention to the adaptive contribution and evolutionary potential of traits that cause these mediated distributions, in addition to incorporating a traditional ecological focus on functional traits per se. More broadly, it expands the scope for evolutionary investigation to include the possible selective feedbacks of an organism's niche-constructing effects on all of its traits and on those of co-occurring species. These potential feedbacks can be empirically tested by comparing phenotypic or genetic change in either natural or experimental populations that differ in niche-constructing effects or in evolutionary potential (e.g., Yoshida et al. 2003; Palkovacs and Post 2008). Theoretical investigations of selective feedbacks can be approached as a generalization of density dependence (Travis et al. 2013) or a two-locus model of feedback dynamics between niche-constructing and "recipient" traits (Odling-Smee et al. 2003; see Chapter 7, Section 7.4.3). The recent research on Trinidadian guppy populations in different community contexts exemplifies a productive, multidimensional approach to these eco-evolutionary feedback dynamics. It incorporates in situ and mesocosm experiments designed to test specific potential environmental impacts; comparative norm of reaction experiments to assess selective response; a tissue bank that provides DNA records of selective change; long-term monitoring of natural populations; and mathematical models to examine potential outcomes (Travis et al. 2014 and references therein; see Chapter 7, Section 7.4.2).

Incorporating niche-constructing effects into coevolutionary studies may be an important way to maintain realism. The impact of a particular cohabiting species on a shared resource or other environmental factor can be included in experimental design, to maintain a community context for studies of pairwise coevolution. Diffuse coevolutionary effects of a particular niche-constructing species can be examined by comparing real or experimental communities in which its presence or abundance varies (Scott-Phillips et al. 2013; e.g., Agrawal et al. 2012). Within populations, testing for genotype–environment associations generated by genetic diversity for niche-constructing traits may provide insight to sympatric speciation and, more generally, to trait heritabilities and complex sources of phenotypic variation (Saltz and Nuzhdin 2014). Evolutionary feedbacks due to niche-constructing effects can also be addressed at a macroevolutionary level, for instance, by using phylogenetic data to test whether a particular niche-constructing trait repeatedly preceded the evolution of related, "recipient" traits (e.g., Schwilk and Ackerly 2001 on the evolutionary association of tree flammability and fire-tolerance traits; see Odling-Smee et al. 2003 for a comprehensive research agenda).

E.4 Practical applications of eco-devo and niche construction principles: environmental and biomedical research

An eco-devo approach emphasizes the breadth of environmental influences on organisms as well as their potential heritability. This expanded awareness has led to the recognition of complex types of environmental impact that may hold important implications for environmental and medical research.

One such implication is the need for more sophisticated protocols to test the safety of agricultural chemicals and other bioactive substances introduced to ecosystems. First, because environmental effects on organisms may be synergistic, test exposures should not be limited to specified amounts of these substances in otherwise favorable lab conditions but should also include combinatorial treatments with naturally occurring variables. Synergistic effects of natural stresses with agrochemicals have been implicated in the ongoing decline of both bee and amphibian populations. A widely used broad-spectrum insecticide thought to be in itself only mildly toxic to bees was found to significantly elevate mortality due to a common bee pathogen (Doublet et al. 2015). Similarly, the purportedly vertebrate-safe herbicide "Roundup" was found to be twice as toxic to certain amphibian species in the presence of stress-inducing predator cues that are found in many habitats (Relyea 2005). The seasonal and therefore developmental timing of chemical exposures may also affect their impact in natural systems (Relyea 2005). Second—and perhaps even more sobering—knowing the full impacts of these substances on organisms (possibly including humans) may require multigeneration tests (Nilsson and Skinner 2015). In laboratory rodents, common environmental contaminants (such as fungicides, insecticides, plastics, and jet fuel) have been found to induce epigenetic changes associated with the expression of tumors, obesity, and other disease states in subsequent generations (Manikkam et al. 2012). These initial studies argue for far more comprehensive testing of the potential effects that chemical contaminants and other novel aspects of real-world environments may have on organisms.

In medicine, an eco-devo approach argues for a greater focus on the interactions that physiological and cellular systems have with environmental factors such as diet, gut biome, physical and emotional stresses, and exposure to pollutants or other toxins. Indeed, a recent *Science* paper estimated that 90% of chronic disease risk "is more likely to be understood in the context of human environmental exposures such as diet and toxicants" than directly linked to genetic factors (S. P. Carroll et al. 2014). Such a shift in focus could be of immense benefit to human well-being and public health, because in many cases these factors can be directly and often inexpensively remediated (S. P. Carroll et al. 2014).

One eco-devo-based avenue of medical inquiry has shown how a uterine environment influenced by poor maternal nutrition interacts with fetal signaling pathways (largely via epigenetic modifications) to initiate a host of profound developmental, behavioral, and health effects that unfold from childhood to adulthood (Low et al. 2014 and references; see Chapter 1, Section 1.2.2). These effects include obesity, cardiovascular disease, and type II diabetes, adult health problems that are often assumed to result from genetic predispositions combined with poor diet and sedentary lifestyle. If, instead, these problems result from early developmental environment, they will call for very different types of intervention (Hanson and Gluckman 2014). In some cases, for example, epigenetic states induced by poor prenatal conditions can be altered simply by maternal nutritional or folic acid supplements (Burdge and Lillycrop 2010 and references therein). Pharmacological or nutritional manipulation of epigenetic states may also prove effective in the prevention and treatment of cancer and thus is a very active area of current research (Schnekenburger et al. 2014; Shukla et al. 2014).

Another exciting medical research focus is the gut microbiome, an internal aspect of the environment that may have extraordinarily important effects on multiple areas of human health. This symbiont community, which is strongly influenced by host diet, regulates metabolic and signaling pathways between the gut, liver, muscle, and brain (Nicholson et al. 2012; David et al. 2014; and references therein). Increasingly, the presence and composition of these internal communities are being investigated directly in human subjects, for instance, through supervised dietary changes. One such study revealed a previously unsuspected effect of certain artificial sweeteners on aspects of metabolic function linked to obesity and diabetes, an effect that was evidently due to specific microbiome changes caused by the sweeteners (Suez et al. 2014). The effects of microbial components on epigenetic regulation of brain pathways are another promising research area (Stilling et al. 2014). Current work suggests that microbiome manipulation could offer entirely new therapeutic tools for the treatment of complex central nervous system disorders such as autism and depression (see Cryan and Dinan 2012). An expanded, mechanistically precise understanding of environmental factors in human health may lead to fundamentally different ideas about causation and treatment. Eventually, these investigations may lead to a fully contextual "personalized medicine," in which not just individual DNA sequence but also inherited epigenetic and maternal factors, and direct environmental inputs such as diet, microbial symbionts, and stress history, are considered as causal elements and potential targets for remediation.

Along with expanded studies of environmental effects on organisms, a focus on organismic impacts on environments can also be a source of solutions for significant practical needs. With respect to predicting global wildfire patterns, for example, it is increasingly recognized that climate alone does not explain fire regimes, so attention is now being paid to the flammability-enhancing traits of particular vegetation types and taxa, including invasive plants (Pausas and Keeley 2014). Habitat-constructing effects may provide guidance for choosing or developing agricultural, arboricultural, and aquacultural varieties, by indicating traits that enhance group productivity, minimize environmental damage, or both (S. P. Carroll et al. 2014). In wheat, for instance, cultivar selection could target growth traits that, in conjunction with specified planting configurations, would cause increased shading of weeds, an approach that could maximize wheat yield while reducing herbicide use (Weiner et al. 2010). High-protein crop varieties being engineered for increased nitrogen: carbon allocation (e.g., L. Li and Wurtele 2015) could be tested to insure that (carbon-rich) root exudates critical to rhizosphere function are maintained. Niche-constructing effects of organisms can even suggest new technologies: inspired by the way bivalves collect calcium from seawater to build their shells, researchers working to develop more efficient lithium-air batteries have genetically altered viruses to collect molecules of manganese oxide from an aqueous solution and build rough accretions that provide an enhanced surface area for electrochemical reactions (Oh et al. 2013).

E.5 Organisms and environments going forward: the question of future adaptation

As natural habitats are increasingly altered, contaminated, and fragmented; as global climatic and atmospheric conditions shift in novel ways; as macro- and microorganisms of all kinds are introduced to new ranges via distribution shifts and human transport; as populations increase, decrease, or disappear from communities due to these changes; the critical question arises: to what extent will organisms be able to persist under these new terms? Will present-day taxa be able to form adaptive relationships with their changing environments? For any type of organism, this question has two parts: can currently evolved genomes express viable phenotypes under new and imminently approaching conditions, and is there potential for selective evolution of reaction norms to accommodate further, and presumably more extreme, changes to come? At the community and ecosystem levels, future stability and productivity will be influenced by changes to niche-constructing impacts under altered conditions, and by resulting selective and coevolutionary feedbacks.

If existing norms of reaction allow for appropriate eco-devo responses to altered conditions (e.g., shifts in life-history timing to match changes in seasonal onset; Visser 2008), such individual plasticity can buffer the risk of extinction and allow populations with such norms to persist (S. P. Carroll et al. 2007; Sultan 2007; Nicotra et al. 2010). Genotypic norms of reaction that accommodate sudden environmental changes may be especially critical for the survival of species that evolve more slowly, such as those with longer generation times (see simulation model by Vedder et al. 2013). However, future conditions may instead elicit maladaptive phenotypes from existing developmental systems (Ghalambor et al. 2007). Because temperature can have a direct effect on growth and metabolism in many taxa, predicted rapid temperature changes are of particular concern. Similarly, rapid increases in atmospheric carbon dioxide and ocean acidity may have especially widespread phenotypic effects, respectively, on photosynthetic and marine organisms. Norm of reaction experiments to characterize phenotypic outcomes under expected future conditions will be of critical value to determine the functional and fitness tolerance ranges of present-day genotypes and thus identify those biological systems at greatest risk.

An important caveat must be raised regarding the potential role of existing plastic responses as a source of future adaptive phenotypes. As natural habitats are disrupted and altered, so too are the environmental elements that serve as eco-devo response cues. It is possible that previously evolved cue perception and transduction pathways will permit organisms to use changed environmental signals to guide appropriate phenotypic expression. However, it seems likely that many of these cue changes will derail selectively evolved patterns of phenotypic response. A dramatic case is that of mayflies (*Ephemeroptera* spp.), aquatic insects that often lay their eggs—fatally—on asphalt roads rather than in mountain streams because the highly polarized light reflected from the road surfaces mimics the kind of reflected light that they evolved to use as their oviposition cue (Kriska et al. 1998). Other water insects can be similarly "deceived" into maladaptive behaviors by reflectance miscues from car roofs or windshields, panes of glass, or the shiny black plastic sheets used to suppress weeds (references in Kriska et al. 1998). Disruption of both abiotic and biotic cues may be a major consequence of anthropogenic change, although in general this has received little attention. The main exception is climate change, which has been studied as a likely disruptor of seasonal transition cues in many animal and plant taxa (Wilczek et al. 2010). To investigate the impact of disrupted cues on phenotypic outcomes, field measurements or predicted future levels of relevant environmental factors (or factor combinations) can be incorporated into norm of reaction experiments.

The ability of eco-devo response systems to produce phenotypes that are appropriate to future environments depends on the developmental influence of parental environment, as well as on plastic responses to immediate conditions. Initial findings provide some encouraging news about the potential adaptive contribution of inherited environmental effects, at least in some taxa. Several studies have found that parental exposures to both higher temperatures and elevated ambient carbon dioxide levels cause offspring to express phenotypes that are adaptive to these predicted future conditions. The effect of higher water temperatures on offspring performance was tested in sheepshead minnow, *Cyprinodon variegatus*, a small Atlantic fish. When parent fish were held for 30 days at one of several temperatures, the growth rate, body size, and expected fecundity of their progeny were highest at whatever temperature the parents had experienced. As a result of this parental effect, offspring growth rates were over 30% higher within a high-temperature environment when parents had encountered that same environment, a far greater advantage than could be achieved in a single generation of selective change (Salinas and Munch 2012). Similarly adaptive transgenerational thermal effects have been found in several insect species and in *Arabidopsis*, where it persisted across two generations and therefore must be epigenetically mediated (Whittle et al. 2009). Adaptive transgenerational plasticity may contribute to plant tolerance of drought stress, another aspect of predicted climate change. As noted previously, in the widespread colonizing annual *Polygonum persicaria*, seedling offspring of drought-stressed parents had larger root systems and greater survivorship in dry soil than offspring of isogenic, undroughted parent plants did (Herman et al. 2012). However, this adaptive inherited effect was lacking in a congener limited to moist soils, *Polygonum hydropiper*; this observation shows that transgenerational effects (like other aspects of eco-devo systems) may vary even among closely related taxa (Sultan et al. 2009). Parental effects may also mediate the acidification risks of higher dissolved carbon dioxide levels to certain marine organisms (e.g., anemonefish; Miller et al. 2012; see Chapter 4, Section 4.4). The potential contribution of transgenerational plasticity to individual tolerance, and hence to population persistence, can only be determined through a greatly expanded effort to test for inherited effects of simulated future conditions on individual eco-devo outcomes.

As the conditions that a population encounters deviate further from that population's previous selective history, evolved environmental inputs and associated phenotypic outcomes may become increasingly dysfunctional. At that point, the population's persistence will depend on its potential for selective change in norms of reaction. This evolutionary potential depends on both the presence of $G \times E$ variation and precisely how such variation is expressed in the future conditions (see Chapter 7, Section 7.2.2). As previously explained, a novel environment can either release variation to allow selection or buffer a population from selective change, depending on whether the genotypic norms are different or similar in that environment (see Chapter 7, Section 7.2.3). If all genotypes of a population converge on low absolute fitness in a novel, stressful environment, there is no opportunity for selective change. If, instead, certain genotypes are less phenotypically disrupted in the novel conditions, selection can promote the evolution of a homeostatic response to the changed conditions (Sultan 2007). It is not yet known whether any general prediction can be made regarding convergent versus variable expression of genotypes in the new and stressful conditions expected to predominate in future habitats (Rice and Emery 2003; Gienapp et al. 2008; Merilä and Hendry 2014).

Evolutionary potential due to $G \times E$ variance (and its allelic components) will inevitably vary among taxa and populations, influencing their relative abilities to evolve new norms of reaction. Variation available to selection may also depend on the trait in question: for instance, if genotypes are more likely to differ in life-history expression than in norms of reaction for metabolic or physiological traits, phenological shifts are more likely to evolve under climate change than altered thermal tolerances are (Sultan 2007). Studies of reptiles have examined the relative evolutionary potential of distinct traits that could influence adaptation to climate change. Because these animals have temperature-based sex determination, they are considered to be at risk of "demographic collapse" as global temperatures increase, due to skewing of sex ratios (N. Mitchell and Janzen 2010). Hatchling sex ratio is shaped by two different traits: the mother's choice of nest site in a given habitat, as this choice influences incubation temperature (Bull et al. 1988), and the embryo's developmental threshold for response to that temperature. Can either of these traits evolve to keep pace with climate warming? Initial findings suggest that thermal norms of reaction may be more likely to change under future selection than maternal nesting behaviors, at least in certain taxa (Morjan 2003; Janes and Wayne 2006). To assess the risk of extinction and devise possible interventions, more comprehensive data are required regarding norm of reaction variation within natural reptile populations for both types of trait (N. Mitchell and Janzen 2010). To predict the evolutionary potential of any organism under future conditions, experiments are needed that test for variation in genotypic norms of reaction for key traits, as expressed in predicted future environmental states (Ghalambor et al. 2007; see Chapter 7, Section 7.2).

An organism–environment perspective raises a final, critical question about the future prospects for biological systems: as abiotic and biotic conditions are increasingly transformed, how will the niche-constructing impacts of organisms change, and how will these changes affect community interactions, ecosystem processes, and evolutionary feedbacks? Organismic impacts will be altered in part due to eco-devo responses to future conditions. Increased ocean acidification, for instance, will promote the growth of marine algae and cyanobacteria, but it will negatively affect organisms that build acid-vulnerable shells, including reef corals that provide critical habitat structure (see Chapter 6, Section 6.3.1). Habitat-constructing effects will also be modified as a result of changes in species distributions, for instance, because of range shifts in response to climate change (Bailey et al. 2014). Species introductions and local extinctions will likewise result in the addition and loss of particular niche-constructing effects, possibly leading to complex functional and coevolutionary consequences. Displacement of native plants by invasive taxa, for example, can alter ecosystem properties such as nutrient cycling, soil biota, and fire regime (D'Antonio and Vitousek 1992; Kourtev et al. 2002; see Chapter 6, Section 6.3.2).

Understanding the prospects for future persistence will require information about the functional and selective consequences of predicted changes in organismic impacts. The Earth's smallest and most abundant photosynthetic organisms, the marine cyanobacteria comprising the genus *Prochlorococcus*, provide a case in point. Despite their small size, the collective habitat-constructing impact of these minute, free-living organisms is enormous. Distributed throughout the shallow, sunlit zones of tropical and subtropical oceans, the primary productivity of *Prochlorococcus* is estimated to equal that of the Earth's combined croplands (Biller et al. 2015). In addition to providing a massive input of carbon and other nutrients to marine ecosystems and releasing substantial quantities of atmospheric oxygen, *Prochlorococcus* appears to play a role in transferring carbon from the atmosphere to deep ocean waters. Curiously, these photosynthetic organisms are also able to *take up* carbon in the form of glucose (Biller et al. 2015). As the oceans warm, *Prochlorococcus* cyanobacteria are predicted to both increase in abundance and shift their distribution, thus causing changes to biogeochemical inputs that may "shift selection pressures in the entire ecosystem" (Biller et al. 2015, 24; also see Flombaum et al. 2013). Informed predictions about organism-mediated aspects of anthropogenic change will require a research program that focuses on environmental impacts in community and ecosystem context.

The research program described here raises only some of the countless possible questions that emerge from a focus on the organism–environment relationship. As we seek to understand biological processes and ecological systems, studying the mutual influences that constitute this relationship could provide crucial insights. It is true that both empirical and theoretical approaches to these complex phenomena are likely to be demanding. Yet this expanded view of causation is exciting as well. An organism–environment research program will, of necessity, integrate studies of gene expression and developmental pathways, ecological conditions, and evolutionary trajectories, in ways that promise to illuminate and enrich these formerly separate disciplines.

References

Abouheif, A., and G. A. Wray. 2002. Evolution of the gene network underlying wing polyphenism in ants. *Science* **297**: 249–252.

Ackerly, D. D. 2003. Community assembly, niche conservatism, and adaptive evolution in changing environments. *International Journal of Plant Sciences* **164**: S165–S184.

Ackerly, D. D., and S. E. Sultan. 2006. Mind the gap: The emerging synthesis of plant "eco-devo." *New Phytologist* **170**: 648–653.

Ackerman, M., and M. Doebeli. 2004. Evolution of niche width and adaptive diversification. *Evolution* **58**: 2599–2612.

Acosta, C. A., and M. J. Butler IV. 1997. Role of mangrove habitat as a nursery for juvenile spiny lobster, *Panulirus argus*, in Belize. *Marine and Freshwater Research* **48**: 721–727.

Agrawal, A. A. 1998a. Induced responses to herbivory and increased plant performance. *Science* **279**: 1201–1202.

Agrawal, A. A. 1998b. Leaf damage and associated cues induce aggressive ant recruitment in a Neotropical ant-plant. *Ecology* **79**: 2100–2112.

Agrawal, A. A. 1999. Induced responses to herbivory in wild radish: Effects on several herbivores and plant fitness. *Ecology* **80**: 1713–1723.

Agrawal, A. A. 2001. Phenotypic plasticity in the interactions and evolution of species. *Science* **294**: 321–326.

Agrawal, A. A., and J. A. Fordyce. 2000. Induced indirect defence in a lycaenid-ant association: The regulation of a resource in a mutualism. *Proceedings of the Royal Society B* **267**: 1857–1861.

Agrawal, A. A., A. P. Hastings, M. T. J. Johnson, J. L. Maron, and J.-P. Salminen. 2012. Insect herbivores drive real-time ecological and evolutionary change in plant populations. *Science* **338**: 113–116.

Agrawal, A. A., M. T. J. Johnson, A. P. Hastings, and J. L. Maron. 2013. A field experiment demonstrating plant life-history evolution and its eco-evolutionary feedback to seed predator populations. *American Naturalist* **181**: S35–S45.

Agrawal, A. A., C. Laforsch, and R. Tollrian. 1999. Transgenerational induction of defenses in animals and plants. *Nature* **401**: 60–63.

Agrawal, A. A., and M. T. Rutter. 1998. Dynamic antiherbivore defense in ant-plants: The role of induced responses. *Oikos* **83**: 227–236.

Alborn T., T. C. J. Turlings, T. H. Jones, G. Stenhagen, J. H. Loughrin, and J. H. Tumlinson. 1997. An elicitor of plant volatiles from beet armyworm oral secretion. *Science* **276**: 945–949.

Albrecht M., B. Padrón, I. Bartomeus, and A. Traveset. 2014. Consequences of plant invasions on compartmentalization and species' roles in plant-pollinator networks. *Proceedings of the Royal Society B* **281**: 20140773.

Allen, R. E., ed. 1990. *Oxford Dictionary of Current English*, 8th ed. Clarendon Press, Oxford.

Amasino, R. M., and S. D. Michaels. 2010. The timing of flowering. *Plant Physiology* **154**: 516–520.

Amundsen, R. 2001. Adaptation and development: On the lack of common ground. In S. Orzack and E. Sober, eds. *Adaptationism and Optimality*, pp. 303–334. Cambridge University Press, Cambridge, UK.

Anderson, J. T., C.-R. Lee, C. Rushworth, R. Colautti, and T. Mitchell-Olds. 2013. Genetic trade-offs and conditional neutrality contribute to local adaptation. *Molecular Ecology* **22**: 699–708.

Anderson, J. T., M. R. Wagner, C. A. Rushworth, K. V. S. K. Prasad, and T. Mitchell-Olds. 2014. The evolution of quantitative traits in complex environments. *Heredity* **112**: 4–12.

Andersson, J., F. Johansson, and T. Söderlund. 2006. Interactions between predator- and diet-induced phenotypic changes in body shape of crucian carp. *Proceedings of the Royal Society B* **273**: 431–437.

Angier, N. 2003. "Not just genes: Moving beyond nature vs. nurture," *New York Times Science Times*, February 25.

Anna, J. M., G. D. Scholes, and R. van Grondelle. 2014. A little coherence in photosynthetic light harvesting. *BioScience* **64**: 14–25.

Anthony, K. R. N., J. A. Maynard, G. Diaz-Pulido, P. J. Mumby, P. A. Marshall, L. Cao, and O. Hoegh-Guldberg.

2011. Ocean acidification and warming will lower coral reef resilience. *Global Change Biology* **17**: 1798–1808.

Ashton, I. W., A. E. Miller, W. D. Bowman, and K. N. Suding. 2010. Niche complementarity due to plasticity in resource use: Plant partitioning of chemical N forms. *Ecology* **91**: 3252–3260.

Atkinson, D., and M. Thorndyke. 2001. The nature of animal developmental ecology: Integrating the "how" and the "why." In D. Atkinson and M. Thorndyke, eds. *Environment and Animal Development: Genes, Life Histories and Plasticity*, pp. 1–28. BIOS Scientific Publishers, Oxford.

Aubin-Horth, N., C. R. Landry, B. H. Letcher, and H. A. Hofmann. 2005. Alternative life histories shape brain gene expression profiles in males of the same population. *Proceedings of the Royal Society B* **272**: 1655–1662.

Auld, J. R., A. A. Agrawal, and R. A. Relyea. 2010. Re-evaluating the costs and limits of adaptive phenotypic plasticity. *Proceedings of the Royal Society B* **277**: 503–511.

Badeaux, A., and Y. Shi. 2013. Emerging roles for chromatin as a signal integration and storage platform. *Nature Reviews Molecular Cell Biology* **14**: 211–224.

Badri, D. V., and J. M. Vivanco. 2009. Regulation and function of root exudates. *Plant, Cell and Environment* **32**: 666–681.

Badyaev, A. V. 2005. Role of stress in evolution: From individual adaptability to evolutionary adaptation. In B. Hallgrímsson and B. Hall, eds. *Variation: A Hierarchical Examination of a Central Concept in Biology*, pp. 277–302. Academic Press, New York.

Badyaev, A. V. 2009. Evolutionary significance of phenotypic accommodation in novel environments: An empirical test of the Baldwin effect. *Philosophical Transactions of the Royal Society B* **364**: 1125–1141.

Badyaev, A. V., and T. Uller. 2009. Parental effects in ecology and evolution: Mechanisms, processes and implications. *Philosophical Transactions of the Royal Society B* **364**: 1169–1177.

Bailey, J. K., M. A. Genung, I. Ware, et al. 2014. Indirect genetic effects: An evolutionary mechanism linking feedbacks, genotypic diversity and coadaptation in a climate change context. *Functional Ecology* **28**: 87–95.

Bailey-Serres, J., and L. A. C. J. Voesenek. 2010. Life in the balance: A signaling network controlling survival of flooding. *Current Opinion in Plant Biology* **13**: 489–494.

Bais, H. P., R. Vepachedu, S. Gilroy, R. M. Callaway, and J. M. Vivanco. 2003. Allelopathy and exotic plant invasion: From molecules and genes to species interactions. *Science* **301**: 1377–1380.

Bais, H. P., T. S. Walker, F. R. Stermitz, R. A. Hufbauer, and J. M. Vivanco. 2002. Enantiomeric dependent phytotoxic and antimicrobial activity of (±)-catechin; a rhizosecreted racemic mixture from *Centaurea maculosa* (spotted knapweed). *Plant Physiology* **128**: 1173–1179.

Baker, H. G. 1965. Characteristics and modes of origin of weeds. In H. G. Baker and G. L. Stebbins, eds. *The Genetics of Colonizing Species*, pp. 147–172. Academic Press, New York.

Balaban, N. Q., J. Merrin, R. Chait, L. Kowalik, and S. Leibler. 2004. Bacterial persistence as a phenotypic switch. *Science* **305**: 1622–1625.

Balbyshev, N. F., and J. H. Lorenzen. 1997. Hypersensitivity and egg drop: A novel mechanism of host plant resistance to Colorado potato beetle (Coleoptera: Chrysomelidae). *Journal of Economic Entomology* **90**: 652–657.

Baldwin, I. T. 2010. Plant volatiles. *Current Biology* **20**: R392–R397.

Baldwin, I. T., R. Halitschke, A. Paschold, C. C. von Dahl, and C. A. Preston. 2006. Volatile signaling in plant-plant interactions: "Talking trees" in the genomics era. *Science* **311**: 812–815.

Balhoff, J. P., and G. A. Wray. 2005. Evolutionary analysis of the well characterized *endo16* promoter reveals substantial variation within functional sites. *Proceedings of the National Academy of Sciences USA* **102**: 8591–8596.

Ballaré, C. L. 2009. Illuminated behaviour: Phytochrome as a key regulator of light foraging and plant anti-herbivore defense. *Plant, Cell and Environment* **32**: 713–725.

Ballaré, C. L., A. L. Scopel, and R. A. Sánchez. 1990. Far-red radiation reflected from adjacent leaves: An early signal of competition in plant canopies. *Science* **247**: 329–332.

Bannister, A. J., and T. Kouzarides. 2011. Regulation of chromatin by histone modifications. *Cell Reports* **21**: 381–395.

Barker, N. P. 2005. A review and survey of basicarpy, geocarpy, and amphicarpy in the African and Madagascar flora. *Annals of the Missouri Botanical Garden* **92**: 445–462.

Barrett, R. D. H., and H. E. Hoekstra. 2011. Molecular spandrels: Tests of adaptation at the genetic level. *Nature Reviews Genetics* **12**: 767–780.

Barrett, R. D. H., S. M. Rogers, and D. Schluter. 2009. Environment specific pleiotropy facilitates divergence at the *ECTODYSPLASIN* locus in threespine stickleback. *Evolution* **63**: 2831–2837.

Bartemucci, P., C. Messier, and C. D. Canham. 2006. Overstory influences on light attenuation patterns and understory plant community diversity and composition in southern boreal forests of Quebec. *Canadian Journal of Forest Research* **36**: 2065–2079.

Barton, N. H., and M. Turelli. 1989. Evolutionary quantitative genetics: How little do we know? *Annual Review of Genetics* **23**: 337–370.

Basille, M., B. Van Moorter, I. Herfindal, et al. 2013. Selecting habitat to survive: The impact of road density on survival in a large carnivore. *Public Library of Science ONE* **7**: e65493.

Baskin, C. C., and J. M. Baskin. 1998. *Seeds: Ecology, Biogeography and Evolution of Dormancy and Germination*. Academic Press, London.

Bassar, R. D., M. C. Marshall, A. López-Sepulcre, et al. 2010. Local adaptation in Trinidadian guppies alters ecosystem processes. *Proceedings of the National Academy of Sciences USA* **107**: 3616–3621.

Bateson, P., D. Barker, T. Clutton-Brock, et al. 2004. Developmental plasticity and human health. *Nature* **430**: 419–421.

Bateson, P., and P. D. Gluckman. 2011. *Plasticity, Robustness, Development and Evolution*. Cambridge University Press, New York.

Bateson, P., P. D. Gluckman, and M. Hanson. 2014. The biology of developmental plasticity and the predictive adaptive response hypothesis. *Journal of Physiology* **592**: 2357–2368C.

Baythavong, B. S. 2011. Linking the spatial scale of environmental variation and the evolution of phenotypic plasticity: Selection favors adaptive plasticity in fine-grained environments. *American Naturalist* **178**: 75–87.

Bazzaz, F. A. 1979. The physiological ecology of plant succession. *Annual Review of Ecology and Systematics* **10**: 351–371.

Bazzaz, F. A. 1996. *Plants in Changing Environments*. Cambridge University Press, Cambridge, UK.

Becker, C., J. Hagmann, J. Müller, et al. 2011. Spontaneous epigenetic variation in the *Arabidopsis thaliana* methylome. *Nature* **480**: 245–249.

Beckman, B. R., and W. W. Dickhoff. 1998. Plasticity of smolting in spring chinook salmon: Relation to growth and insulin-like growth factor-I. *Journal of Fish Biology* **53**: 808–826.

Beerling, D. J., and R. A. Berner. 2005. Feedbacks and the coevolution of plants and atmospheric CO_2. *Proceedings of the National Academy of Sciences USA* **102**: 1302–1305.

Begum, G., A. Davies, A. Stevens, et al. 2013. Maternal undernutrition programs tissue-specific epigenetic changes in the glucucorticoid receptor in adult offspring. *Endocrinology* **154**: 4560–4569.

Bell, D. L., and S. E. Sultan. 1999. Dynamic phenotypic plasticity for root growth in *Polygonum*: A comparative study. *American Journal of Botany* **86**: 807–819.

Bennett, A. F., R. B. Huey, H. John-Alder, and K. A. Nagy. 1984. The parasol tail and thermoregulatory behavior of the cape ground squirrel *Xerus inauris*. *Physiological Zoology* **57**: 57–62.

Bennetts, R. E., G. C. White, F. G. Hawksworth, and S. E. Severs. 1996. The influence of dwarf mistletoe on bird communities in Colorado ponderosa pine forests. *Ecological Applications* **6**: 899–909.

Benschop, J. J., M. B. Jackson, K. Gühl, et al. 2005. Contrasting interactions between ethylene and abscisic acid in *Rumex* species differing in submergence tolerance. *Plant Journal* **44**: 756–768.

Bergmann, D. 2006. Stomatal development: From neighborly to global communication. *Current Opinion in Plant Biology* **9**: 478–483.

Bergmann, S., J. Ihmels, and N. Barkai. 2004. Similarities and differences in genome-wide expression data of six organisms. *Public Library of Science Biology* **2**: e9.

Bernstein, E., and C. D. Allis. 2005. RNA meets chromatin. *Genes and Development* **19**: 1635–1655.

Berrigan, D., and S. M. Scheiner. 2004. Modeling the evolution of phenotypic plasticity. In T. M. DeWitt and S. M. Scheiner, eds. *Phenotypic Plasticity: Functional and Conceptual Approaches*, pp. 82–97. Oxford University Press, New York.

Bertness, M. D. 1989. Intraspecific competition and facilitation in a northern acorn barnacle population. *Ecology* **70**: 257 –268.

Bertness, M. D., and R. Callaway. 1994. Positive interactions in communities. *Trends in Ecology and Evolution* **9**: 191–193.

Bertness, M. D., G. H. Leonard, J. M. Levine, P. R. Schmidt, and A. O. Ingraham. 1999. Testing the relative contribution of positive and negative interactions in rocky intertidal communities. *Ecology* **80**: 2711–2726.

Bevins, C. L., and N. H. Salzman. 2011. The potter's wheel: The host's role in sculpting its microbiota. *Cell Molecular Life Sciences* **68**: 3675–3685.

Bhatia, A., A. Yadav, C. Zhu, et al. 2014. Yeast growth plasticity is regulated by environment-specific multi-QTL interactions. *G3: Genes, Genomes, Genetics* **4**: 769–777.

Bhattacharya, D., K. N. Pelletreau, D. C. Price, K. E. Sarver, and M. E. Rumpho. 2013. Genome analysis of *Elysia chlorotica* egg DNA provides no evidence for horizontal gene transfer into the germ line of this kleptoplastic mollusc. *Molecular Biology and Evolution* **30**: 1843–1852.

Biddle, F. G., and B. A. Eales. 1999. Mouse genetic model for left-right hand usage: Context, direction, norms of reaction, and memory. *Genome* **42**: 1150–1166.

Biedrzycki, M. M. L., T. A. Jilany, S. A. Dudley, and H. P. Bais. 2010. Root exudates mediate kin recognition in plants. *Community Integrated Biology* **1**: 28–35.

Biere, A. 1995. Genotypic and plastic variation in plant size: Effects on fecundity and allocation patterns in *Lychnis flos-cuculi* along a gradient of natural soil fertility. *Journal of Ecology* **83**: 629–642.

Biesmeijer, J. C., S. P. M. Roberts, M. Reemer, et al. 2006. Parallel declines in pollinators and insect-pollinated plants in Britain and the Netherlands. *Science* **313**: 351–354.

Biller, S. J., P. M. Berube, D. Lindell, and S. W. Chisholm. 2015. *Prochlorococcus*: The structure and function of collective diversity. *Nature Reviews Microbiology* **13**: 13–27.

Binkley, D. A. N., and C. Giardina.1998. Why do tree species affect soils? The warp and woof of tree–soil interactions. *Biogeochemistry* **42**: 89–106.

Bird, A. 2007. Perceptions of epigenetics. *Nature* **447**: 396–398.

Blackstone, N. W., and D. M. Bridge. 2005. Model systems for environmental signaling. *Integrative and Comparative Biology* **45**: 605–614.

Bleuler-Martínez, S., A. Butschi, M. Garbani, et al. 2011. A lectin-mediated resistance of higher fungi against predators and parasites. *Molecular Ecology* **20**: 3056–3070.

Bloom, J. S., I. M. Ehrenreich, W. T. Loo, T.-L. Võ Lite, and L. Kruglyak. 2013. Finding the sources of missing heritability in a yeast cross. *Nature* **494**: 234–237.

Blossfeld, S., and D. Gansert. 2007. A novel non-invasive optical method for quantitative visualization of pH dynamics in the rhizosphere of plants. *Plant Cell and Environment* **30**: 176–186.

Blum, A. 2005. Drought resistance, water-use efficiency, and yield potential: Are they compatible, dissonant, or mutually exclusive? *Crop and Pasture Science* **56**: 1159–1168.

Boland, M. J., K. L. Nazor, and J. F. Loring. 2014. Epigenetic regulation of pluripotency and differentiation. *Circulation Research* **115**: 311–324.

Bolker, B., M. Holyoak, V. Krivan, L. Rowe, and O. Schmitz. 2003. Connecting theoretical and empirical studies of trait-mediated interactions. *Ecology* **84**: 1101–1114.

Bolker, J. A. 1995. Model systems in developmental biology. *BioEssays* **17**: 451–455.

Bollati, V., and A. Baccarelli. 2010. Environmental epigenetics. *Heredity* **105**: 1105–1112.

Bolnick, D. I., R. Svanbäck, J. A. Fordyce, et al. 2003. The ecology of individuals: Incidence and implications of individual specialization. *American Naturalist* **161**: 1–28.

Bolter, C. J., and M. A. Jongsma. 1995. Colorado potato beetles (*Leptinotarsa decemlineata*) adapt to protein inhibitors induced in potato leaves by methyl jasmonate. *Journal of Insect Physiology* **41**: 1071–1078.

Bonaventure, G., A. VanDoorn, and I. T. Baldwin. 2011. Herbivore-associated elicitors: FAC signaling and metabolism. *Trends in Plant Science* **16**: 294–299.

Bonduriansky, R. 2012. Rethinking heredity, again. *Trends in Ecology and Evolution* **27**: 330–336.

Bonduriansky, R., A. J. Crean, and T. Day. 2012. The implications of non-genetic inheritance for evolution in changing environments. *Evolutionary Applications* **5**: 192–201.

Bonner, J. T. 1988. *The Evolution of Complexity*. Princeton University Press, Princeton.

Boogert N. J., D. M. Paterson, and K. N. Laland. 2006. The implications of niche construction and ecosystem engineering for conservation biology. *BioScience* **56**: 570–578.

Boose, D. L. 1997. Sources of variation in floral nectar production rate in *Epilobium canum*: Implications for natural selection. *Oecologia* **110**: 493–500.

Bossan, B., P. Hammerstein, and A. Koehncke. 2013. We were all young once: An intragenomic perspective on parent–offspring conflict. *Proceedings of the Royal Society B* **280**: 20122637.

Bossuyt, H., J. Six, and P. F. Hendrix. 2005. Protection of soil carbon by micro-aggregates within earthworm casts. *Soil Biology and Biochemistry* **37**: 251–258.

Both, C., M. Van Asch, R. G. Bijlsma, A. B. Van Den Burg, and M. E. Visser. 2009. Climate change and unequal phenological changes across four trophic levels: Constraints or adaptations? *Journal of Animal Ecology* **78**: 73–83.

Bourdeau, P. E. 2010. An inducible morphological defense is a passive by-product of behaviour in a marine snail. *Proceedings of the Royal Society B* **277**: 455–462.

Bourdeau, P. E., and F. Johansson. 2012. Predator-induced morphological defenses as by-products of prey behaviour: A review and prospectus. *Oikos* **1221**: 1175–1190.

Bouton, N., F. Witte, and J. J. M. van Alphen. 2002. Experimental evidence for adaptive phenotypic plasticity in a rock-dwelling cichlid fish from Lake Victoria. *Biological Journal of the Linnean Society* **77**: 185–192.

Braam, J. 2005. In touch: Plant responses to mechanical stimuli. *New Phytologist* **165**: 373–389.

Braendle, C., G. K. Davis, J. A. Brisson, and D. L. Stern. 2006. Wing dimorphism in aphids. *Heredity* **97**: 192–199.

Brandon, R. N. 1990. *Adaptation and Environment*. Princeton University Press, Princeton.

Bray, S. R., K. Kitajima, and D. M. Silvia. 2003. Mycorrhizae differentially alter growth, physiology and competitive ability of an invasive shrub. *Ecological Applications* **13**: 565–574.

Brem, R. B., J. D. Storey, J. Whittle, and L. Kruglyak. 2005. Genetic interactions between polymorphisms that affect gene expression in yeast. *Nature* **436**: 701–703.

Brenner, E. D., R. Stahlberg, S. Mancuso, J. Vivanco, F. Baluska, and E. Van Volkenburgh. 2006. Plant neurobiology: An integrated view of plant signaling. *Trends in Plant Science* **11**: 1360–1385.

Bridge, L. J., K. A. Franklin, and M. E. Homer. 2013. Impact of plant shoot architecture on leaf cooling: A coupled heat and mass transfer model. *Journal of the Royal Society Interface* **10**: 20130326.

Briggs, D., and S. M. Walters. 1997. *Plant Variation and Evolution*, 3rd ed. Cambridge University Press, Cambridge, UK.

Brock, M. T., J. R. Stinchcombe, and C. Weinig. 2009. Indirect effects of *FRIGIDA*: Floral trait (co)variances are altered by seasonally variable abiotic factors associated with flowering time. *Journal of Evolutionary Biology* **22**: 1826–1838.

Brodie, E. D., III. 2005. Caution: Niche construction ahead. *Evolution* **59**: 249–251.

Brönmark, C., and L. B. Pettersson. 1994. Chemical cues from piscivores induce a change in morphology in crucian carp. *Oikos* **70**: 396–402.

Brookes, J. I., and R. Rochette. 2007. Mechanism of a plastic phenotypic response: Predator-induced shell thickening in the intertidal gastropod *Littorina obtusata*. *Journal of Evolutionary Biology* **20**: 1015–1027.

Brosi, B. J., and H. M. Briggs. 2013. Single pollinator species losses reduce floral fidelity and plant reproductive function. *Proceedings of the National Academy of Sciences USA* **110**: 13044–13048.

Brotosudarmo, T. H. P., A. M. Collins, A. Gall, et al. 2011. The light intensity under which cells are grown controls the type of peripheral light-harvesting complexes that are assembled in a purple photosynthetic bacterium. *Biochemical Journal* **440**: 51–61.

Brown, D. D., and L. Cai. 2007. Amphibian metamorphosis. *Developmental Biology* **306**: 20–33.

Broz, A. K., D. K. Manter, and J. M. Vivanco. 2007. Soil fungal abundance and diversity: Another victim of the invasive plant *Centaurea maculosa*. *International Society for Microbial Ecology Journal* **1**: 763–765.

Bruin, J., M. Dicke, and M. W. Sabelis. 1992. Plants are better protected against spider-mites after exposure to volatiles from infested conspecifics. *Experientia* **48**: 525–529.

Brundrett, M. C. 2009. Mycorrhizal associations and other means of nutrition of vascular plants: Understanding the global diversity of host plants by resolving conflicting information. *Plant and Soil* **320**: 37–77.

Bruno, J. F., and C. W. Kennedy. 2000. Patch-size dependent habitat modification and facilitation on New England cobble beaches by *Spartina alterniflora*. *Oecologia* **122**: 98–108.

Bruno, J. F., J. J. Stachowicz, and M. D. Bertness. 2003. Inclusion of facilitation into ecological theory. *Trends in Ecology and Evolution* **18**: 119–125.

Buchholz, D. R., S.-C. V. Hsia, L. Fu, and Y.-B. Shi. 2003. A dominant-negative thyroid hormone receptor blocks amphibian metamorphosis by retaining corepressors at target genes. *Molecular and Cellular Biology* **23**: 6750–6758.

Bull, J. J., W. H. N. Guizke, and M. G. Bulmer. 1988. Nest choice in a captive lizard with temperature-dependent sex determination. *Journal of Evolutionary Biology* **2**: 177–184.

Bundt, M., F. Widmer, M. Pesaro, J. Zeyer, and P. Blaser. 2001. Preferential flow paths: Biological "hot spots" in soil. *Soil Biology and Biochemistry* **33**: 729–738.

Burdge, G. C., and K. A. Lillycrop. 2010. Nutrition, epigenetics, and developmental plasticity: Implications for understanding human disease. *Annual Review of Nutrition* **30**: 315–339.

Burkle, L. A., and R. Alarcón. 2011. The future of plant–pollinator diversity: Understanding interaction networks across time, space, and global change. *American Journal of Botany* **98**: 528–538.

Burmeister, S. S., E. D. Jarvis, and R. D. Fernald. 2005. Rapid behavioral and genomic responses to social opportunity. *Public Library of Science Biology* **3**: e383.

Burns, J. H., and S. Y. Strauss. 2012. Effects of competition on phylogenetic signal and phenotypic plasticity in plant functional traits. *Ecology* **93**: S126–S137.

Burton, N. O., K. B. Burkhart, and S. Kennedy. 2011. Nuclear RNAi maintains heritable gene silencing in *Caenorhabditis elegans*. *Proceedings of the National Academy of Sciences USA* **108**: 19683–19688.

Byrnes, W. M., and W. R. Eckberg. 2006. E. E. Just: An early ecological developmental biologist. *Developmental Biology* **296**: 1–11.

Caldwell, M. M., T. E. Dawson, and J. H. Richards. 1998. Hydraulic lift: Consequences of water efflux from the roots of plants. *Oecologia* **113**: 151–161.

Callahan, H. S., and M. Pigliucci. 2002. Shade-induced plasticity and its ecological significance in wild populations of *Arabidopsis thaliana*. *Ecology* **83**: 1965–1980.

Callaway, R. M. 1990. Effects of soil water distribution on the lateral root development of three species of California oaks. *American Journal of Botany* **77**: 1469–1474.

Callaway, R. M., R. W. Brooker, P. Choler, et al. 2002. Positive interactions among alpine plants increase with stress. *Nature* **417**: 844–848.

Callaway, R. M., T. H. DeLuca, and W. M. Belliveau. 1999. Biological-control herbivores may increase competitive ability of the noxious weed *Centaurea maculosa*. *Ecology* **80**: 1196–1201.

Callaway, R. M., S. C. Pennings, and C. L. Richards. 2003. Phenotypic plasticity and interactions among plants. *Ecology* **84**: 1115–1128.

Callaway, R. M., G. C. Thelen, S. Barth, P. W. Ramsey, and J. E. Gannon. 2004. Soil fungi alter interactions between the invader *Centaurea maculosa* and North American natives. *Ecology* **85**: 1062–1071.

Cameron, S. L., N. Lo, T. Bourguignon, G. J. Svenson, and T. A. Evans. 2012. A mitochondrial genome phylogeny of termites (Blattodea: Termitoidae): Robust support for interfamilial relationships and molecular synapomorphies define major clades. *Molecular Phylogenetics and Evolution* **65**: 163–173.

Cano, L. M., S. Raffaele, R. H. Haugen, et al. 2013. Major transcriptome reprogramming underlies floral mimicry induced by the rust fungus *Puccinia monoica* in *Boechera stricta*. *Public Library of Science ONE* **8**: e75293.

Cantor, A., A. Hale, J. Aaron, M. B. Traw, and S. Kalisz. 2011. Low allelochemical concentrations detected in garlic mustard-invaded forest soils inhibit fungal growth and AMF spore germination. *Biological Invasions* **13**: 3015–3025.

Cardinale, B. J. 2011. Biodiversity improves water quality through niche partitioning. *Nature* **472**: 86–89.

Cardinale, B. J., C. T. Harvey, K. Gross, and A. R. Ives. 2003. Biodiversity and biocontrol: Emergent impacts of a multi-enemy assemblage on pest suppression and crop yield in an agroecosystem. *Ecology Letters* **6**: 857–865.

Cardinale, B. J., J. P. Wright, M. W. Cadotte, et al. 2007. Impacts of plant diversity on biomass production increase through time because of species complementarity. *Proceedings of the National Academy of Science USA* **104**: 18123–18128.

Carroll, S. B. 2008. Evo-devo and an expanding evolutionary synthesis: A genetic theory of morphological evolution. *Cell* **134**: 25–36.

Carroll, S. B., J. K. Grenier, and S. D. Weatherbee. 2005. *From DNA to Diversity: Molecular Genetics and the Evolution of Animal Design*, 2nd ed. Blackwell Publishing, Malden.

Carroll, S. P., A. P. Hendry, D. N. Reznick, and C. W. Fox. 2007. Evolution on ecological time-scales. *Functional Ecology* **21**: 387–393.

Carroll, S. P., P. Søgaard Jørgensen, M. T. Kinnison, et al. 2014. Applying evolutionary biology to address global challenges. *Science* **346**: 1245993.

Casal, J. J., and Smith, H. 1989. The function, action and adaptive significance of phytochrome in light-grown plants. *Plant, Cell and Environment* **12**: 855–862.

Casson, S. A., and A. M. Hetherington. 2010. Environmental regulation of stomatal development. *Current Opinion in Plant Biology* **13**: 90–95.

Cedar, H., and Y. Bergman. 2009. Linking DNA methylation and histone modification: Patterns and paradigms. *Nature Reviews Genetics* **10**: 295–304.

Chakir, M., A. Chafik, P. Gibert, and J. R. David. 2002. Phenotypic plasticity of adult size and pigmentation in *Drosophila* thermosensitive periods during development in two sibling species. *Journal of Thermal Biology* **27**: 61–70.

Champagne, F. A. 2008. Epigenetic mechanisms and the transgenerational effects of maternal care. *Frontiers in Neuroendocrinology* **29**: 386–397.

Chandler, V. L. 2007. Paramutation: From maize to mice. *Cell* **128**: 641–645.

Chang, S. S., B. B. Prézelin, and R. K. Trench. 1983. Mechanisms of photoadaptation in three strains of the symbiotic dinoflagellate *Symbiodinium microadriaticum*. *Marine Biology* **76**: 219–229.

Chapman, L. J., J. Albert, and F. Galis. 2008. Developmental plasticity, genetic differentiation, and hypoxia-induced trade-offs in an African cichlid fish. *The Open Evolution Journal* **2**: 75–88.

Chapman, L. J., F. Galis, and J. Shinn. 2000. Phenotypic plasticity and the possible role of genetic assimilation: Hypoxia-induced trade-offs in the morphological traits of an African cichlid. *Ecology Letters* **3**: 387–393.

Chapman, N. C., B. P. Oldroyd, and W. O. H. Hughes. 2007. Differential responses of honeybee (*Apis mellifera*) patrilines to changes in stimuli for the generalist tasks of nursing and foraging. *Behavioral Ecology and Sociobiology* **61**: 1185–1194.

Charlesworth, D., and J. H. Willis. 2009. Fundamental concepts in genetics: The genetics of inbreeding depression. *Nature Reviews Genetics* **10**: 783–796.

Chaves-Fonnegra, A., and S. Zea. 2007. Observations on reef coral undermining by the Caribbean excaving sponge *Cliona delitrix*. In M. R. Custódio, E. Lobo-Hajdu, and G. Muricy, eds. *Porifera Research: Biodiversity, Innovation and Sustainability*, pp. 247–264. Museu Nacional, Rio de Janeiro.

Chen, Y., F. Ding, H. Nie, A. W. Serohijos, S. Sharma, K. C. Wilcox, S. Yin, and N. V. Dokholyan. 2008. Protein folding: Then and now. *Archives of Biochemistry and Biophysics* **469**: 4–19.

Chevin, L.-M., S. Collins, and F. Lefèvre. 2013. Phenotypic plasticity and evolutionary demographic responses to climate change: Taking theory out to the field. *Functional Ecology* **27**: 967–979.

Chevin, L.-M., and R. Lande. 2009. When do adaptive plasticity and genetic evolution prevent extinction of a density-regulated population? *Evolution* **64**: 1143–1150.

Chevin, L.-M., R. Lande, and G. M. Mace. 2010. Adaptation, plasticity, and extinction in a changing environment: Towards a predictive theory. *Public Library of Science Biology* **8**: e1000357.

Chinnusamy, V., and J. K. Zhu. 2009. Epigenetic regulation of stress responses in plants. *Current Opinion in Plant Biology* **12**: 133–139.

Clements, K. D., E. R. Angert, W. L. Montgomery, and J. H. Choat. 2014. Intestinal microbiota in fishes: What's known and what's not. *Molecular Ecology* **23**: 1891–1898.

Clutton-Brock, T. H. 1991. *The Evolution of Parental Care*. Princeton University Press, Princeton.

Cody, M. L. 1985. An introduction to habitat selection in birds. In M. L. Cody, ed. *Habitat Selection in Birds*, pp. 4–58. Academic Press, New York.

Cody, M. L., and J. J. M. Overton. 1996. Short-term evolution of reduced dispersal in island plant populations. *Journal of Ecology* **84**: 53–61.

Coe, M. 2004. Orientation, movement and thermoregulation in the giant tortoises (*Testudo*) of Aldabra Atoll, Seychelles. *Transactions of the Royal Society of South Africa: Proceedings of a Colloquium on Adaptations in Desert Fauna and Flora* **59**: 73–77.

Coggan, N., F. J. Clissold, and S. J. Simpson. 2011. Locusts use dynamic thermoregulatory behaviour to optimize nutritional outcomes. *Proceedings of the Royal Society B* **278**: 2745–2752.

Coleman, F. C., and S. L. Williams. 2002. Overexploiting marine ecosystem engineers: Potential consequences for biodiversity. *Trends in Ecology and Evolution* **17**: 40–44.

Coley, P. D., and J. A. Barone. 1996. Herbivory and plant defenses in tropical forests. *Annual Review of Ecology and Systematics* **27**: 305–335.

Colicchio, J. M., P. J. Monnahan, J. K. Kelly, and L. C. Hileman. 2015. Gene expression plasticity resulting from parental leaf damage in *Mimulus guttatus*. *New Phytologist* **205**: 894–906.

Colwell, R. K. 1992. Niche: A bifurcation in the conceptual lineage of the term. In E. F. Keller and E. A. Lloyd, eds. *Keywords in Evolutionary Biology*, pp. 241–248. Harvard University Press, Cambridge.

Conner, J. K., and D. L. Hartl. 2004. *A Primer of Ecological Genetics*. Sinauer Associates Inc., Sunderland, MA.

Cortijo, S., R. Wardenaar, M. Colomé-Tatché, et al. 2014. Mapping the epigenetic basis of complex traits. *Science* **343**: 1145–1148.

Cossetti, C., L. Lugini, L. Astrologo, I. Saggio, S. Fais, and C. Spadafora. 2014. Soma-to-germline transmission of RNA in mice xenografted with human tumour cells: Possible transport by exosomes. *Public Library of Science ONE* **9**: e101629.

Coupe, S. A., B. G. Palmer, J. A. Lake, et al. 2006. Systemic signaling of environmental cues in *Arabidopsis* leaves. *Journal of Experimental Botany* **57**: 329–341.

Covello, P. S., and M. W. Gray. 1989. RNA editing in plant mitochondria. *Nature* **341**: 662–666.

Crain, C. M., and M. D. Bertness. 2005. Community impacts of a tussock sedge: Is ecosystem engineering important in benign habitats? *Ecology* **86**: 2695–2704.

Crain, C. M., and M. D. Bertness. 2006. Ecosystem engineering across environmental gradients: Implications for conservation and management. *Bioscience* **56**: 211–218.

Cramer, G. R., K. Urano, M. Pezzotti, and K. Shinozaki. 2011. Effects of abiotic stress on plants: A systems biology perspective. *BioMed Central Plant Biology* **11**: 163.

Cramer, M. J., M. R. Willig, and C. Jones. 2001. *Trachops cirrhosus*. *Mammalian Species* **656**: 1–6.

Crawford, A. J., D. H. McLachlan, A. M. Hetherington, and K. A. Franklin. 2012. High temperature exposure increases plant cooling capacity. *Current Biology* **22**:R396.

Crayton, M. E., III, B. C. Powell, T. J. Vision, and M. C. Giddings. 2006. Tracking the evolution of alternatively spliced exons within the *Dscam* family. *BioMed Central Evolutionary Biology* **6**: 16.

Crews, D. 2003. Sex determination: Where environment and genetics meet. *Evolution and Development* **5**: 50–55.

Crews, D., J. M. Bergeron, J. J. Bull, et al. 1994. Temperature-dependent sex determination in reptiles: Proximate mechanisms, ultimate outcomes, and practical applications. *Developmental Genetics* **15**: 297–312.

Crews, D., R. Gillette, S. V. Scarpino, M. Manikkam, M. I. Savenkova, and M. K. Skinner. 2012. Epigenetic transgenerational inheritance of altered stress responses. *Proceedings of the National Academy of Sciences USA* **109**: 9143–9148.

Crooks, J. A. 2002. Characterizing ecosystem-level consequences of biological invasions: The role of ecosystem engineers. *Oikos* **97**: 153–166.

Cropley, J. E., C. M. Suter, K. B. Beckman, and D. I. K. Martin. 2006. Germ-line epigenetic modification of the murine A^{vy} allele by nutritional supplementation. *Proceedings of the National Academy of Sciences USA* **103**: 17308–17312.

Cryan, J. F., and T. G. Dinan. 2012. Mind-altering microorganisms: The impact of the gut microbiota on brain and behaviour. *Nature Reviews Neuroscience* **13**: 701–712.

Cubas, P., C., Vincent, and E. Coen. 1999. An epigenetic mutation responsible for natural variation in floral symmetry. *Nature* **401**: 157–161.

Dalin, P., J. Ågren, C. Björkman, P. Huttunen, and K. Kärkkäinen. 2008. Leaf trichome formation and plant resistance to herbivory. In A. Schaller, ed. *Induced Plant Resistance to Herbivory*, pp. 89–105. Springer Science and Business Media, New York.

Dalin, P., and C. Björkman. 2003. Adult beetle grazing induces willow trichome defence against subsequent larval feeding. *Oecologia* **134**: 112–118.

Danchin, É., A. Charmentier, F. A. Champagne, A. Mesoudi, B. Pujol, and S. Blanchet. 2011. Beyond DNA: Integrating inclusive inheritance into an extended theory of evolution. *Nature Reviews Genetics* **12**: 475–486.

Daniel, O., and J. M. Anderson. 1992. Microbial biomass and activity in contrasting soil materials after passage through the gut of the earthworm *Lumbricus rubellus*. *Soil Biology and Biochemistry* **24**: 465–470.

D'Antonio, C. M., and P. M. Vitousek. 1992. Biological invasions by exotic grasses, the grass/fire cycle, and global change. *Annual Review of Ecology and Systematics* **23**: 63–87.

Darwin, C. 1859. *The Origin of Species*. Murray, London. (Ed. by J. W. Burrow and reprinted by Penguin Classics, 1985.)

Darwin, C. 1881. *The Formation of Vegetable Mould through the Action of Worms with Obervations on their Habits*. Murray, London.

Darwin, C., and F. Darwin. 1880. *The Power of Movement in Plants*. Murray, London.

Daugherty, R. L., and C. J. Gottardi. 2007. Phospho-regulation of ß-catenin adhesion and signalling functions. *Physiology* **22**: 303–309.

David, L. A., C. F. Maurice, R. N. Carmody, et al. 2014. Diet rapidly and reproducibly alters the human gut microbiome. *Nature* **505**: 559–563.

Davidson, E. H. 2006. *The Regulatory Genome: Gene Regulatory Networks in Development and Evolution*. Elsevier Academic Press, London.

Davidson, E. H. 2010. Emerging properties of animal gene regulatory networks. *Nature* **468**: 911–920.

Davidson, E. H., J. P. Rast, P. Olivieri, et al. 2002. A genomic regulatory network for development. *Science* **295**: 1669–1678.

Davis, P. A., and R. P. Hangarter. 2012. Chloroplast movement provides photoprotection to plants by redistributing PSII damage within leaves. *Photosynthesis Research* **112**: 153–161.

Dawkins, R. 1982. *The Extended Phenotype: The Long Reach of the Gene*. Oxford University Press, New York.

Dawkins, R. 2004. Extended phenotype—but not *too* extended. A reply to Laland, Turner and Jablonka. *Biology and Philosophy* **19**: 377–396.

Day, T., and R. Bonduriansky. 2011. A unified approach to the evolutionary consequences of genetic and nongenetic inheritance. *American Naturalist* **178**: E18–E36.

DeBlasio, S. L., J. L. Mullen, D. R. Luesse, and R. P. Hangarter. 2003. Phytochrome modulation of blue light-induced chloroplast movements in *Arabidopsis*. *Plant Physiology* **133**: 1471–1479.

de Bruyn, L. A., and A. J. Conacher. 1990. The role of termites and ants in soil modification: A review. *Australian Journal of Soil Research* **28**: 55–93.

De Jong, G. 1994. The fitness of fitness concepts and the description of natural selection. *Quarterly Review of Biology* **69**: 3–29.

De Jong, G. 1995. Phenotypic plasticity as a product of selection in a variable environment. American Naturalist **145**: 493–512.

de Kroon, H. 2007. How do roots interact? *Science* **318**: 1562–1563.

de Kroon, H., E. J. W. Visser, H. Huber, L. Mommer, and M. J. Hutchings. 2009. A modular concept of plant foraging behaviour: The interplay between local responses and systemic control. *Plant, Cell and Environment* **32**: 704–712.

De Leo, F., and R. Gallerani. 2002. The mustard trypsin inhibitor 2 affects the fertility of *Spodoptera littoralis* larvae fed on transgenic plants. *Insect Biochemistry and Molecular Biology* **32**: 489–496.

De Moraes, C. M., W. J. Lewis, P. W. Pare, H. T. Alborn, and J. H. Tumlinson. 1998. Herbivore-infested plants selectively attract parasitoids. *Nature* **393**: 570–573.

Dempsey, M. A., M. C. Fisk, J. B. Yavitt, T. J. Fahey, and T. C. Balser. 2013. Exotic earthworms alter soil microbial community composition and function. *Soil Biology and Biochemistry* **67**: 263–270.

Dennis, S. R., G. A. LeBlanc, and A. P. Beckerman. 2014. Endocrine regulation of predator-induced phenotypic plasticity. *Oecologia* **176**: 625–635.

Denver, R. J. 1997a. Proximate mechanisms of phenotypic plasticity in amphibian metamorphosis. *American Zoologist* **37**: 172–184.

Denver, R. J. 1997b. Environmental stress as a developmental cue: Corticotropin-releasing hormone is a proximate mediator of adaptive phenotypic plasticity in amphibian metamorphosis. *Hormones and Behavior* **31**: 169–179.

Denver, R. J. 1998. Hormonal correlates of environmentally induced metamorphosis in the Western Spadefoot Toad, *Scaphiopus hammondii*. *General and Comparative Endocrinology* **110**: 326–336.

Denver, R. J. 2013. Neuroendocrinology of amphibian metamorphosis. *Current Topics in Developmental Biology* **103**: 195–227.

Denver, R. J., N. Mirhadi, and M. Phillips. 1998. Adaptive plasticity in amphibian metamorphosis: Response of *Scaphiopus hammondii* tadpoles to habitat dessication. *Ecology* **79**: 1859–1873.

Depew, D. J., and B. H. Weber. 2011. The fate of Darwinism: Evolution after the modern synthesis. *Biological Theory* **6**: 89–102.

de Sampaio, C., P. Camilo-Alves, and G. de Miranda Mourao. 2006. Responses of a specialized insectivorous mammal to variation in ambient temperature. *Biotropica* **38**: 52–56.

Des Marais, D. L., K. M. Hernandez, and T. E. Juenger. 2013. Genotype-by-environment interaction and plasticity: Exploring genomic responses of plants to the abiotic environment. *Annual Review of Ecology, Evolution and Systematics* **44**: 5–29.

de Vos, M., and G. Jander. 2009. *Myzus persicae* (green peach aphid) salivary components induce defence responses in *Arabidopsis thaliana*. *Plant Cell Environment* **32**: 1548–1560.

de Vos, M., and G. Jander. 2010. Volatile communication in plant–aphid interactions. *Current Opinion in Plant Biology* **13**: 366–371.

de Vries, J., J. Habicht, C. Woehle, et al. 2013. Is *ftsH* the key to plastic longevity in sacoglossan slugs? *Genome Biology and Evolution* **5**: 2540–2548.

DeWitt, T. J., and S. M. Scheiner, eds. 2004. *Phenotypic Plasticity: Functional and Conceptual Approaches*. Oxford University Press, Oxford.

DeWitt, T. J., A. Sih, and D. S. Wilson. 1998. Costs and limits of phenotypic plasticity. *Trends in Ecology and Evolution* **13**: 77–81.

Dicke, M., and M. W. Sabelis. 1987. How plants obtain predatory mites as bodyguards. *Netherlands Journal of Zoology* **38**: 148–165.

Dickens, W. T., and J. R. Flynn. 2001. Heritability estimates versus large environmental effects: The IQ paradox resolved. *Psychological Review* **108**: 346–349.

Djakovic-Petrovic, T., M. de Wit, L. A. C. J. Voesenek, and R. Pierik. 2007. DELLA protein function in growth responses to canopy signals. *Plant Journal* **51**: 117–126.

Dobkin, D. S. 1985. Heterogeneity of tropical floral microclimates and the response of hummingbird flower mites. *Ecology* **66**: 536–543.

Dobzhansky, T. H. 1937. *Genetics and the Origin of Species*. Columbia University Press, New York.

Dolinoy, D. C. 2008. The agouti mouse model: An epigenetic biosensor for nutritional and environmental alterations on the fetal epigenome. *Nutrition Reviews* **66**: S7–S11.

Dolinoy, D. C., J. R. Weidman, R. A. Waterland, and R. L. Jirtle. 2006. Maternal genistein alters coat color and protects A^{vy} mouse offspring from obesity by modifying the fetal epigenome. *Environmental Health Perspectives* **114**: 567–572.

Donohue, K. 2003. Setting the stage: Phenotypic plasticity as habitat selection. *International Journal of Plant Science* **164**: S79–S92.

Donohue, K. 2005. Niche construction through phenological plasticity: Life history dynamics and ecological consequences. *New Phytologist* **166**: 83–92.

Donohue, K. 2009. Completing the cycle: Maternal effects as the missing link in plant life histories. *Philosophical Transactions of the Royal Society B* **364**: 1059–1074.

Donohue, K., L. Dorn, C. Griffith, et al. 2005. Niche construction through germination cueing: Life-history responses to timing of germination in *Arabidopsis thaliana*. *Evolution* **59**: 771–785.

Donohue, K., D. Messiqua, E. H. Pyle, M. S. Heschel, and J. Schmitt. 2000. Evidence of adaptive divergence in plasticity: Density- and site-dependent selection on shade-avoidance responses in *Impatiens capensis*. *Evolution* **54**: 1956–1968.

Donohue, K., and J. Schmitt. 1998. Maternal environmental effects in plants: Adaptive plasticity? In T. A. Mousseau and C. W. Fox, eds. *Maternal Effects as Adaptations*, pp. 137–158. Oxford University Press, Oxford.

Dorn, L. A., E. H. Pyle, and J. Schmitt. 2000. Plasticity to light cues and resources in *Arabidopsis rhaliana:* Testing for adaptive value and costs. *Evolution* **54**: 1982–1994.

Doublet, V., M. Labarussias, J. R. Miranda, R. F. A. Moritz, and R. J. Paxton. 2015. Bees under stress: Sublethal doses of a neonicotinoid pesticide and pathogens interact to elevate honey bee mortality across the life cycle. *Environmental Microbiology* **17**: 969–983.

Draghi, J. A., and M. C. Whitlock. 2012. Phenotypic plasticity facilitates mutational variance, genetic variance, and evolvability along the major axis of environmental variation. *Evolution* **66**: 2891–2902.

Drepper, T., U. Krauss, S. Meyer zu Berstenhorst, J. Pietruszka, and K.-E. Jaeger. 2011. Lights on and action! Controlling microbial gene expression by light. *Applied Microbiology and Biotechnology* **90**: 23–40.

Drew, M. C. 1975. Comparison of the effects of a localized supply of phosphate, nitrate, ammonium and potassium on the growth of the seminal root system, and the shoot, in barley. *New Phytologist* **75**: 479–490.

Drew, M. C., L. R. Saker, and T. W. Ashley. 1973. Nutrient supply and the growth of the seminal root system in barley. I. The effect of nitrate concentration on the grown of axes and laterals. *Journal of Experimental Botany* **24**: 1189–1202.

Drong, A. W., C. M. Lindgren, and M. I. McCarthy. 2012. The genetic and epigenetic basis of type 2 diabetes and obesity. *Clinical Pharmacology and Therapeutics* **92**: 707–715.

Du, Y., G. M. Poppy, W. Powell, J. A. Pickett, L. J. Wadhams, and C. M. Woodcock. 1998. Identification of semiochemicals released during aphid feeding that attract parasitoid *Aphidius ervi*. *Journal of Chemical Ecology* **24**: 1355–1368.

Dudley, S. A., and J. Schmitt. 1996. Testing the adaptive plasticity hypothesis: Density-dependent selection on manipulated stem length in *Impatiens capensis*. *American Naturalist* **147**: 445–465.

Dufty, A. M., Jr., J. Clobert, and A. P. Møller. 2002. Hormones, developmental plasticity and adaptation. *Trends in Ecology and Evolution* **17**: 190–196.

Duncan, E. J., P. D. Gluckman, and P. K. Dearden. 2014. Epigenetics, plasticity, and evolution: How do we link epigenetic change to phenotype? *Journal of Experimental Zoology* **322B**: 208–220.

Dusek, J. A., H. H. Otu, A. L. Wohlhueter, et al. 2008. Genomic counter-stress changes induced by the relaxation response. *Public Library of Science ONE* **3**: e2576.

Ehleringer, J., and H. A. Mooney. 1978. Leaf hairs: Effects on physiological activity and adaptive value to a desert shrub. *Oecologia* **37**: 183–200.

Ehret, G., and R. Fisher. 1991. Neuronal activity and tonotopy in the auditory system visualized by *c-fos* gene expression. *Brain Research* **567**: 350–354.

Emery, N. C., and D. D. Ackerly. 2014. Ecological release exposes genetically based niche variation. *Ecology Letters* **17**: 1149–1157.

Emery, N. C., E. J. Forrestel, G. Jui, S. Park, B. G. Baldwin, and D. D. Ackerly. 2012. Niche evolution across spatial scales: Climate and habitat specialization in California *Lasthenia* (Asteraceae). *Ecology* **93**: S151–S166.

Emery, N. C., K. J. Rice, and M. L. Stanton. 2011. Fitness variation and local distribution limits in an annual plant population. *Evolution* **65**: 1011–1020.

ENCODE Project Consortium. 2012. An integrated encyclopedia of DNA elements in the human genome. *Nature* **489**: 57–74.

Endler, J. A. 1986. *Natural Selection in the Wild*. Princeton University Press, Princeton.

Engel, G. S., T. R. Calhoun, E. L. Read, et al. 2007. Evidence for wavelike energy transfer through quantum coherence in photosynthetic systems. *Nature* **446**: 782–786.

Engelberth, J., H. T. Alborn, E. A. Schmelz, and J. H. Tunlinson. 2004. Airborne signals prime plants against herbivore attack. *Proceedings of the National Academy of Sciences USA* **101**: 1781–1785.

Engen, S. R., R. Lande, and B. E. Saether. 2011. Evolution of a plastic quantitative trait in an age-structured population in a fluctuating environment. *Evolution* **65**: 2893–2906.

Ernest, S. K. M., and J. H. Brown. 2001. Delayed compensation for missing keystone species by colonization. *Science* **292**: 101–103.

Estes, J. A., J. Terborgh, J. S. Brashares, et al. 2011. Trophic downgrading of planet Earth. *Science* **333**: 301–306.

Evans, A. S., and R. J. Cabin. 1995. Can dormancy affect the evolution of post-germination traits? The case of *Lesquerella fendleri*. *Ecology* **76**: 344–356.

Eveland, A. L., and D. P. Jackson. 2012. Sugars, signalling, and plant development. *Journal of Experimental Botany* **63**: 3367–3377.

Ewart, J. C. 1899. Is natural selection all metaphor? *Nature* **59**: 369.

Facelli, J. M. 1994. Multiple indirect effects of plant litter affect the establishment of woody seedlings in old fields. *Ecology* **75**: 1727–1735.

Fagiolini, M., C. L. Jensen, and F. A. Champagne. 2009. Epigenetic influences on brain development and plasticity. *Current Opinion in Neurobiology* **19**: 207–212.

Falconer, D. S. 1989. *Introduction to Quantitative Genetics*, 3rd ed. Longman, London.

Falconer, D. S., and T. F. C. Mackay. 1996. *Introduction to Quantitative Genetics*, 4th ed. Longman, Harlowe.

Falster, D. S., and M. Westoby. 2003. Leaf size and angle vary widely across species: What consequences for light interception? *New Phytologist* **158**: 509–525.

Farmer, C. G. 2000. Parental care: The key to understanding endothermy and other convergent features in birds and mammals. *American Naturalist* **155**: 326–334.

Farnsworth, E. 2004. Hormones and shifting ecology throughout plant development. *Ecology* **85**: 5–15.

Fellous, S., O. Duron, and F. Rousset. 2011. Adaptation due to symbionts and conflicts between heritable agents of biological information. *Nature Review Genetics* **12**: 663.

Fenchel, T. 2002. Microbial behavior in a heterogeneous world. *Science* **296**: 1068–1070.

Fenner, M., and K. Thompson. 2005. *The Ecology of Seeds*. Cambridge University Press, Cambridge, UK.

Fenster, C. B., W. S. Armbruster, P. Wilson, M. R. Dudash, and J. D. Thompson. 2004. Pollination syndromes and floral specialization. *Annual Reviews of Ecology, Evolution and Systematics* **35**: 375–403.

Fernald, R. D., and K. P. Maruska. 2012. Social information changes the brain. *Proceedings of the National Academy of Sciences USA* **109**: 17194–17199.

Fierer, N., and R. B. Jackson. 2006. The diversity and biogeography of soil bacterial communities. *Proceedings of the National Academy of Sciences USA* **103**: 626–631.

Filchak, K. E., J. B. Roethele, and J. L. Feder. 2000. Natural selection and sympatric divergence in the apple maggot *Rhagoletis pomonella*. *Nature* **407**: 739–742.

Fischer, R. C., A. Richter, W. Wanek, and V. Mayer. 2002. Plants feed ants: Food bodies of myrmecophytic *Piper* and their significance for the interaction with *Pheidole bicornis* ants. *Oecologia* **133**: 186–192.

Fisher, R. A. 1930. *The Genetical Theory of Natural Selection*. Oxford University Press, London.

Fisher, R. A., and W. A. Mackenzie. 1923. Studies in crop variation. II. The manurial response of different potato varieties. *Journal of Agricultural Science* **13**: 311–320.

Fitter, A. H. 1994. Architecture and biomass allocation as components of the plastic response of root systems to soil heterogeneity. In M. M. Caldwell and R. W. Pearcy, eds. *Exploitation of Environmental Heterogeneity by Plants*, pp. 305–323. Academic Press, New York.

Fitter, A. H., and R. K. M. Hay. 2002. *Environmental Physiology of Plants*. Academic Press, London.

Fitzpatrick, B. M. 2014. Symbiote transmission and maintenance of extra-genomic associations. *Frontiers in Microbiology* **5**: 46.

Fleck, S., Ü. Ninimets, A. Cescatti, and J. D. Tenhunen. 2003. Three-dimensional lamina architecture alters light-harvesting efficiency in *Fagus*: A leaf-scale analysis. *Tree Physiology* **23**: 577–589.

Flombaum, P., J. L. Gallegos, R. A. Gordillo, et al. 2013. Present and future global distributions of the marine cyanobacteria *Prochlorococcus* and *Synechococcus*. *Proceedings of the National Academy of Sciences USA* **110**: 9824–9829.

Foret, S., R. Kucharski, M. Pellegrini, et al. 2012. DNA methylation dynamics, metabolic fluxes, gene splicing, and alternative phenotypes in honey bees. *Proceedings of the National Academy of Sciences USA* **109**: 4968–4973.

Forrester, T. E., A. V. Badaloo, M. S. Boyne, et al. 2012. Prenatal factors contribute to the emergence of kwashiorkor or marasmus in severe undernutrition: Evidence for the predictive adaptation model. *Public Library of Science ONE* **7**: e0035907.

Forsman, A. 2014. Rethinking phenotypic plasticity and its consequences for individuals, populations and species. *Heredity* doi: 10.1038/hdy.2014.92.

Foster, B. L., and K. L. Gross. 1998. Species richness in a successional grassland: Effects of nitrogen enrichment and plant litter. *Ecology* **79**: 2593–2602.

Fournier-Level, A., A. Korte, M. D. Cooper, M. Nordberg, J. Schmitt, and A. M. Wilczek. 2011. A map of local adaptation in *Arabidopsis thaliana*. *Science* **334**: 86–89.

Fraker, M. E., F. Hu, V. Cuddapah, S. A. McCollum, R. A. Relyea, J. Hempel, and R. J. Denver. 2009. Characterization of an alarm pheromone secreted by amphibian tadpoles that induces behavioral inhibition and suppression of the neuroendocrine stress axis. *Hormones and Behavior* **55**: 520–529.

Francis, D., J. Diorio, D. Liu, and M. J. Meaney. 1999. Nongenomic transmission across generations of maternal behavior and stress responses in the rat. *Science* **286**: 1155–1158.

Frank, S. A. 2013. Natural selection. VI. Partitioning the information in fitness and characters by path analysis. *Journal of Evolutionary Biology* **26**: 457–471.

Franklin, K. A. 2008. Shade avoidance. *New Phytologist* **179**: 930–944.

Franklin, T. B., and I. M. Mansuy. 2010. Epigenetic inheritance in mammals: Evidence for the impact of adverse environmental effects. *Neurobiology of Disease* **39**: 61–65.

Franklin, T. B., H. Russig, I. C. Weiss, et al. 2010. Epigenetic transmission of the impact of early stress across generations. *Biological Psychiatry* **68**: 408–415.

Frederickson, M. E., and D. M. Gordon. 2007. The devil to pay: The cost of mutualism with *Myrmelachista schumanni* ants in "devil's gardens" is increased herbivory on *Duroia hirsuta* trees. *Proceedings of the Royal Society B* **274**: 1117–1123.

Frederickson, M. E., M. J. Greene, and D. M. Gordon. 2005. "Devil's gardens" bedeviled by ants. *Nature* **437**: 495–496.

Freedberg, S., T. J. Greives, M. A. Ewert, G. E. Demas, N. Beecher, and C. E. Nelson. 2008. Incubation environment affects immune system development in a turtle with environmental sex determination. *Journal of Herpetology* **42**: 536–541.

Freund, J., A. M. Brandmaier, L. Lewejohann, et al. 2013. Emergence of individuality in genetically identical mice. *Science* **340**: 756–759.

Frommen, J. G., F. Herder, L. Engqvist, M. Mehlis, T. C. M. Bakker, J. Schwarzer, and T. Thünken. 2011. Costly plastic morphological responses to predator specific odour cues in three-spined sticklebacks (*Gastaerosteus aculeatus*). *Evolutionary Ecology* **25**: 641–656.

Fründ, J., C. F. Dormann, A. Holzschuh, and T. Tscharntke. 2013. Bee diversity effects on pollination depend on functional complementarity and niche shifts. *Ecology* **94**: 2042–2054.

Fry, J. D. 1996. The evolution of host specialization: Are trade-offs overrated? *American Naturalist* **148**: S84–S107.

Fujita M., Y. Fujita, Y. Noutoshi, et al. 2006. Crosstalk between abiotic and biotic stress responses: A current view from the points of convergence in the stress signaling networks. *Current Opinion in Plant Biology* **9**: 436–442.

Fuller, T., S. Sarkar, and D. Crews. 2005. The use of norms of reaction to analyze genotypic and environmental influences on behavior in mice and rats. *Neuroscience and Biobehavioral Reviews* **29**: 445–456.

Furrow, R. E., and M. W. Feldman. 2014. Genetic variation, environmental variability, and the evolution of epigenetic regulation. *Evolution* **68**: 673–683.

Futuyma, D. J. 2009. *Evolution*, 2nd ed. Sinauer Associates, Sunderland.

Gabet, E. J., J. T. Perron, and D. L. Johnson. 2014. Biotic origin for Mima mounds supported by numerical modeling. *Geomorphology* **206**: 58–66.

Gabryś, H. 2012. Blue light-activated chloroplast movements: Progress in the last decade. *Progress in Botany* **73**: 189–205.

Galant, R., and S. B. Carroll. 2002. Evolution of a transcriptional repression domain in an insect *Hox* protein. *Nature* **415**: 910–913.

Galen, C., J. Huddle, and E. Liscum. 2004. An experimental test of the adaptive evolution of phototropins: Blue-light photoreceptors controlling phototropism in *Arabidopsis thaliana*. *Evolution* **58**: 515–523.

Gallet, C., M.-C. Nilsson, and O. Zackrisson. 1999. Phenolic metabolites of ecological significance in *Empetrum hermaphroditum* leaves and associated humus. *Plant and Soil* **210**: 1–9.

Galloway, L. F. 2005. Maternal effects provide phenotypic adaptation to local environmental conditions. *New Phytologist* **166**: 93–100.

Galloway, L. F., and J. R. Etterson. 2007. Transgenerational plasticity is adaptive in the wild. *Science* **318**: 1134–1136.

Galloway, L. F., and J. R. Etterson. 2009. Plasticity to canopy shade in a monocarpic herb: Within- and between-generation effects. *New Phytologist* **182**: 1003–1012.

Gamon, J. A., and R. W. Pearcy. 1990. Photoinhibition in *Vitis californica*: The role of temperature during high-light treatment. *Plant Physiology* **92**: 487–494.

Gapp, K., A. Jawaid, P. Sarkies, et al. 2014. Implication of sperm RNAs in transgenerational inheritance of the effects of early trauma in mice. *Nature Neuroscience* **17**: 667–669.

Garfield, D. A., and G. A. Wray. 2010. The evolution of gene regulatory interactions. *BioScience* **60**: 15–23.

Gaston, K. J. 2009. Geographic range limits: Achieving synthesis. *Proceedings of the Royal Society B* **276**: 1395–1406.

Gatehouse, J. A. 2002. Plant resistance towards insect herbivores: A dynamic interaction. *New Phytologist* **156**: 145–169.

Gates, D. M. 1968. Transpiration and leaf temperature. *Annual Review of Plant Physiology* **19**: 211–238.

Gaut, B. 2012. *Arabidopsis thaliana* as a model for the genetics of local adaptation. *Nature Genetics* **44**: 115–116.

Geervliet, J. B. F., L. E. M. Vet, and M. Dicke. 1994. Volatiles from damaged plants as major cues in long-range host searching by the specialist parasitoid *Cotesia rubecula*. *Entomologia Experimentalis et Applicata* **73**: 289–297.

Geoghegan, J. L., and H. G. Spencer. 2012. Population-epigenetic models of selection. *Theoretical Population Biology* **81**: 232–242.

Geoghegan, J. L., and H. G. Spencer. 2013a. Exploring epiallele stability in a population-epigenetic model. *Theoretical Population Biology* **83**: 136–144.

Geoghegan, J. L., and H. G. Spencer. 2013b. The adaptive invasion of epialleles in a heterogeneous environment. *Theoretical Population Biology* **88**: 1–8.

Gersani, M., Z. Abramsky, and O. Falik. 1998. Density-dependent habitat selection in plants. *Evolutionary Ecology* **12**: 223–234.

Gerten, D., S. Schaphoff, U. Haberlandt, W. Lucht, and S. Sitch. 2004. Terrestrial vegetation and water balance: Hydrological evaluation of a dynamic global vegetation model. *Journal of Hydrology* **286**: 249–270.

Ghalambor, C. K., J. S. McKay, S. P. Carroll, and D. N. Reznick. 2007. Adaptive versus non-adaptive phenotypic plasticity and the potential for contemporary adaptation in new environments. *Functional Ecology* **21**: 394–407.

Gibson, G., and D. S. Hogness. 1996. Effect of polymorphism in the *Drosophila* regulatory gene *Ultrabithorax* on homeotic stability. *Science* **271**: 200–203.

Gienapp, P., C. Teplitsky, J. S. Alho, J. A. Mills, and J. Merilä. 2008. Climate change and evolution: Disentangling environmental and genetic responses. *Molecular Ecology* **17**: 167–178.

Gilbert, C., Y. Le Maho, M. Perret, and A. Ancel. 2007. Body temperature changes induced by huddling in breeding male emperor penguins. *American Journal of Physiology: Regulatory, Integrative and Comparative Physiology* **292**: R176–R185.

Gilbert, S. F. 2001. Ecological developmental biology: Developmental biology meets the real world. *Developmental Biology* **233**: 1–12.

Gilbert, S. F. 2005. Mechanisms for the environmental regulation of gene expression: Ecological aspects of animal development. *Journal of Biosciences* **30**: 65–74.

Gilbert, S. F. 2011. Expanding the temporal dimensions of developmental biology: The role of environmental agents in establishing adult-onset phenotypes. *Biological Theory* **6**: 65–72.

Gilbert, S. F. 2012. Ecological developmental biology: Environmental signals for normal animal development. *Evolution and Development* **14**: 20–28.

Gilbert, S. F., and J. A. Bolker. 2003. Ecological developmental biology: Preface to the symposium. *Evolution and Development* **5**: 3–8.

Gilbert, S. F., and D. Epel. 2009. *Ecological Developmental Biology: Integrating Epigenetics, Medicine, and Evolution*. Sinauer Associates, Sunderland.

Gill, M. E., S. Erkek, and A. H. F. M. Peters. 2012. Parental epigenetic control of embryogenesis: A balance between inheritance and reprogramming? *Current Opinion in Cell Biology* **24**: 387–396.

Gillespie, J. H., and Turelli, M. 1989. Genotype–environment interactions and the maintenance of polygenic variation. *Genetics* **121**: 129–138.

Gilroy, S., and A. Trewavas. 2001. Signal processing and transduction in plant cells: The end of the beginning? *Nature Reviews Molecular Cell Biology* **2**: 307–314.

Giri, A. P., H. Wunsche, S. Mitra, et al. 2006. Molecular interactions between the specialist herbivore *Manduca sexta* (Lepidoptera, Sphingidae) and its natural host *Nicotiana attenuata*. VII. Changes in the plant's proteome. *Plant Physiology* **142**: 1621–1641.

Glaser, R., S. Kennedy, W. P. Lafuse, R. H. Bonneau, C. Speicher, and J. K. Kiecolt-Glaser. 1990. Psychological stress-induced modulation of IL-2 receptor gene expression and IL-2 production in peripheral blood leukocytes. *Archives of General Psychiatry* **47**: 707–712.

Glennemeier, K. A., and R. J. Denver. 2002. Role for corticoids in mediating the response of *Rana pipiens* tadpoles to intraspecific competition. *Journal of Experimental Zoology* **292**: 32–40.

Gluckman, P. D., and M. Hanson. 2005. *The Fetal Matrix: Evolution, Development and Disease*. Cambridge University Press, Cambridge, UK.

Gluckman, P. D., M. A. Hanson, and A. S. Beedle. 2007. Non-genomic transgenerational inheritance of disease risk. *BioEssays* **29**: 145–154.

Gluckman, P. D., M. A. Hanson, A. S. Beedle, T. Buklijas, and F. M. Low. 2011. Epigenetics of human disease. In B. Hallgrímsson and B. K. Hall, eds. *Epigenetics: Linking Genotype and Phenotype in Development and Evolution*, pp. 398–423. University of California Press, Berkeley.

Gluckman, P. D., M. A. Hanson, T. Buklijas, F. M. Low, and A. S. Beedle. 2009. Epigenetic mechanisms that underpin metabolic and cardiovascular diseases. *Nature Reviews Endocrinology* **5**: 401–408.

Gluckman, P. D., K. A. Lillycrop, M. H. Vickers, et al. 2007. Metabolic plasticity during mammalian development is directionally dependent on early nutritional status. *Proceedings of the National Academy of Sciences USA* **104**: 12796–12800.

Godfrey, K. M., A. Sheppard, P. D. Gluckman, et al. 2011. Epigenetic gene promoter methylation at birth is associated with child's later adiposity. *Diabetes* **60**: 1528–1534.

Gomahr, A., M. Palzenberger, and K. Kotrschal. 1992. Density and distribution of external taste buds in cyprinids. *Environmental Biology of Fishes* **33**: 125–134.

Gomez-Mestre, I., and R. Jovani. 2013. A heuristic model on the role of plasticity in adaptive evolution: Plasticity increases adaptation, population viability and genetic variation. *Proceedings of the Royal Society B* **280**: 20131869.

Gomulkiewicz, R., and M. Kirkpatrick. 1992. Quantitative genetics and the evolution of reaction norms. *Evolution* **46**: 390–411.

González-Rivero, M., L. Yakob, and P. J. Mumby. 2011. The role of sponge competition on coral reef alternative steady states. *Ecological Modeling* **222**: 1847–1853.

Gottlieb, G. 2004. Normally occurring environmental and behavioral influences on gene activity: From central dogma to probabilistic epigenesis. In C. G. Coll, E. L. Bearer, and R. M. Lerner, eds. *Nature and Nurture: The Complex Interplay of Genetic and Environmental Influences on Human Behavior and Development*, pp. 85–106. Lawrence Erlbaum Associates, London.

Gould, S. J., and R. C. Lewontin. 1979. The Spandrels of San Marco and the Panglossian paradigm: A critique of the adaptationist programme. *Proceedings of the Royal Society B* **205**: 581–598.

Grafen, A. 1988. On the uses of data on lifetime reproductive success. In T. H. Clutton-Brock, ed. *Reproductive Success*, pp. 454–471. University of Chicago Press, Chicago.

Grant, B. S. 2004. Allelic melanism in American and British peppered moths. *Journal of Heredity* **95**: 97–102.

Gray, R. 1992. Death of the gene: Developmental systems strike back. In R. D. Gray and P. Griffiths, eds. *Trees of Life: Essays in the Philosophy of Biology*, pp. 165–209. Kluwer Publishing, Dordrecht.

Grenier, J. K., T. Garber, R. Warren, P. M. Whitington, and S. Carroll. 1997. Evolution of the entire arthropod *Hox* gene set predated the origin and radiation of the onycophoran/arthropod clade. *Current Biology* **7**: 547–553.

Griesemer, J. R. 1992. Niche: Historical perspectives. In E. F. Keller and E. A. Lloyd, eds. *Keywords in Evolutionary Biology*, pp. 231–240. Harvard University Press, Cambridge.

Griffith, M., and M. W. F. Yaish. 2004. Antifreeze proteins in overwintering plants: A tale of two activities. *Trends in Plant Science* **9**: 399–405.

Griffith, T. M., and S. E. Sultan. 2005. Shade tolerance plasticity in response to neutral vs green shade cues in *Polygonum* species of contrasting ecological breadth. *New Phytologist* **166**: 141–148.

Griffith, T. M., and S. E. Sultan. 2006. Plastic and constant developmental traits contribute to adaptive differences in co-occurring *Polygonum* species. *Oikos* **114**: 5–14.

Griffith, T. M., and S. E. Sultan. 2012. Field-based insights to the evolution of specialization: Plasticity and fitness across habitats in a specialist/generalist species pair. *Ecology and Evolution* **2**: 778–791.

Griffiths, P. E. 2006. Philip Kitcher, genetic determinism, and the informational gene. In E. M. Neumann-Held and C. Rehmann-Sutter, eds. *Genes in Development: Rereading the Molecular Paradigm*, pp. 175–198. Duke University Press, Durham.

Griffiths, P. E., and R. D. Gray. 1994. Developmental systems and evolutionary explanation. *Journal of Philosophy* **91**: 277–304.

Grimaldi, A. G., N. Buisine, P. Bilesimo, and L. M. Sachs. 2013. High-throughput sequencing will metamorphose the analysis of thyroid hormone receptor function during amphibian development. *Current Topics in Developmental Biology* **103**: 277–290.

Grözinger, F., T. H. Jürgen, H. Feldhaar, and M.-O. Rödel. 2014. Giants, dwarfs and the environment: Metamorphic trait plasticity in the common frog. *Public Library of Science ONE* **9**: e89982.

Gruntman, M., and A. Novoplansky. 2004. Physiologically mediated self/non-self discrimination in roots. *Proceedings of the National Academy of Sciences USA* **101**: 3863–3867.

Gubernick, D. J., and T. Teferi. 2000. Adaptive significance of male parental care in a monogamous mammal. *Proceedings of the Royal Society B* **267**: 147–150.

Guisan, A., and W. Thuiller. 2005. Predicting species distribution: Offering more than simple habitat models. *Ecology Letters* **8**: 993–1009.

Gupta, A. P., and R. C. Lewontin. 1982. A study of reaction norms in natural populations of *Drosophila pseudoobscura*. *Evolution* **36**: 934–948.

Gutierrez-Arcelus, M., T. Lappalainen, S. B. Montgomery, et al. 2013. Passive and active DNA methylation and the interplay with genetic variation in gene regulation. *Elife* **2**:e00523.

Gutteling, E. W., J. A. G. Riksen, J. Bakker, and J. E. Kammenga. 2007. Mapping phenotypic plasticity and genotype–environment interactions affecting life-history traits in *Caenorhabditis elegans*. *Heredity* **98**: 28–37.

Ha, M., and V. N. Kim. 2014. Regulation of microRNA biogenesis. *Nature Reviews Molecular Cell Biology* **15**: 509–524.

Hacker, S. D., and S. D. Gaines. 1997. Some implications of direct positive interactions for community species diversity. *Ecology* **78**: 1990–2003.

Haig, D. 2007. Weismann rules! Ok? Epigenetics and the Lamarckian temptation. *Biology and Philosophy* **22**: 415–428.

Hairston, N. G. 1995. Editor's note. *Ecology* **76**: 1371.

Haldane, J. B. S. 1946. The interaction of nature and nurture. *Annals of Eugenics* **13**: 197–205.

Halitschke, R., U. Schittko, G. Pohnert, W. Boland, and I. T. Baldwin. 2001. Molecular interactions between the specialist herbivore *Manduca sexta* (Lepidoptera, Sphingidae) and its natural host *Nicotiana attenuata*. III. Fatty acid-amino acid conjugates in herbivore oral secretions are necessary and sufficient for herbivore- specific plant responses. *Plant Physiology* **125**: 711–717.

Hallgrímsson, B., and B. K. Hall, eds. 2011. *Epigenetics: Linking Genotype and Phenotype in Development and Evolution*. University of California Press, Berkeley.

Hamburger, V. 1980. Embryology and the modern synthesis in evolutionary theory. In E. Mayr and W. Provine, eds. *The Evolutionary Synthesis: Perspectives in the Unification of Biology*, pp. 97–1112. Cambridge University Press, New York.

Hammerstein, P., and R. Boyd. 2012. Learning, cognitive limitations, and the modeling of social behavior. In P. Hammerstein and J. R. Stevens, eds. *Evolution and the Mechanisms of Decision Making*, pp. 311–334. MIT Press, Cambridge.

Hammond, K. A., J. Szewczak, and E. Król. 2001. Effects of altitude and temperature on organ phenotypic plasticity along an altitudinal gradient. *Journal of Experimental Biology* **204**: 1991–2000.

Hammoud, S. S., D. A. Nix, H. Zhang, J. Purwar, D. T. Carrell, and B. R. Cairns. 2009. Distinctive chromatin in human sperm packages genes for embryo development. *Nature* **460**: 473–478.

Han, X., C. Hui, and Y. Zhang. 2009. Effects of time-lagged niche construction on metapopulation dynamics and environmental heterogeneity. *Applied Mathematics and Computation* **215**: 449–458.

Hanson, M. A., and P. D. Gluckman. 2014. Early developmental conditioning of later health and disease: Physiology or pathophysiology? *Physiological Reviews* **94**: 1027–1076.

Hardoim, P. R., L. S. van Overbeek, and J. D. van Elsas. 2008. Properties of bacterial endophytes and their proposed role in plant growth. *Trends in Microbiology* **16**: 463–471.

Hare, J. D. 2011. Ecological role of volatiles produced by plants in response to damage by herbivorous insects. *Annual Review of Entomology* **56**: 161–180.

Hargreaves, A. L., K. E. Samis, and C. G. Eckert. Are species' range limits simply niche limits writ large? A review of transplant experiments beyond the range. *The American Naturalist* **183**: 157–173.

Hartmann, A., M. Rothballer, and M. Schmid. 2008. Lorenz Hiltner, a pioneer in rhizosphere microbial ecology and soil bacteriology research. *Plant and Soil* **312**: 7–14.

Hartnett, D. C., and F. A. Bazzaz. 1983. Physiological integration among intraclonal ramets in *Solidago canadensis*. *Ecology* **64**: 779–788.

Harvey, P. H., and M. D. Pagel. 1991. *The Comparative Method in Evolutionary Biology*. Oxford University Press, New York.

Hastings, A., J. E. Byers, J. A. Crooks, et al. 2007. Ecosystem engineering in space and time. *Ecology Letters* **10**: 153–164.

Haupt, W., and R. Scheuerlein. 1990. Chloroplast movement. *Plant, Cell and Environment* **13**: 595–614.

Heard, E., ed. 2013. *Epigenetics and Development*, Current Topics in Development **Vol. 104**. Elsevier Academic Press, San Diego.

Heard, E., and R. A. Martienssen. 2014. Transgenerational epigenetic inheritance: Myths and mechanisms. *Cell* **157**: 95–109.

Heath, J. J., and J. O. Stireman III. 2010. Dissecting the association between a gall midge, *Asteromyia carbonifera*, and its symbiotic fungus, *Botryosphaeria dothidea*. *Entomologia Experimentalis et Applicata* **137**: 36–49.

Hegrenes, S. 2001. Diet-induced phenotypic plasticity of feeding morphology in the orangespotted sunfish, *Lepimis humilis*. *Ecology of Freshwater Fish* **10**: 35–42.

Heintzman, N. D., R. K. Stuart, G. Hon, et al. 2007. Distinct and predictive chromatin signatures of transcriptional promoters and enhancers in the human genome. *Nature Genetics* **39**: 311–318.

Heldmaier, G., and D. Werner. 2003. Environmental signal processing and adaptation. In G. Heldmaier and D. Werner, eds. *Environmental Signal Processing and Adaptation*, pp. 1–9. Springer-Verlag, Berlin.

Hemberg, M., J. M. Gray, N. Cloonan, S. Kuersten, S. Grimmond, M. E. Greenberg, and G. Kreiman. 2012. Integrated genome analysis suggests that most conserved non-coding sequences are regulatory factor binding sites. *Nucleic Acids Research* **40**: 7858–7869.

Henderson, I. R., and S. E. Jacobsen. 2007. Epigenetic inheritance in plants. *Nature* **447**: 418–424.

Herbert, R. A., A. Gall, T. Maoka, et al. 2008. Phototrophic purple sulfur bacteria as heat engines in the South Andros Black Hole. *Photosynthesis Research* **95**: 261–268.

Herbold, B., and P. B. Moyle. 1986. Introduced species and vacant niches. *The American Naturalist* **128**: 751–760.

Herman, J. J., H. G. Spencer, K. Donohue, and S. E. Sultan. 2014. How stable "should" epigenetic modifications be? Insights from adaptive plasticity and bet hedging. *Evolution* **68**: 632–643.

Herman, J. J., and S. E. Sultan. 2011. Adaptive transgenerational plasticity: Case studies, mechanisms, and implications for natural populations. *Frontiers in Plant Genetics and Genomics* **2**: 102.

Herman, J. J., S. E. Sultan, T. Horgan-Kobelski, and C. E. Riggs. 2012. Adaptive transgenerational plasticity in an annual plant: Grandparental and parental drought stress enhance performance of seedlings in dry soil. *Integrative and Comparative Biology* **52**: 1–12.

Hermsmeier, D., U. Schittko, and I. T. Baldwin. 2001. Molecular interactions between the specialist herbivore *Manduca sexta* (Lepidoptera, Sphingidae) and its natural host *Nicotiana attenuata*. I. Large-scale changes in the accumulation of growth- and defense-related plant mRNAs. *Plant Physiology* **125**: 683–700.

Herrera, C. M., and P. Bazaga. 2010. Epigenetic differentiation and relationship to adaptive genetic divergence in discrete populations of the violet *Viola cazorlensis*. *New Phytologist* **187**: 867–876.

Herrera, C. M., and P. Bazaga. 2011. Untangling individual variation in natural populations: Ecological, genetic, and epigenetic correlates of long-term inequality in herbivory. *Molecular Ecology* **20**: 1675–1688.

Herrera, C. M., M. Medrano, and P. Bazaga. 2014. Variation in DNA methylation transmissibility, genetic heterogeneity and fecundity-related traits in natural populations of the perennial herb *Helleborus foetidus*. *Molecular Ecology* **23**: 1085–1095.

Herrera, C. M., M. I. Pozo, and P. Bazaga. 2012. Jack of all nectars, master of most: DNA methylation and the epigenetic basis of niche width in a flower-living yeast. *Molecular Ecology* **21**: 2602–2616.

Heschel, M. S., C. M. Butler, D. Barua, et al. 2008. New roles of phytochromes during seed germination. *International Journal of Plant Science* **169**: 531–540.

Heschel, M. S., J. Selby, C. Butler, G. C. Whitelam, R. A. Sharrock, and K. Donohue. 2007. A new role for phytochromes in temperature-dependent germination. *New Phytologist* **174**: 735–741.

Heschel, M. S., S. E. Sultan, D. Sloan, and S. Glover. 2004. Population differentiation and plastic responses to drought stress in the generalist annual *Polygonum persicaria*. *International Journal of Plant Sciences* **165**: 817–824.

Hetherington, A. M., and F. I. Woodward. 2003. The role of stomata in sensing and driving environmental change. *Nature* **424**: 901–908.

Hicks, J. W., and T. Wang. 2004. Hypometabolism in reptiles: Behavioural and physiological mechanisms that reduce aerobic demands. *Respiratory Physiology and Neurobiology* **141**: 261–271.

Himanen, S. J., J. D. Blande, T. Klemola, J. Pulkkinen, J. Heijari, and J. Holopainen. 2010. Birch (*Betula* spp.) leaves adsorb and re-release volatiles specific to neighbouring plants: A mechanism for associational herbivore resistance? *New Phytologist* **186**: 722–732.

Hinsinger, P., A. G. Bengough, D. Vetterlein, and I. M. Young. 2009. Rhizosphere: Biophysics, biogeochemistry and ecological relevance. *Plant Soil* **321**: 117–152.

Hodge, A. 2004. The plastic plant: Root responses to heterogeneous supplies of nutrients. *New Phytologist* **162**: 9–24.

Hodge, A. 2009. Root decisions. *Plant, Cell and Environment* **32**: 628–640.

Hodge, A. 2010. Roots: The acquisition of water and nutrients from the heterogeneous soil environment. *Progress in Botany* **71**: 307–337.

Hodge, A., and A. H. Fitter. 2013. Microbial mediation of plant competition and community structure. *Functional Ecology* **27**: 865–875.

Hofmann, H. A. 2006. Gonadotropin-releasing hormone signaling in behavioral plasticity. *Current Opinions in Neurobiology* **16**: 343–350.

Hofmann, H. A., P. Y. Le Bail, and R. D. Fernald. 1999. Social control of growth and growth hormone levels in an African cichlid fish. *Society for Neurosicence Abstracts* **29**: 348.

Hoffmann, M. H. 2002. Biogeography of *Arabidopsis thaliana* (L.). *Journal of Biogeography* **29**: 125–134.

Hogenhout, S. A., R. A. L. Van der Hoorn, R. Terauchi, and S. Kamoun. 2009. Emerging concepts in effector biology of plant-associated organisms. *Molecular Plant–Microbe Interactions* **22**: 115–122.

Holeski, L. M. 2007. Within and between generation phenotypic plasticity in trichome density of *Mimulus guttatus*. *Journal of Evolutionary Biology* **20**: 2092–2100.

Holeski, L. M., G. Jander, and A. A. Agrawal. 2012. Transgenerational defense induction and epigenetic inheritance in plants. *Trends in Ecology and Evolution* **27**: 618–626.

Holliday, R. 1990. Mechanisms for the control of gene activity during development. *Biological Reviews of the Cambridge Philosophical Society* **65**: 431–471.

Holliday, R., and J. E. Pugh. 1975. DNA modification mechanisms and gene activity during development. *Science* **187**: 226–232.

Holt, R. D. 2006. Asymmetry and stability. *Nature* **442**: 252–253.

Holt, R. D. 2009. Bringing the Hutchinsonian niche into the 21st century: Ecological and evolutionary perspectives. *Proceedings of the Natural Academy of Sciences* **106**: 19659–19665.

Holt, R. D., and M. S. Gaines. 1992. Analysis of adaptation in heterogeneous landscapes: Implications for the evolution of fundamental niches. *Evolutionary Ecology* **6**: 433–447.

Hom, E. F. Y., and A. W. Murray. 2014. Niche engineering demonstrates a latent capacity for fungal–algal mutualism. *Science* **345**: 94–98.

Hooper, L. V. 2004. Bacterial contributions to mammalian gut development. *Trends in Microbiology* **12**: 129–134.

Hooper, D. U., E. C. Adair, B. J. Cardinale, et al. 2012. A global synthesis reveals biodiversity loss as a major driver of ecosystem change. *Nature* **486**: 105–108.

Hooper, D. U., F. S. Chapin III, J. J. Ewel, et al. 2005. Effects of biodiversity on ecosystem functioning: A consensus of current knowledge. *Ecological Monographs* **75**: 3–35.

Howe, G. A., and G. Jander. 2008. Plant immunity to insect herbivores. *Annual Review of Plant Biology* **59**: 41–66.

Hsia, C. C. W., J. J. Polo Carbayo, X. Yan, and D. J. Bellotto. 2005. Enhanced alveolar growth and remodeling in guinea pigs raised at high altitude. *Respiratory Physiology and Neurobiology* **147**: 105–115.

Huber, H., J. de Brouwer, H. de Caluwe, J. Wijschedé, and N. P. R. Anten. 2008. Shade induced changes in biochemical petiole properties in the stoloniferous herb *Trifolium repens*. *Evolutionary Ecology* **22**: 399–416.

Huber, H., N. C. Kane, M. S. Heschel, et al. 2004. Frequency and microenvironmental pattern of selection on plastic shade-avoidance traits in a natural population of *Impatiens capensis*. *American Naturalist* **163**: 548–563.

Huey, R. B. 1982. Temperature, physiology and the ecology of reptiles. In C. Gans and F. H. Pough, eds. *Biology of the Reptilia*, Vol. **12**, pp. 25–91. Academic Press, London.

Huey, R. B. 1991. Physiological consequences of habitat selection. *American Naturalist* **137**: S91–S115.

Huey, R. B., M. Carlson, L. Crozier, et al. 2002. Plants versus animals: Do they deal with stress in different ways? *Integrative Comparative Biology* **42**: 415–423.

Hughes, L., M. Westoby, and E. Jurado. 1994. Convergence of elaiosomes and insect prey: Evidence from ant foraging behavior and fatty acid composition. *Functional Ecology* **8**: 358–365.

Hughes, T. P., N. A. J. Graham, J. B. C. Jackson, P. J. Mumby, and R. S. Steneck. 2010. Rising to the challenge of sustaining coral reef resilience. *Trends in Ecology and Evolution* **25**: 633–642.

Hughes, W. O. H., and J. J. Boomsma. 2007. Genetic polymorphism in leaf-cutting ants is phenotypically plastic. *Proceedings of the Royal Society B* **274**: 1625–1630.

Huhta, A., K. Hellstrom, P. Rautio, and J. Tuomi. 2000. A test of the compensatory continuum: Fertilization increases and below-ground competition decreases the grazing tolerance of tall wormseed mustard (*Erysimum strictum*). *Evolutionary Ecology* **14**: 353–372.

Humphries, S. N., A. G. Bengough, B. S. Griffiths, et al. 2005. Root cap influences root colonization by *Pseudomonas fluorescens SBW25* on maize. *Federation of European Microbiological Societies Microbiology Ecology* **54**: 123–130.

Hurst, L. D., and C. Pál. 2008. Genomic redundancy and dispensability. In M. Pagel and A. Pomiankowski, eds. *Evolutionary Genomics and Proteomics*, pp. 141–165. Sinauer Associated, Sunderland.

Husby, A., D. H. Nussey, M. E. Visser, A. J. Wilson, B. C. Sheldon, and L. E. B. Kruuk. 2010. Contrasting patterns of phenotypic plasticity in reproductive traits in two great tit (*Parus major*) populations. *Evolution* **64**: 2221–2237.

Husby, A., M. E. Visser, and L. E. Kruuk. 2011. Speeding up microevolution: The effects of increasing temperature on selection and genetic variance in a wild bird population. *Public Library of Science Biology* **9**: e1000585.

Hutchings, J. A. 2004. Evolutionary biology: The cod that got away. *Nature* **428**: 899–900.

Hutchings, J. A. 2011. Old wine in new bottles: Reaction norms in salmonid fishes. *Heredity* **106**: 421–437.

Hutchings, M. J., and H. de Kroon. 1994. Foraging in plants: The role of morphological plasticity in resource acquisition. *Advances in Ecological Research* **25**: 159–238.

Hutchinson, G. E. 1957. Concluding remarks. *Cold Spring Harbor Symposium* **22**: 415–427.

Hutchison, M. L., M. A. Tester, and D. C. Gross. 1995. Role of biosurfactant and ion channel-forming activities of syringomycin in transmembrane ion flux: A model for the mechanism of action in the plant–pathogen interaction. *Molecular Plant–Microbe Interactions* **8**: 610–620.

Hyman, S. E. 2004. Foreword: Behavioral genetics in the post-genomic era. In C. G. Coll, E. L. Bearer, and R. M. Lerner, eds. *Nature and Nurture: The Complex Interplay of Genetic and Environmental Influences on Human Behavior and Development*, pp. i–xiii. Lawrence Erlbaum Associates, London.

Ihara, M., M. L. Meyer-Ficca, N. A. Leu, et al. 2014. Paternal poly(ADP-ribose) metabolism modulates retention of inheritable sperm histones and early embryonic gene expression. *Public Library of Science Genetics* **10**: e1004317.

Inouye, B., and J. R. Stinchcombe. 2001. Relationships between ecological interaction modifications and diffuse coevolution: Similarities, differences, and causal links. *Oikos* **95**: 353–360.

Jablonka, E., and M. J. Lamb. 1998. Epigenetic inheritance in evolution. *Journal of Evolutionary Biology* **11**: 159–183.

Jablonka, E., and M. J. Lamb. 2002. The changing concept of epigenetics. *Annals of the New York Academy of Sciences* **981**: 82–96.

Jablonka, E., and M. J. Lamb. 2005. *Evolution in Four Dimensions: Genetic, Epigenetic, Behavioral, and Symbolic Variation in the History of Life*. MIT Press, Cambridge.

Jablonka, E., and G. Raz. 2009. Transgenerational epigenetic inheritance: Prevalence, mechanisms, and implications for the study of heredity and evolution. *Quarterly Review of Biology* **84**: 1331–1176.

Jacquemyn, H., R. Brys, K. Vandepitte, O. Honnay, I. Roldan-Ruiz, and T. Wiegand. 2007. A spatially explicit analysis of seedling recruitment in the terrestrial orchid *Orchis purpurea*. *New Phytologist* **176**: 448–459.

Jaenisch, R., and A. Bird. 2003. Epigenetic regulation of gene expression: How the genome integrates intrinsic and environmental signals. *Nature Genetics* **33**: 245–254.

Jain, S. K. 1979. Adaptive strategies: Polymorphism, plasticity, and homeostasis. In O. Solbrig, G. B. Johnson, P. Raven, and S. K. Jain, eds. *Topics in Plant Population Biology*, pp. 160–187. Columbia University Press, New York.

Jain, S. K. 1990. Variation and selection in plant populations. In K. Wöhrmann and S. K. Jain, eds. *Population Biology: Ecological and Evolutionary Viewpoints*, pp. 199–230. Springer-Verlag, Berlin.

Janes, D. E., and M. L. Wayne. 2006. Evidence for a genotype × environment interaction in sex-determining response to incubation temperation in the leopard gecko, *Eublepharis macularius*. *Herpetologica* **62**: 56–62.

Jeong, M.-J., C.-K. Shim, J.-O. Lee, et al. 2008. Plant gene responses to frequency-specific sound signals. *Molecular Breeding* **21**: 217–226.

Jiménez-Chillarón, J. C., R. Díaz, D. Martínez, et al. 2012. The role of nutrition on epigenetic modifications and their implications on health. *Biochimie* **94**: 2242–2263.

Johannes, F., V. Colot, and R. C. Jansen. 2008. Epigenome dynamics: A quantitative genetics perspective. *Nature Reviews Genetics* **9**: 883–890.

Johannes, F., E. Porcher, F. K. Teixeira, et al. 2009. Assessing the impact of transgenerational epigenetic variation on complex traits. *Public Library of Science Genetics* **5**: e1000530.

Johannsen, W. 1911. The genotype concept of heredity. *American Naturalist* **45**: 129–159.

Johnson, J. E., S. M. Webb, K. Thomas, S. Ono, J. L. Kirschvink, and W. W. Fischer. 2013. Manganese-oxidizing photosynthesis before the rise of cyanobacteria. *Proceedings of the National Academy of Sciences USA* **110**: 11238–11243.

Johnston, A., and E. Reekie. 2008. Regardless of whether rising atmospheric carbon dioxide levels increase air temperature, flowering phenology will be affected. *International Journal of Plant Sciences* **169**: 1210–1218.

Jones, C. G., J. H. Lawton, and M. Shachak. 1994. Organisms as ecosystem engineers. *Oikos* **69**: 373–386.

Jones, C. G., J. H. Lawton, and M. Shachak. 1997. Positive and negative effects of organisms as physical ecosystem engineers. *Ecology* **778**: 1946–1957.

Jones, P. A. 2012. Functions of DNA methylation: Islands, start sites, gene bodies and beyond. *Nature Reviews Genetics* **13**: 484–492.

Jongsma, M. A., P. L. Bakker, J. Peters, D. Bosch, and W. J. Stiekema. 1995. Adaptation of *Spodoptera exigua* larvae to plant proteinase inhibitors by induction of gut proteinase activity insensitive to inhibition. *Proceedings of the National Academy of Sciences USA* **92**: 8041–8045.

Joos, B., T. M. Casey, T. D. Fitzgerald, and W. A. Buttemer. 1988. Roles of the tent in behavioral thermoregulation of eastern tent caterpillars. *Ecology* **69**: 2004–2011.

Judd, W. S., C. S. Campbell, E. A. Kellogg, P. F. Stevens, and M. J. Donoghue. 2002. *Plant Systematics: A Phylogenetic Approach*. Sinauer Associates, Sunderland.

Jung, V., C. Violle, C. Mondy, L. Hoffmann, and S. Muller. 2010. Intraspecific variability and trait-based community assembly. *Journal of Ecology* **98**: 1134–1140.

Juracka, P. J., C. Laforsch, and A. Petrusek. 2011. Neckteeth formation in two species of the *Daphnia curvirostris* complex (Crustacea: Cladocera). *Journal of Limnology* **70**: 359–368.

Kalisz, S., and M. D. Purugganan. 2004. Epialleles via DNA methylation: Consequences for plant evolution. *Trends in Ecology and Evolution* **19**: 309–314.

Kanai, Y., and S. Hirohashi. 2007. Alterations of DNA methylation associated with abnormalities of DNA methyltransferases in human cancers during transition from a precancerous to a malignant state. *Carcinogenesis* **28**: 2434–2442.

Kanchisa, M., S. Goto, Y. Sato, M. Kawashima, M. Furunichi, and M. Tanabe. 2014. Data, information, knowledge and principle: Back to metabolism in KEGG (Kyoto Encyclopedia of Genes and Genomes). *Nucleic Acids Research* **42**: D199–D205.

Kao, W. Y., and I. N. Forseth. 1992. Diurnal leaf movement, chlorophyll fluorescence and carbon assimilation in a soybean grown under different nitrogen and water availabilities. *Plant Cell and Environment* **21**: 1055–1062.

Karan, D., J. P. Morin, P. Gibert, B. Moreteau, S. M. Scheiner, and J. R. David. 2000. The genetics of phenotypic plasticity. IX. Genetic architecture, temperature, and sex differences in *Drosohila melanogaster*. *Evolution* **54**: 1035–1040.

Karban, R., and I. T. Baldwin. 1997. *Induced Responses to Herbivory*. University of Chicago Press, Chicago.

Karban, R., C. A. Black, and S. A. Weinbaum. 2000. How 17-year cicadas keep track of time. *Ecology Letters* **3**: 253–256.

Karban, R., K. Shiojiri, M. Huntzinger, and A. C. McCall. 2006. Damage-induced resistance in sagebrush: Volatiles are key to intra- and inter-plant communication. *Ecology* **87**: 922–930.

Karl, D., R. Letelier, L. Tupas, J. Dore, J. Christian, and D. Hebel. 1997. The role of nitrogen fixation in biogeochemical cycling in the subtropical North Pacific Ocean. *Nature* **388**: 533–538.

Kasahara, M., T. Kagawa, K. Oikawa, N. Suetsugu, M. Miyao, and M. Wada. 2002. Chloroplast avoidance movement reduces photodamage in plants. *Nature* **420**: 829–832.

Kasahara, M., T. Kagawa, Y. Sato, T. Kiyosue, and M. Wada. 2004. Phototropins mediate blue and red light-induced chloroplast movements in *Physcomitrella patens*. *Plant Physiology* **135**: 1388–1397.

Kasting, J. F., and J. L. Siefert. 2002. Life and the evolution of earth's atmosphere. *Science* **296**: 1066–1067.

Kawecki, T. J. 1994. Accumulation of deleterious mutations and the evolutionary cost of being a generalist. *American Naturalist* **144**: 833–838.

Kawecki, T. J., N. H. Barton, and J. D. Fry. 1997. Mutational collapse of fitness in marginal habitats and the evolution of ecological specialization. *Journal of Evolutionary Biology* **10**: 407–429.

Kearney, M., and W. P. Porter. 2004. Mapping the fundamental niche: Physiology, climate, and the distribution of a nocturnal lizard. *Ecology* **85**: 3119–3131.

Keeley, J. E., and C. J. Fotheringham. 2000. Role of fire in regeneration from seed. In M. Fenner, ed. *Seeds: The Ecology of Regeneration in Plant Communities*, pp. 311–330. CABI Publishing, Wallingford.

Keller, E. F. 2000. *The Century of the Gene*. Harvard University Press, Cambridge.

Kelly, M. W., E. Sanford, and R. K. Grosberg. 2012. Limited potential for adaptation to climate change in a broadly distributed marine crustacean. *Proceedings of the Royal Society B* **279**: 349–356.

Kemp, D. J., and A. K. Krockenberger. 2004. Behavioural thermoregulation in butterflies: The interacting effects of body size and basking posture in *Hypolimnas bolina* (L.). *Australian Journal of Zoology* **52**: 229–236.

Kempken, F., and M. Rohlfs. 2010. Fungal secondary metabolite biosynthesis: A chemical defence strategy against antagonistic animals? *Fungal Ecology* **3**: 107–114.

Kerney, R., E. Kim, R. P. Hangarter, A. A. Heiss, C. D. Bishop, and B. K. Hall. 2011. Intracellular invasion of green algae in a salamander host. *Proceedings of the National Academy of Sciences USA* **108**: 6497–6502.

Kessler, A., and I. T. Baldwin. 2001. Defensive function of herbivore-induced plant volatile emissions in nature. *Science* **291**: 2141–2144.

Kessler, A., R. Halitschke, C. Diezel, and I. T. Baldwin. 2006. Priming of plant defense responses in nature by airborne signaling between *Artemisia tridentata* and *Nicotiana attenuata*. *Oecologia* **148**: 280–292.

Kessler, A., and M. Heil. 2011. The multiple faces of indirect defences and their agents of natural selection. *Functional Ecology* **25**: 348–357.

Kessler, S. A., H. Shimosato-Asano, N. F. Keinath, et al. 2010. Conserved molecular components for pollen tube reception and fungal invasion. *Science* **330**: 968–971.

Kevles, D., and L. Hood, eds. 1992. *The Code of Codes: Scientific and Social Issues in the Human Genome Project*. Harvard University Press, Cambridge.

Kim, T.-H., and B. Ć. Maik. 2010. Guard cell signal transduction network: Advances in understanding abscisic acid, CO_2, and Ca^{2+} signaling. *Annual Review of Plant Biology* **61**: 561–591.

Kingsolver, J. G., S. E. Diamond, A. M. Siepielski, and S. M. Carlson. 2012. Synthetic analyses of phenotypic selection in natural populations: Lessons, limitations and future directions. *Evolutionary Ecology* **26**: 1101–1118.

Kingsolver, J. G., and R. B. Huey. 1998. Evolutionary analyses of morphological and physiological plasticity in thermally variable environments. *American Zoologist* **38**: 545–560.

Kingsolver, J. G., and D. W. Pfennig. 2007. Patterns and power of phenotypic selection in nature. *BioScience* **57**: 561–572.

Kingsolver, J. G., G. J. Ragland, and J. G. Shlichta. 2004. Quantitative genetics of continuous reaction norms: Thermal sensitivity of caterpillar growth rates. *Evolution* **58**: 1521–1529.

Kingsolver, J. G., and D. W. Schemske. 1991. Path analyses of selection. *Trends in Ecology and Evolution* **6**: 276–280.

Kinnison, M. T., and A. P. Hendry. 2001. The pace of modern life II: From rates of contemporary microevolution to pattern and process. *Genetica* 112–113: 145–164.

Kirchhoff, H. 2013. Architectural switches in plant thylakoid membranes. *Photosynthesis Research* **116**: 481–487.

Kirkpatrick, M., and R. Lande. 1989. The evolution of maternal characters. *Evolution* **43**: 484–503.

Kirschner, M., and J. Gerhart. 1998. Evolvability. *Proceedings of the National Academy of Sciences USA* **95**: 8420–8427.

Klironomos F. D., J. Berg, and S. Collins. 2013. How epigenetic mutations can affect genetic evolution: Model and mechanism. *BioEssays* **35**: 571–578.

Knight, H., and M. R. Knight. 2001. Abiotic stress signaling pathways: Specificity and cross-talk. *Trends in Plant Science* **6**: 262–267.

Knight, T. M., M. W. McCoy, J. M. Chase, K. A. McCoy, and R. D. Holt. 2005. Trophic cascades across ecosystems. *Nature* **437**: 880–883.

Kniskern, J. M., and M. D. Rausher. 2006. Environmental variation mediates the deleterious effects of *Coleosporium ipomoeae* on *Ipomoea purpurea*. *Ecology* **87**: 675–685.

Knouft, J. H., J. B. Losos, R. E. Glor, and J. J. Kolbe. 2006. Phylogenetic analysis of the evolution of the niche in lizards of the *Anolis sagrei* group. *Ecology* **87**: S29–S38.

Koch, P. B., and D. Buckmann. 1987. Hormonal control of seasonal morphs by the timing of ecdysteroid release in *Araschnia levana* L. (Nymphalidae, Lepidoptera). *Journal of Insect Physiology* **33**: 823–829.

Koiwa, H., R. A. Bressan, and P. M. Hasegawa. 1997. Regulation of protease inhibitors and plant defense. *Trends in Plant Science* **2**: 379–387.

Kokko, H., and A. López-Sepulcre. 2007. The ecogenetic link between demography and evolution: Can we bridge the gap between theory and data? *Ecology Letters* **10**: 773–782.

Koonin, E. V., and Y. I. Wolf. 2008. Evolutionary systems biology. In M. Pagel and A. Pomiankowski, eds. *Evolutionary Genomics and Proteomics*, pp. 11–25. Sinauer Associates, Sunderland.

Kopp, M., and R. Tollrian. 2003a. Reciprocal phenotypic plasticity in a predator–prey system: Inducible offenses against inducible defenses? *Ecology Letters* **6**: 742–748.

Kopp, M., and R. Tollrian. 2003b. Trophic size polyphenism in *Lembadion bullinum*: Cost and benefits of an inducible offense. *Ecology* **84**: 641–651.

Körner, C. 2013. Plant–environment interactions. In A. Bresinsky, C. Korner, J. Kadereit, eds. *Strasburger's Plant Sciences*, pp. 1065–1166. Springer Berlin, Heidelberg.

Kouril, R., E. Wientjes, J. B. Bultema, R. Croce, and E. J. Boekema. 2013. High-light vs. low-light: Effect of light acclimation on photosystem II composition and organization in *Arabidopsis thaliana*. *Biochimica et Biophysica Acta* **1827**: 411–419.

Kourtev, P. S., J. G. Ehrenfeld, and M. Häggblom. 2002. Exotic plants alter the microbial community structure and function in the soil. *Ecology* **83**: 3152–3166.

Koza, A., O. Moshynets, W. Otten, and A. J. Spiers. 2011. Environmental modification and niche construction: Developing O_2 gradients drive the evolution of the Wrinkly Spreader. *The International Society for Microbial Ecology Journal* **5**: 665–673.

Koziol, M. J., and J. L. Rinn. 2010. RNA traffic control of chromatin complexes. *Current Opinion in Genetics and Development* **20**: 142–148.

Kriska, G., G. Horváth, and S. Andrikovics. 1998. Why do mayflies lay their eggs en masse on dry asphalt roads? Water-imitating polarized light reflected from asphalt attracts Ephemeroptera. *Journal of Experimental Biology* **201**: 2273–2286.

Krivan, V. 2003. Competitive co-existence caused by adaptive predators. *Evolutionary Ecology Research* **5**: 1163–1182.

Kruuk, L. E. B., J. Merilä, and B. C. Sheldon. 2003. When environmental variation short-circuits natural selection. *Trends in Ecology and Evolution* **18**: 207–209.

Kruuk, L. E. B., J. Slate, and A. J. Wilson. 2008. New answers for old questions: The evolutionary quantitative genetics of wild animal populations. *Annual Review of Ecology, Evolution and Systematics* **39**: 525–548.

Kucharski, R., J. Maleszka, S. Foret, and R. Maleszka. 2008. Nutritional control of reproductive status in honeybees via DNA methylation. *Science* **319**: 1827–1830.

Kuhlmann, H.-W., J. Kusch,and K. Heckmann. 1999. Predator-induced defences in ciliated protozoa. In R. Tollrian and C. D. Harvell, eds. *The Evolution of Inducible Defences*, pp. 142–159. Princeton University Press, Princeton.

Kulkarni, S. S., and D. R. Buchholz. 2014. Corticosteroid signaling in frog metamorphosis. *General and Comparative Endocrinology* **203**: 225–231.

Kussell, E., R. Kishony, N. Q. Balaban, and S. Leibler. 2005. Bacterial persistence: A model of survival in changing environments. *Genetics* **169**: 1807–1814.

Kwiatkowski, D. P. 2005. How malaria has affected the human genome and what human genetics can teach us about malaria. *American Journal of Human Genetics* **77**: 171–192.

Kylafis, G., and M. Loreau. 2008. Ecological and evolutionary consequences of niche construction for its agent. *Ecology Letters* **11**: 1072–1081.

Lacey, E. P., and D. Herr. 2005. Phenotypic plasticity, parental effects and parental care in plants? I. An examination of spike reflectance in *Plantago lanceolata*. *American Journal of Botany* **92**: 920–930.

Lake, J., F. I. Woodward, and W. P. Quick. 2002. Long-distance CO_2 signalling in plants. *Journal of Experimental Botany* **53**: 183–193.

Laland, K., T. Uller, M. Feldman, et al. 2014. Does evolutionary theory need a rethink? Yes, urgently. *Nature* **514**: 161–164.

Laland, K. N., and N. J. Boogert. 2008. Niche construction, co-evolution and biodiversity. *Ecological Economics* **69**: 731–736.

Laland, K. N., and M. J. O'Brien. 2012. Cultural niche construction: An introduction. *Biological Theory* **6**: 191–202.

Laland, K. N., J. Odling-Smee, and M. W. Feldman. 1999. Evolutionary consequences of niche construction and their implications for ecology. *Proceedings of the National Academy of Sciences USA* **96**: 10242–10247.

Laland, K. N., J. Odling-Smee, and S. F. Gilbert. 2008. EvoDevo and niche construction: Building bridges. *Journal of Experimental Zoology* **310B**: 549–566.

Laland, K. N., and K. Sterelny. 2006. Seven reasons not to neglect niche construction. *Evolution* **60**: 1751–1762.

Lambers, H., F. S. Chapin III, and T. H. L. Poons. 2008. *Plant Physiological Ecology*, 2nd ed. Springer Science, New York.

Lambrot, R., C. Xu, S. Saint-Phar, et al. 2013. Low paternal dietary folate alters the mouse sperm epigenome and is associated with negative pregnancy outcomes. *Nature Communications* **4**: Article no. 2889.

Lance, V. A. 2009. Is regulation of aromatase expression in reptiles the key to understanding temperature-dependent sex determination? *Journal of Experimental Zoology* **311A**: 314–322.

Lande, R. 2009. Adaptation to an extraordinary environment by evolution of phenotypic plasticity and genetic assimilation. *Journal of Evolutionary Biology* **22**: 1435–1446.

Langerhans, R. B., L. J. Chapman, and T. J. DeWitt. 2007. Complex phenotype–environment associations revealed in an East African cyprinid. *Journal of Evolutionary Biology* **20**:1171–1181.

Larios, E., A. Búrquez, J. X. Becerra, and D. L. Venable. 2014. Natural selection on seed size through the life cycle of a desert annual plant. *Ecology* **95**: 3213–3220.

Latzel, V., Y. Zhang, K. K. Moritz, M. Fischer, and O. Bossdorf. 2012. Epigenetic variation in plant responses to defense hormones. *Annals of Botany* **110**: 1423–1428.

Laudet, V. 2011. The origins and evolution of vertebrate metamorphosis. *Current Biology* **21**: R726–R737.

Lavelle, P. 1988. Earthworm activities and the soil system. *Biology and Fertility of Soils* **6**: 237–251.

Leck, M. A., V. T. Parker, and R. I. Simpson. 2008. *Seedling Ecology and Evolution*. Cambridge University Press, Cambridge, UK.

Ledón-Rettig, C. C., and D. W. Pfennig. 2011. Emerging model systems in eco-evo-devo: The environmentally responsive spadefoot toad. *Evolution and Development* **13**: 391–400.

Ledón-Rettig, C., D. W. Pfennig, C. Chunco, and I. Dworkin. 2014. Cryptic genetic variation in natural populations: A predictive framework. *Integrative and Comparative Biology* **54**: 783–793.

Ledón-Rettig, C. C., D. W. Pfennig, and N. Nascone-Yoder. 2008. Ancestral variation and the potential for genetic accommodation in larval amphibians: Implications for the evolution of novel feeding strategies. *Evolution and Development* **10**: 316–325.

Lee, T. M., L. Smale, I. Zucker, and J. Dark. 1987. Influence of daylength experienced by dams on post-natal development of young meadow voles (*Microtus pennsylvanicus*). *Journal of Reproduction and Fertility* **81**: 337–342.

Lefevre, T. H., L. Oliver, M. D. Hunter, and J. C. de Roode. 2010. Evidence for trans-generational medication in nature. *Ecology Letters* **13**: 1485–1493.

Leibold, M. A. 1995. The niche concept revisited: Mechanistic models and community context. *Ecology* **76**: 1371–1382.

Leigh, E. G. 2010. The evolution of mutualism. *Journal of Evolutionary Biology* **23**: 2507–2528.

Leimar, O. 2007. Environmental and genetic cues in the evolution of phenotypic polymorphism. *Evolutionary Ecology* **23**: 125–135.

Leimar, O., P. Hammerstein, and T. J. M. Van Dooren. 2006. A new perspective on developmental plasticity and the principles of adaptive morph determination. *American Naturalist* **167**: 367–376.

Leinonen, P. H., D. L. Remington, J. Leppälä, and O. Savolainen. 2013. Genetic basis of local adaptation and flowering time variation in *Arabidopsis lyrata*. *Molecular Ecology* **22**: 709–723.

Leips, J., and T. F. C. Mackay. 2000. Quantitative trait loci for life span in *Drosophila melanogaster*: Interactions with genetic background and larval density. *Genetics* **155**: 1773–1788.

Leishman, M. R., I. J. Wright, A. T. Moles, and M. Westoby. 2000. The evolutionary ecology of seed size. In M. Fenner, ed. *Seeds: The Ecology of Regeneration in Plant Communities*, 2nd ed., pp. 31–57. CABI Publishing, Wallingford.

Lema, S. C., and J. Kitano. 2013. Hormones and phenotypic plasticity: Implications for the evolution of integrated adaptive phenotypes. *Current Zoology* **59**: 506–525.

Lemos, B., C. R. Landry, P. Fontanillas, et al. 2008. Evolution of genomic expression. In M. Pagel and A. Pomiankowski, eds. *Evolutionary Genomics and Proteomics*, pp. 81–118. Sinauer Associates, Sunderland.

Lengyel, S., A. D. Gove, A. M. Latimer, J. D. Majer, and R. R. Dunn. 2009. Ants sow the seeds of global diversification in flowering plants. *Public Library of Science ONE* **4**: e5480.

Lenhard, B., A. Sandelin, and P. Carninci. 2012. Metazoan promoters: Emerging characteristics and insights into transcriptional regulation. *Nature Reviews Genetics* **13**: 233–244.

Leonard, G. H., M. D. Bertness, and P. O. Yund. 1999. Crab predation, waterborne cues, and inducible defenses in the blue mussel, *Mytilus edulis*. *Ecology* **80**: 1–14.

Leveau, J. H. J., and S. E. Lindow. 2001. Appetite of an epiphyte: Quantitative monitoring of bacterial sugar consumption in the phyllosphere. *Proceedings of the National Academy of Sciences USA* **98**: 3446–3453.

Levin, D. A. 2009. Flowering-time plasticity facilitates niche shifts in adjacent populations. *New Phytologist* **183**: 661–666.

Lev-Yadun, S., G. Katzir, and G. Ne`eman. 2009. *Rheum palaestinum*, a self-irrigating desert plant. *Naturwissenschaften* **96**: 393–397.

Lewontin, R. C. 1957. The adaptation of populations to varying environments. *Cold Spring Harbor Symposium in Quantitative Biology* **22**: 395–408.

Lewontin, R. C. 1978. Adaptation. *Scientific American* **239**: 212–230.

Lewontin, R. C. 1983. Gene, organism and environment. In D. S. Bendall, ed. *Evolution: From Molecules to Men*, pp. 273–285. Cambridge University Press, Cambridge, UK.

Lewontin, R. C. 1985. The organism as the subject and object of evolution. In R. Levins and R. C. Lewontin, eds. *The Dialectical Biologist*, pp. 85–106. Harvard University Press, Cambridge.

Lewontin, R. C. 2000. *The Triple Helix: Gene, Organism and Environment*. Harvard University Press, Cambridge.

Lewontin, R. C. 2001. Gene, organism and environment: A new introduction. In S. G. Oyama, P. E. Griffiths, and R. D. Gray, eds. *Cycles of Contingency*, pp. 55–66. MIT Press, Cambridge.

Li, J., E. J. Helmerhorst, C. W. Leone, R. F. Troxler, T. Yaskell, A. D. Haffajee, S. S. Socransky, and F. G. Oppenheim. 2004. Identification of early microbial colonizers in human dental biofilm. *Journal of Applied Microbiology* **97**: 1311–1318.

Li, L., and E. S. Wurtele. 2015. The *QQS* orphan gene of *Arabidopsis* modulates carbon and nitrogen allocation in soybean. *Plant Biotechnology Journal* **13**: 177–187.

Lilien, J., J. Balsamo, C. Arregui, and G. Xu. 2002. Functional state switching of cadherins. *Developmental Dynamics* **224**: 18–29.

Lindow, S. E., and M. T. Brandl. 2003. Microbiology of the phyllosphere. *Applied and Environmental Microbiology* **69**: 1875–1883.

Lindroth, R. L., S. Roth, and E. V. Nordheim. 2001. Genotypic variation in response of quaking aspen to atmospheric CO_2 enrichment. *Oecologia* **126**: 371–379.

Linn, C. E., W. L. Yee, S. B. Sim, et al. 2012. Behavioral evidence for fruit odor discrimination and sympatric host races of *Rhagoletis pomonella* flies in the western united states. *Evolution* **66**: 3632–3641.

Lira-Medeiros, C. F., C. Parisod, R. A. Fernandes, C. S. Mata, M. A. Cardoso, and P. C. G. Ferreira. 2010. Epigenetic variation in mangrove plants occurring in contrasting natural environments. *Public Library of Science ONE* **5**: e10326.

Liu, D., J. Diorio, B. Tannenbaum, et al. 1997. Maternal care, hippocampal glucocorticoid receptors, and hypothalamic–pituitary adrenal responses to stress. *Science* **277**: 1659–1662.

Lively, C. M. 1986. Predator-induced shell dimorphism in the acorn barnacle *Chthamalus anisopoma*. *Evolution* **40**: 232–242.

Lively, C. M., W. N. Hazel, M. J. Schellenberger, and K. S. Michelson. 2000. Predator-induced defense: Variation for inducibility in an intertidal barnacle. *Ecology* **81**: 1240–1247.

Lolle, S. J., J. L. Victor, J. M. Young, and R. E. Pruitt. 2005. Genome-wide non-Mendelian inheritance of extra-genomic information in *Arabidopsis*. *Nature* **434**: 505–506.

Long, H. H., D. D. Schmidt, and I. T. Baldwin. 2008. Native bacterial endophytes promote host growth in a species-specific manner. *Public Library of Science ONE* **3**: e2702.

López-Juez, E. 2009. Steering the solar panel: Plastids influence development. *New Phytologist* **182**: 287–290.

Lorrain, S., T. Allen, P. D. Duek, G. C. Whitelam, and C. Fankhauser. 2008. Phytochrome-mediated inhibition of shade avoidance involves degredation of growth-promoting bHLH transcription factors. *Plant Journal* **53**: 312–323.

Low, F. M., P. D. Gluckman, and M. A. Hanson. 2014. Epigenetic and developmental basis of risk of obesity and metabolic disease. In A. Ulloa-Aguirre and P. M. Conn, eds. *Cellular Endocrinology in Health and Disease*, pp. pp. 111–132. Elsevier Academic Press, London.

Lucassen, P. J., E. F. G. Naninck, J. B. van Goudoever, C. Fitsimons, M. Joels, and A. Korosi. 2013. Perinatal programming of adult hippocampal structure and function; emerging roles of stress, nutrition and epigenetics. *Trends in Neurosciences* **36**: 621–631.

Ludwig, M. Z., N. H. Patel, and M. Kreitman. 1998. Functional analysis of *eve stripe* 2 enhancer evolution in *Drosophila*: Rules governing conservation and change. *Development* **125**: 949–958.

Lundgren, M. R., and S. E. Sultan. 2005. Seedling expression of cross-generational plasticity depends on reproductive architecture. *American Journal of Botany* **92**: 377–381.

Luttbeg, B., L. Rowe, and M. Mangel. 2003. Prey state and experimental design affect relative size of trait- and density-mediated indirect effects. *Ecology* **84**: 1140–1150.

Luzar, N., and G. Gottsberger. 2001. Flower heliotropism and floral heating of five alpine plant species and the effect on flower visiting in *Ranunculus montanus* in the Austrian Alps. *Arctic, Antarctic, and Alpine Research* **33**: 93–99.

Lynch, K. E., and D. J. Kemp. 2013. Nature-via-nurture and unravelling causality in evolutionary genetics. *Trends in Ecology and Evolution* **29**: 2–4.

Lynch, V., and G. P. Wagner 2008. Resurrecting the role of transcription factor change in developmental evolution. *Evolution* **62**: 2131–2154.

Lyons, T. W., C. T. Reinhard, and N. J. Planavsky. 2014. The rise of oxygen in Earth's early ocean and atmosphere. *Nature* **506**: 307–315.

MacColl A. D. 2011. The ecological causes of evolution. *Trends in Ecology and Evolution* **26**: 514–522.

MacDonald, W. A. 2012. Epigenetic mechanisms of genomic imprinting: Common themes in the regulation of imprinted regions in mammals, plants, and insects. *Genetics Research International*: Article ID 585024.

Mack, K. J., and P. A. Mack. 1992. Induction of transcription factors in somatosensory cortex after tactile stimulation. *Molecular Brain Research* **12**: 141–147.

Mackay, T. F. C. 2013. Epistasis and quantitative traits: Using model organisms to study gene–gene interactions. *Nature Reviews Genetics* **15**: 22–33.

Madritch, M. D., and M. D. Hunter. 2005. Phenotypic variation in oak litter influences short- and long-term nutrient cycling through litter chemistry. *Soil Biology and Biochemistry* **37**: 319–327.

Maeno, K. O., C. Piou, M. A. Ould Babah, and S. Nakamura. 2013. Eggs and hatchlings variations in desert locusts: Phase related characteristics and starvation tolerance. *Frontiers in Physiology* **4**: 345–359.

Maestripieri, D., and J. M. Mateo, eds. 2009. *Maternal Effects in Mammals*. University of Chicago Press, Chicago.

Maffei, M. E., A. Mithofer, G. Arimura, et al. 2006. Effects of feeding *Spodoptera littoralis* on lima bean leaves. III. Membrane depolarization and involvement of hydrogen peroxide. *Plant Physiology* **140**: 1022–1035.

Mahan, J. R., and D. R. Upchurch. 1988. Maintenance of constant leaf temperature by plants. I. Hypothesis-limited homeothermy. *Environmental and Experimental Botany* **28**: 351–357.

Maher, J. M., E. E. Werner, and R. J. Denver. 2013. Stress hormones mediate predator-induced phenotypic plasticity in amphibian tadpoles. *Proceedings of the Royal Society B* **280**: 20123075.

Majetic, C. J., R. A. Raguso, and T.-L. Ashman. 2009. Sources of floral scent variation. *Plant Signaling and Behavior* **4**: 129–131.

Maniatis, T. 1991. Mechanisms of alternative pre-mRNA splicing. *Science* **251**: 33–34.

Manikkam, M., C. Guerrero-Bosagna, R. Tracey, M. M. Haque, and M. K. Skinner. 2012. Transgenerational actions of environmental compounds on reproductive disease and epigenetic biomarkers of ancestral exposures. *Public Library of Science ONE* **7**: e31901.

Manolio, T. A., F. S. Collins, N. J. Cox, et al. 2009. Finding the missing heritability of complex diseases. *Nature* **461**: 747–753.

March, W. A., and D. M. Watson. 2007. Parasites boost productivity: Effects of mistletoe on litterfall dynamics in a temperate Australian forest. *Oecologia* **154**: 339–347.

March, W. A., and D. M. Watson. 2010. The contribution of mistletoes to nutrient returns: Evidence for a critical role in nutrient cycling. *Austral Ecology* **35**: 713–721.

Marco, D. 2008. Metagenomics and the niche concept. *Theory in Biosciences* **127**: 241–247.

Maron, J. L., and P. G. Connors. 1996. A native nitrogen-fixing shrub facilitates weed invasion. *Oecologia* **105**: 302–312.

Marschner, H. 1995. *Mineral Nutrition of Higher Plants*, 2nd ed. Academic Press,London.

Maruska, K. P., R. E. Carpenter, and R. D. Fernald. 2012. Characterization of cell proliferation throughout the brain of the African cichlid fish *Astatotilapia burtoni* and its regulation by social status. *Journal of Comparative Neurology* **520**: 3471–3491.

Massa, G. D., and S. Gilroy. 2003. Touch modulates gravity sensing to regulate the growth of primary roots in *Arabidopsis thaliana*. *The Plant Journal* **33**: 435–445.

Matesanz, S., T. Horgan-Kobelski, and S. E. Sultan. 2012. Phenotypic plasticity and population differentiation in an ongoing species invasion. *Public Library of Science ONE* **9**: e44955.

Matesanz, S., T. Horgan-Kobelski, and S. E. Sultan. 2014. Contrasting levels of evolutionary potential in populations of the invasive plant *Polygonum cespitosum*. *Biological Invasions* **16**: 455–468.

Matesanz, S., T. Horgan-Kobelski, and S. E. Sultan. 2015. Evidence for rapid ecological range expansion in a newly invasive plant. *AoB Plants* **7**: 038.

Mathis, A., and R. J. F. Smith. 1993. Fathead minnows, *Pimephales promelas*, learn to recognize northern pike *Esox lucius* as predators on the basis of chemical stimuli from minnows in the pike's diet. *Animal Behaviour* **46**: 645–656.

Mattick, J. S. 2012. Rocking the foundations of molecular genetics. *Proceedings of the National Academy of Sciences USA* **109**: 16400–16401.

Mattick, J. S., and M. F. Mehler. 2008. RNA editing, DNA recoding and the evolution of human cognition. *Trends in Neurosciences* **31**: 227–233.

Maynard Smith, J. 1989. *Evolutionary Genetics*. Oxford University Press, Oxford.

Mayntz, D., D. Raubenheimer, M. Salomon, S. Toft, and S. J. Simpson. 2005. Nutrient-specific foraging in invertebrate predators. *Science* **307**: 111–113.

Mayr, E. 1970. *Populations, Species, and Evolution: An Abridgment of "Animal Species and Evolution."* Harvard University Press, Cambridge.

Mayr, E. 1983. How to carry out the adaptationist program? *American Naturalist* **121**: 324–334.

Mayr, E., and W. Provine, eds. 1980. *The Evolutionary Synthesis: Perspectives in the Unification of Biology*. Cambridge University Press, New York.

Mazin, P., J. Xiong, X. Liu, et al. 2013. Widespread splicing changes in human brain development and aging. *Molecular Systems Biology* **9** doi: 10.1038/msb.2012.67.

McCully, M. E. 1999. Roots in soil: Unearthing the complexities of roots and their rhizospheres. *Annual Review of Plant Physiology and Plant Molecular Biology* **50**: 695–718.

McFall-Ngai, M. J. 2002. Unseen forces: The influence of bacteria on animal development. *Developmental Biology* **242**: 1–14.

Meaney, M. J., and A. C. Ferguson-Smith. 2010. Epigenetic regulation of the neural transcriptome: The meaning of the marks. *Nature Neuroscience* **13**: 1313–1318.

Mello, C. V., D. S. Vicario, and D. F. Clayton. 1992. Song presentation induces gene expression in the songbird forebrain. *Proceedings of the National Academy of Sciences USA* **89**: 6818–6822.

Melotto, M., W. Underwood, and S. Y. He. 2008. Role of stomata in plant innate immunity and foliar bacterial diseases. *Annual Review of Phytopathology* **46**: 101–122.

Memmott, J., P. G. Craze, N. M. Waser, and M. V. Price. 2007. Global warming and the disruption of plant-pollinator interactions. *Ecology Letters* **10**: 710–717.

Memmott, J., and N. M. Waser. 2002. Integration of alien plants into a native flower-pollinator visitation web. *Proceedings of the Royal Society B* **269**: 2395–2399.

Merilä, J., and A. P. Hendry. 2014. Climate change, adaptation, and phenotypic plasticity: The problem and the evidence. *Evolutionary Applications* **7**: 1–14.

Merilä, J., B. C. Sheldon, and L. E. Kruuk. 2001. Explaining stasis: Microevolutionary studies in natural populations. *Genetica* 112–113: 199–222.

Messier, J., B. J. McGill, and M. J. Lechowicz. 2010. How do traits vary across ecological scales? A case for trait-based ecology. *Ecology Letters* **13**: 838–848.

Messmer, V., G. P. Jones, L. L. Herwerden, and P. L. Munday. 2005. Genetic and ecological characterisation of colour dimorphism in a coral reef fish. *Environmental Biology of Fish* **74**: 175–183.

Messmer, V., G. P. Jones, P. L. Munday, S. J. Holbrook, R. J. Schmitt, and A. J. Brooks. 2011. Habitat biodiversity as a determinant of fish community structure on coral reefs. *Ecology* **92**: 2285–2298.

Meyer, A. 1987. Phenotypic plasticity and heterochrony in *Cichlasoma managuense* and their implications for speciation in cichlid fishes. *Evolution* **41**: 1357–1369.

Meyer, A. 1990. Ecological and evolutionary consequences of the trophic polymorphism in *Cichlasoma citrinellum*. *Biological Journal of the Linnean Society* **39**: 279–299.

Miklos, G. L. G. 2005. The Human Cancer Genome Project: One more misstep in the war on cancer. *Nature Biotechnology* **23**: 535–537.

Milewski, A. V., T. P. Young, and D. Madden. 1991. Thorns as induced defenses: Experimental evidence. *Oecologia* **86**: 70–75.

Miller, G. M., S.-A. Watson, J. M. Donelson, M. I. McCormick, and P. L. Munday. 2012. Parental environment mediates impacts of increased carbon dioxide on a coral reef fish. *Nature Climate Change* **2**: 858–861.

Miner, B. G., S. E. Sultan, S. G. Morgan, D. K. Padilla, and R. A. Relyea. 2005. Ecological consequences of phenotypic plasticity. *Trends in Ecology and Evolution* **20**: 685–692.

Mitchell, R. J., R. E. Irwin, R. J. Flanagan, and J. D. Karron. 2009. Ecology and evolution of plant–pollinator interactions. *Annals of Botany* **103**: 1355–1363.

Mitchell, N. J., and F. J. Janzen. 2010. Temperature-dependent sex determination and contemporary climate change. *Sexual Development* **4**: 129–140.

Mitchell-Olds, T., and J. J. Rutledge. 1986. Quantitative genetics in natural plant populations. A review of the theory. *American Naturalist* **127**: 379–402.

Mitchell-Olds, T., and R. G. Shaw. 1987. Regression analysis of natural selection: Statistical inference and biological interpretation. *Evolution* **41**: 1149–1161.

Mithofer, A., G. Wanner, and W. Boland. 2005. Effects of feeding *Spodoptera littoralis* on lima bean leaves. II. Continuous mechanical wounding resembling insect feeding is sufficient to elicit herbivory-related volatile emission. *Plant Physiology* **137**: 1160–1168.

Moberg, F., and C. Folke. 1999. Ecological goods and services of coral reef ecosystems. *Ecological Economics* **29**: 215–233.

Moctezuma, E. 2003. The peanut gynophore: A developmental and physiological perspective. *Canadian Journal of Botany* **81**: 183–190.

Moczek, A. P. 1998. Horn polyphenism in the beetle *Onthophagus taurus*: Larval diet quality and plasticity in parental investment determine adult body size and male horn morphology. *Behavioral Ecology* **9**: 636–641.

Moczek, A. P., and D. J. Emlen. 1999. Proximate determination of male horn dimorphism in the beetle *Onthophagus taurus* (Coleoptera: Scarabaeidae). *Journal of Evolutionary Biology* **12**: 27–37.

Moczek, A. P., and D. J. Emlen. 2000. Male horn dimorphism in the scarab beetle, *Onthophagus taurus*: Do alternative reproductive tactics favour alternative phenotypes? *Animal Behaviour* **59**: 459–466.

Moczek, A. P., and H. F. Nijhout. 2003. Rapid evolution of a polyphenic threshold. *Evolution and Development* **5**: 259–268.

Moczek, A. P., S. E. Sultan, S. Foster, et al. 2011. The role of developmental plasticity in evolutionary innovation. *Proceedings of the Royal Society B* **278**: 2705–2713.

Mokady, O., L. Mizrahi, R. Perl-Treves, and Y. Achituv. 2000. The different morphs of *Chthamalus anisopoma*: A phenotypic response? Direct molecular evidence. *Journal of Experimental Marine Biology and Ecology* **243**: 295–304.

Moles, A. T., D. W. Hodson, and C. J. Webb. 2000. Seed size and shape and persistence in the soil in the New Zealand flora. *Oikos* **89**: 541–545.

Mommer, L., T. L. Pons, and E. J. W. Visser. 2006. Photosynthetic consequences of phenotypic plasticity in response to submergence: *Rumex palustris* as a case study. *Journal of Experimental Botany* **57**: 283–290.

Mommer, L., T. L. Pons, M. Wolters-Arts, J. H. Venema, and E. J. W. Visser. 2005. Submergence-induced morphological, anatomical and biochemical responses in a terrestrial species affect gas diffusion resistance and photosynthetic performance. *Plant Physiology* **139**: 497–508.

Mommer, L., and E. J. W. Visser. 2005. Underwater photosynthesis in flooded terrestrial plants: A matter of leaf plasticity. *Annals of Botany* **96**: 581–589.

Mondor, E. B., J. A. Rosenheim, and J. F. Addicott. 2005. Predator-induced transgenerational phenotypic plasticity in the cotton aphid. *Oecologia* **142**: 104–108.

Mopper, S., and A. A. Agrawal. 2004. Phytohormonal ecology. *Ecology* **85**: 3–4.

Moran, N. A. 1992. The evolutionary maintenance of alternative phenotypes. *American Naturalist* **139**: 971–989.

Morjan, C. L. 2003. How rapidly can maternal behavior affecting primary sex ratio evolve in a reptile with environmental sex determination? *The American Naturalist* **162**: 205–219.

Morris, C. E., and L. L. Kinkel. 2002. Fifty years of phyllosphere microbiology: Significant contributions to research in related fields. In S. E. Lindow, E. I. Hecht-Poinar, and V. Elliott, eds. *Phyllosphere Microbiology*, pp. 365–375. APS Press, St. Paul.

Morris, D. W. 2003. Toward an ecological synthesis: A case for habitat selection. *Oecologia* **136**: 1–13.

Morris, K. V., and J. S. Mattick. 2014. The rise of regulatory RNA. *Nature Reviews Genetics* **15**: 423–437.

Mousseau, T. A., and Fox, C. 1998. *Maternal Effects as Adaptations*. Oxford University Press, New York.

Mucci, L. A., S. Wedren, R. M. Tamimi, D. Trichopoulos, and H. -O. Adami. 2001. The role of gene–environment interaction in the aetiology of human cancer: Examples from cancers of the large bowel, lung and breast. *Journal of Internal Medicine* **249**: 477–493.

Müller, G. B., and S. A. Newman, eds. 2003. *Origination of Organismal Form: Beyond the Gene in Developmental and Evolutionary Biology*. MIT Press, Cambridge.

Mumby, P. J., C. P. Dahlgren, A. R. Harborne, et al. 2006. Fishing, trophic cascades, and the process of grazing on coral reefs. *Science* **311**: 98–101.

Muñoz, L. P., G. R. Finke, P. A. Camus, and F. Bozinovic. 2005. Thermoregulatory behavior, heat gain and thermal tolerance in the periwinkle *Echinolittorina peruviana* in central Chile. *Comparative Biochemical Physiology* **142**: 92–98.

Murata, N., and I. Suzuki. 2006. Exploitation of genomic sequences in a systematic analysis to access how cyanobacteria sense environmental stress. *Journal of Experimental Botany* **57**: 235–247.

Muscatine, L. 1990. The role of symbiotic algae in carbon and energy flux in reef corals. In Z. Dubinsky, ed. *Coral Reefs*, pp. 75–87. Ecosystems of the World, Vol. **25**. Elsevier Press, Amsterdam.

Muscatine, L., D. Grossman, and J. Doino. 1991. Release of symbiotic algae by tropical sea anemones and corals after cold shock. *Marine Ecology Progress Series* **77**: 233–243.

Muschick, M., M. Barluenga, W. Salzburger, and A. Meyer. 2011. Adaptive phenotypic plasticity in the Midas cichlid fish pharyngeal jaw and its relevance in adaptive radiation. *BioMed Central Evolutionary Biology* **11**: 116.

Mutti, N. S., A. G. Dolezal, F. Wolschin, J. S. Mutti, K. S. Gill, and G. V. Amdam. 2011. IRS and TOR nutrient-signaling pathways act via juvenile hormone to influence honey bee caste fate. *Journal of Experimental Biology* **214**: 3977–3984.

Myers, M. P., K. Wager-Smith, A. Rothenfluh-Hilfiker, and M. W. Young. 1996. Light-induced degradation of *TIMELESS* and entrainment of the *Drosophila* circadian clock. *Science* **271**: 1736–1740.

Nager, R. G., L. F. Keller, and A. J. Van Noordwijk. 2000. Understanding natural selection on traits that are influenced by environmental conditions. In T. Mousseau, B. Sinervo, and J. A. Endler, eds. *Adaptive Genetic Variation in the Wild*, pp. 95–115. Oxford University Press, New York.

Naiman, R. J., C. A. Johnston, and J. C. Kelley. 1988. Alteration of North American streams by beaver. *BioScience* **38**: 753–762.

Nakagawa, S., M. Lagisz, K. L. Hector, and H. G. Spencer. 2012. Comparative and meta-analytic insights into life extension via dietary restriction. *Aging Cell* **11**: 401–409.

National Library of Medicine and National Human Genome Institute. 2015. *Gene Ed: Genetics, Education, Discovery*. <http://geneed.nlm.nih.gov>, accessed February 10, 2015.

NCBI. 2010. *A Science Primer*. <www.ncbi.nlm.nih.gov/About/primer/genetics_genome.html>, accessed January 24, 2010.

Nelson, V. R., J. D. Heaney, P. J. Tesar, N. O. Davidson, and J. H. Nadeau. 2012. Transgenerational epigenetic effects of the *Apobec1* cytidine deaminase deficiency on testicular germ cell susceptibility and embryonic viability. *Proceedings of the National Academy of Sciences USA* **109**: e2766–2773.

Neslund-Dudas, C., A. M. Levin, A. Rundle, et al. 2014. Case-only gene–environment interaction between *ALAD* tag SNPs and occupational lead exposure in prostate cancer. *The Prostate* **74**: 637–646.

Ness, J. H., D. F. Morin, and I. Giladi. 2009. Uncommon specialization in a mutualism between a temperate herbaceous plant guild and an ant: Are *Aphaeogaster* ants keystone mutualists? *Oikos* **118**: 1793–1804.

Neumann-Held, E. M. 2006. Genes—causes—codes: Deciphering DNA's ontological privilege. In E. M. Neumann-Held and C. Rehmann-Sutter, eds. *Genes in Development: Re-reading the Molecular Paradigm*, pp. 238–271. Duke University Press, Durham.

Newman, R. A. 1992. Adaptive plasticity in amphibian metamorphosis. *BioScience* **42**: 671–678.

Newman, R. A. 1994. Effects of changing density and food level on metamorphosis of a desert amphibian, *Scaphiopus couchii*. *Ecology* **75**: 1085–1096.

Newman, S. A., and G. B. Müller. 2010. *Morphological Evolution: Epigenetic Mechanisms*. Encyclopedia of Life Sciences. John Wiley and Sons, Chichester.

Nicholson, E. Holmes, J. K., J. Kinross, et al. 2012. Host–gut microbiota metabolic interactions. *Science* **336**: 1262–1267.

Nicotra, A. B., O. K. Atkin, S. P. Bonser, et al. 2010. Plant phenotypic plasticity in a changing climate. *Trends in Plant Science* **15**: 684–692.

Niederhuth, C. E., and R. J. Schmitz. 2014. Covering your bases: Inheritance of DNA methylation in plant genomes. *Molecular Plant* **7**: 472–480.

Nijhout, H. F. 1990. Metaphors and the role of genes in development. *BioEssays* **12**: 441–446.

Nijhout, H. F. 1994. *Insect Hormones*. Princeton University Press, Princeton.

Nijhout, H. F. 1999. Control mechanisms of polyphenic development in insects. *Bioscience* **49**: 181–192.

Nijhout, H. F. 2003. Development and evolution of adaptive polyphenisms. *Evolution and Development* **5**: 9–18.

Nilsson, E. E., and M. K. Skinner. 2015. Environmentally induced transgenerational inheritance of disease susceptibility. *Translational Research* **165**: 12–17.

Nishikura, K. 2010. Functions and regulation of RNA editing by ADAR deaminases. *Annual Review of Biochemistry* **79**: 321–349.

Noble, D., E. Jablonka, M. J. Joyner, G. B. Müller, and S. W. Omholt. 2014. Evolution evolves: Physiology returns to center stage. *Journal of Physiology* **592**: 2237–2244.

Norris, A. L., and T. H. Kunz. 2012. Effects of solar radiation on animal thermoregulation. In E. B. Babatunde, ed. *Solar Radiation*, pp. 195–220. INTech Books, Rijeka.

Nowacki, M., V. Vijayan, Y. Zhou, K. Schotanus, T. G. Doak, and L. F. Landweber. 2008. RNA-mediated epigenetic programming of a genome-rearrangement pathway. *Nature* **451**: 153–158.

Nozue, K., M. F. Covington, P. D. Duek, et al. 2007. Rhythmic growth explained by coincidence between internal and external cues. *Nature* **448**: 358–362.

Nussey, D. H., E. Postma, P. Gienapp, and M. E. Visser. 2005. Selection on heritable phenotypic plasticity in a wild bird population. *Science* **310**: 304–306.

Nussey, D. H., A. J. Wilson, and J. E. Brommer. 2007. The evolutionary ecology of individual phenotypic plasticity in wild populations. *Journal of Evolutionary Biology* **20**: 831–844.

Nye, P. H. 1981. Changes of pH across the rhizosphere induced by roots. *Plant Soil* **61**: 7–26.

O'Connor, B. M., R. K. Colwell, and S. Naeem. 1991. Flower mites of Trinidad. II. The genus *Proctolaelaps* (Acari: Ascidae). *Western North American Naturalist* **51**: 348–376.

Odling-Smee, F. J., K. N. Laland, and M. W. Feldman. 2003. *Niche Construction: The Neglected Process in Evolution*. Princeton University Press, Princeton.

Odling-Smee, J. 2007. Niche inheritance: A possible basis for classifying multiple inheritance systems in evolution. *Biological Theory* **2**: 276–289.

Odling-Smee, J., D. H. Erwin, E. P. Palkovacs, M. W. Feldman, and K. N. Laland. 2013. Niche construction theory: A practical guide for ecologists. *The Quarterly Review of Biology* **88**: 3–28.

Oh, D., J. Qi, Y.-C. Lu, Y. Zhang, Y. Shao-Horn, and A. M. Belcher. 2013. Biologically enhanced cathode design for improved capacity and cycle life for lithium–oxygen batteries. *Nature Communications* **4**: 2756.

Oldroyd, G. E. D., J. D. Murray, P. S. Poole, and J. A. Downie. 2011. The rules of engagement in the legume–rhizobial symbiosis. *Annual Reviews in Genetics* **45**: 119–144.

Oliveri, P., Q. Tu, and E. H. Davidson. 2008. Global regulatory logic for specification of an embryonic cell lineage. *Proceedings of the National Academy of Sciences USA* **105**: 5955–5962.

Orr, H. A. 2009. Fitness and its role in evolutionary genetics. *Nature Reviews Genetics* **10**: 531–539.

Orr, H. A., and J. A. Coyne. 1992. The genetics of adaptation: A reassessment. *American Naturalist* **140**: 725–742.

Ostrander, G. K., K. M. Armstrong, E. T. Knobbe, D. Gerace, and E. P. Scully. 2000. Rapid transition in the structure of a coral reef community: The effects of coral bleaching and physical disturbance. *Proceedings of the National Academy of Sciences USA* **97**: 5297–5302.

Ovaskainen, O., S. Skorokhodova, M. Yakovleva, et al. 2013. Community-level phenological response to climate change. *Proceedings of the National Academy of Sciences USA* **110**: 13434–13439.

Oyama, S. 2006. Boundaries and (constructive) interaction. In E. M. Neumann-Held and C. Rehmann-Sutter, eds. *Genes in Development: Re-reading the Molecular Paradigm*, pp. 272–289. Duke University Press, Durham.

Oyama, S. E., P. E. Griffiths, and R. D. Gray, eds. 2001. *Cycles of Contingency: Developmental Systems and Evolution*. MIT Press, Cambridge.

Paaby, A. B., and M. V. Rockman. 2014. Cryptic genetic variation: Evolution's hidden substrate. *Nature Reviews Genetics* **15**: 247–258.

Pacurar, D. I., H. Thordal-Christensen, M. L. Pacurar, D. Pamfil, C. Botez, and C. Bellini. 2011. *Agrobacterium tumefaciens*: From crown gall tumors to genetic transformation. *Physiological and Molecular Plant Pathology* **76**: 76–81.

Pagel, M., and A. Pomiankowski. 2008. The organismal prospect. In M. Pagel and A. Pomiankowski, eds.

Evolutionary Genomics and Proteomics, pp. 1–10. Sinauer Associates, Sunderland.

Palkovacs, E. P., and D. M. Post. 2008. Eco-evolutionary interactions between predators and prey: Can predator-induced changes to prey communities feed back to shape predator foraging traits? *Evolutionary Ecology Research* **10**: 699–720.

Pan, X., W. Song, and D. Zhang. 2010. Earthworm (*Eisenia foetida*) mucus as a complexing ligand for imidacloprid. *Biology and Fertility of Soils* **46**: 845–850.

Papakostas, S., L. A. Vøllestad, M. Bruneaux, et al. 2014. Gene pleiotropy constrains gene expression changes in fish adapted to different thermal conditions. *Nature Communications* **5**: 4071.

Parcy, F. 2005. Flowering: A time for integration. *International Journal of Developmental Biology* **49**: 585–593.

Parmesan, C. 2006. Ecological and evolutionary responses to recent climate change. *Annual Reviews of Ecology, Evolution and Systematics* **37**: 637–669.

Paszkowski, J., and U. Grossniklaus. 2011. Selected aspects of transgenerational epigenetic inheritance and resetting in plants. *Current Opinion in Plant Biology* **14**: 195–203.

Patel, D., and K. A. Franklin. 2009. Temperature-regulation of plant architecture. *Plant Signaling and Behavior* **4**: 577–579.

Paul, B. D., D. R. Buchholz, L. Fu, and Y.-B. Shi. 2007. SRC-p300 coactivator complex is required for thyroid hormone-induced amphibian metamorphosis. *Journal of Biological Chemistry* **282**: 7472–7481.

Paul, M., T. Pellny, and O. Goddijn. 2001. Enhancing photosynthesis with sugar signals. *Trends in Plant Science* **6**: 197–200.

Pausas, J. G., and J. E. Keeley. 2014. Abrupt climate-independent fire regime changes. *Ecosystems* **17**: 1109–1120.

Pelletier, F., D. Garant, and A. P. Hendry. 2009. Eco-evolutionary dynamics. *Philosophical Transactions of the Royal Society B* **364**: 1483–1489.

Pelletreau, K. N., J. M. Wordul, K. E. Sarver, and M. E. Rumpho. 2012. Laboratory culturing of *Elysia chlorotica* reveals a shift from transient to permanent kleptoplasty. *Symbiosis* **58**: 221–232.

Peña, C. J., Y. D. Neugut, and F. A. Champagne. 2013. Developmental timing of the effects of maternal care on gene expression and epigenetic regulation of hormone receptor levels in female rats. *Endocrinology* **154**: 4340–4351.

Penfield, S. 2008. Temperature perception and signal transduction in plants. *New Phytologist* **179**: 615–628.

Pennekamp, F., K. A. Mitchell, A. Chaine, and N. Schtickzelle. 2014. Dispersal propensity in *Tetrahymena thermophila* ciliates: A reaction norm perspective. *Evolution* **68**: 2319–2330.

Pennings, S. C., and R. M. Callaway. 2000. The advantages of clonal integration under different ecological conditions: A community-wide test. *Ecology* **81**: 709–716.

Peoples, M. B., D. F. Herridge, and J. K. Ladha. 1995. Biological nitrogen fixation: An efficient source of nitrogen for sustainable agricultural production. *Plant Soil* **174**: 3–28.

Peter, I. S., and E. H. Davidson. 2010. The endoderm gene regulatory network in sea urchin embryos up to mid-blastula stage. *Developmental Biology* **340**: 188–199.

Peter, I. S., and E. H. Davidson. 2011. A gene regulatory network controlling the embryonic specification of endoderm. *Nature* **474**: 635–639.

Petrescu, A. S., E. B. Mondor, and B. D. Roitberg. 2001. Subversion of alarm communication: Do plants habituate aphids to their own alarm signals? *Canadian Journal of Zoology* **79**: 737–340.

Petrusek, A., R. Tollrian, K. Schwenk, A. Haas, and C. Laforsch. 2009. A "crown of thorns" is an inducible defense that protects *Daphnia* against an ancient predator. *Proceedings of the National Academy of Sciences USA* **106**: 2248–2252.

Pfennig, D. W. 1992. Proximate and functional causes of polyphenism in an anuran tadpole. *Functional Ecology* **6**: 167–174.

Pfennig, D. W., and P. J. Murphy. 2002. How fluctuating competition and phenotypic plasticity mediate species divergence. *Evolution* **56**: 1217–1228.

Pfennig, D. W., and K. S. Pfennig. 2012. Development and evolution of character displacement. *Annals of the NY Academy of Sciences: The Year in Evolutionary Biology* **1256**: 89–107.

Pfennig, D. W., A. M. Rice, and R. A. Martin. 2006. Ecological opportunity and phenotypic plasticity interact to promote character displacement and species coexistence. *Ecology* **87**: 769–779.

Pfennig, D. W., M. A. Wund, E. C. Snell-Rood, T. Cruickshank, C. D. Schlichting, and A. P. Moczek. 2010. Phenotypic plasticity's impacts on diversification and speciation. *Trends in Ecology and Evolution* **25**: 459–467.

Pfennig, K. S., and D. W. Pfennig. 2009. Character displacement: Ecological and reproductive responses to a common evolutionary problem. *The Quarterly Review of Biology* **84**: 253.

Philippot, L., J. M. Raaijmakers, P. Lemanceau, and W. H. van der Putten. 2013. Going back to the roots: The microbial ecology of the rhizosphere. *Nature Reviews Microbiology* **11**: 789–799.

Phillips, P. C. 2008. Epistasis: The essential role of gene interactions in the structure and evolution of genetic systems. *Nature Reviews Genetics* **9**: 855–867.

Pianka, E. 1994. *Evolutionary Ecology*, 5th ed. HarperCollins College Publishers, New York.

Pickett, S. T. A., M. L. Cadenasso, and C. G. Jones. 2001. Generation of heterogeneity by organisms: Creation, maintenance, and transformation. In M. L. Hutchings, E. A. John, and A. J. A. Stewart, eds. *Ecological Consequences of Habitat Heterogeneity*, pp. 33–52. Blackwell, London.

Pierik, R., and M. de Wit. 2013. Shade avoidance: Phytochrome signalling and other aboveground neighbour detection cues. *Journal of Experimental Botany* **65**: 2815–2824.

Pierik, R., T. Djakovic-Petrovic, D. H. Keuskamp, M. de Wit, and L. A. C. J. Voesenek. 2009. Auxin and ethylene regulate elongation responses to neighbor proximity signals independent of gibberellin and DELLA proteins in *Arabidopsis*. *Plant Physiology* **149**: 1701–1712.

Piersma, T. H., and Å. Lindström. 1997. Rapid reversible changes in organ size as a component of adaptive behaviour. *Trends in Ecology and Evolution* **12**: 134–138.

Pigliucci, M. 2001. *Phenotypic Plasticity: Beyond Nature and Nurture*. Johns Hopkins University Press, Baltimore.

Pigliucci, M., and G. B. Müller, eds. 2010. *Evolution: The Extended Synthesis*. MIT Press, Cambridge.

Plasschaert, M. N., and M. S. Bartolemei. 2014. Genomic imprinting in development, growth, behavior and stem cells. *Development* **141**: 1805–1813.

Plucain, J., T. Hindré, M. Le Gac, et al. 2014. Epistasis and allele specificity in the emergence of a stable polymorphism in *Escherichia coli*. *Science* **343**: 1366–1369.

Podolsky, R. D., and J. S. McAlister. 2005. Developmental plasticity in *Macrophiothrix* brittlestars: Are morphologically convergent larvae also convergently plastic? *Biological Bulletin* **209**: 127–138.

Polatto, L. P., J. Chaud-Netto, and V. V. Alves-Junior. 2014. Influence of abiotic factors and floral resource availability on daily foraging activity of bees. *Journal of Insect Behavior* **27**: 1–20.

Popovkin, A. V., K. G. Mathews, J. C. M. Santos, M. C. Molina, and L. Struwe. 2011. *Spigelia genuflexa* (Loganiaceae), a new geocarpic species from the Atlantic forest of northeastern Bahia, Brazil. *PhytoKeys* **6**: 47–56.

Post, D. M., and E. P. Palkovacs. 2009. Eco-evolutionary feedbacks in community and ecosystem ecology: Interactions between the ecological theatre and the evolutionary play. *Philosophical Transactions of the Royal Society B* **364**: 1620–1640.

Potts, S. G., J. C. Biesmeijer, C. Kremens, P. Neumann, O. Schweiger, and W. E. Kunin. 2010. Global pollinator declines: Trends, impacts and drivers. *Trends in Ecology and Evolution* **25**: 345–353.

Powell, A., D. J. Smith, L. J. Hepburn, T. Jones, J. Berman, J. Jompa, and J. J. Bell. 2014. Reduced diversity and high sponge abundance on a sedimented Indo-Pacific reef system: Implications for future changes in environmental quality. *Public Library of Science ONE* **9**:e85253.

Pradeu, T. 2011. A mixed self: The role of symbiosis in development. *Biological Theory* **6**: 80–88.

Pradeu, T., L. Laplane, M. Morange, A. Nicoglou, and M. Vervoort. 2011. The boundaries of development. *Biological Theory* **6**: 1–3.

Pratchett, M. S., A. S. Hoey, and S. K. Wilson. 2014. Reef degredation and the loss of critical ecosystem goods and services provided by coral reef fishes. *Current Opinion in Environmental Sustainability* **7**: 37–43.

Pratchett, M. S., A. S. Hoey, S. K. Wilson, V. Messmer, and N. A. J. Graham. 2011. Changes in biodiversity and functioning of reef fish assemblages following coral bleaching and coral loss. *Diversity* **3**: 424–452.

Prati, D., and O. Bossdorf. 2004. Allelopathic inhibition of germination by *Alliaria petiolata*. *American Journal of Botany* **91**: 285–288.

Press, M. 1998. Dracula or Robin Hood? A functional role for root hemiparasites in nutrient poor ecosystems. *Oikos* **82**: 609–611.

Prestrud, P. 1991. Adaptations by the Arctic fox (*Alopex lagopus*) to the polar winter. *Arctic* **44**: 132–138.

Pringle, R. M. 2008. Elephants as agents of habitat creation for small vertebrates at the patch scale. *Ecology* **89**: 26–33.

Pringle, R. M., T. P. Young, D. I. Rubenstein, and D. J. McCauley. 2007. Herbivore-initiated interaction cascades and their modulation by productivity in an African savanna. *Proceedings of the National Academy of Sciences USA* **104**: 193–197.

Provine, W. 1971. *The Origins of Theoretical Population Biology*. University of Chicago Press, Chicago.

Qu, H., and X. Fang. 2013. A brief review on the Human Encyclopedia of DNA Elements (ENCODE) Project. *Genomics Proteomics Bioinformatics* **11**: 135–141.

Queitsch, C., T. Sanster, and S. Lindquist. 2002. Hsp90 as a capacitor of phenotypic variation. *Nature* **417**: 618–624.

Queral-Regil, A., and R. B. King. 1998. Evidence for phenotypic plasticity in snake body size and relative head dimensions in response to amount and size of prey. *Copeia* **1998**: 423–429.

Quintero, C., K. E. Barton, and K. Boege. 2013. The ontogeny of plant indirect defenses. *Perspectives in Plant Ecology, Evolution and Systematics* **15**: 245–254.

Raguso, R. A., and B. A. Roy. 1998. "Floral" scent productions by *Puccinia* rust fungi that mimic flowers. *Molecular Ecology* **7**: 1127–1136.

Raimondi, P. T., S. E. Forde, L. F. Delph, and C. M. Lively. 2000. Processes structuring communities: Evidence for trait-mediated indirect effects through induced polymorphisms. *Oikos* **91**: 353–361.

Rainey, P. B., and K. Rainey. 2003. Evolution of cooperation and conflict in experimental bacterial populations. *Nature* **425**: 72–74.

Rainey, P. B., and M. Travisano. 1998. Adaptive radiation in a heterogeneous environment. *Nature* **394**: 69–72.

Rampon, C., C. H. Jiang, H. Dong, et al. 2000. Effects of environmental enrichment on gene expression in the brain. *Proceedings of the National Academy of Sciences USA* **97**: 12880–12884.

Rapp, R. A., and J. F. Wendel. 2005. Epigenetics and plant evolution. *New Phytologist* **168**: 81–91.

Rasmann, S., T. G. Kollner, J. Degenhardt, et al. 2005. Recruitment of entomopathogenic nematodes by insect-damaged maize roots. *Nature* **434**: 732–737.

Rassoulzadegan, M., V. Grandjean, P. Gounon, S. Vincent, I. Gillot, and F. Cuzin. 2006. RNA-mediated non-Mendelian inheritance of an epigenetic change in the mouse. *Nature* **441**: 469–474.

Rausher, M. D. 1992. The measurement of selection on quantitative traits: Biases due to environmental covariances between traits and fitness. *Evolution* **46**: 616–626.

Read, D. B., A. G. Bengough, P. J. Gregory, J. W. Crawford, D. Robinson, C. M. Scrimgeour, I. M. Young, K. Zhang, and X. Zhang. 2003. Plant roots release phospholipid surfactants that modify the physical and chemical properties of soil. *New Phytologist* **157**: 315–326.

Reed, T. E., R. S. Waples, D. E. Schindler, J. J. Hard, and M. T. Kinnison. 2010. Phenotypic plasticity and population viability: The importance of environmental predictability. *Proceedings of the Royal Society B* **277**: 3391–3400.

Reeve, H. K., and P. W. Sherman. 1993. Adaptation and the goals of evolutionary research. *Quarterly Reviews of Biology* **68**: 1–31.

Refardt, D., and P. B. Rainey. 2010. Tuning a genetic switch: Experimental evolution and natural variation of prophage induction. *Evolution* **64**: 1086–1097.

Rehrig, E. M., H. M. Appel, and J. C. Schultz. 2011. Measuring "normalcy" in plant gene expression after herbivore attack. *Molecular Ecology Resources* **11**: 294–304.

Reichman, O. J., and E. W. Seabloom. 2002. The role of pocket gophers as subterranean ecosystem engineers. *Trends in Ecology and Evolution* **17**: 44–49.

Reik, W. 2007. Stability and flexibility of epigenetic gene regulation in mammalian development. *Nature* **447**: 425–432.

Reinhart, K. O., and R. M. Callaway. 2006. Soil biota and invasive plants. *New Phytologist* **170**: 445–457.

Reinhart, K. O., A. Packer, W. H. van der Putten, and K. Clay. 2003. Plant–soil biota interactions and spatial distribution of black cherry in its native and invasive ranges. *Ecology Letters* **6**: 1046–1050.

Relyea, R. A. 2004. Fine-tuned phenotypes: Tadpole plasticity under 16 combinations of predators and competitors. *Ecology* **85**: 172–179.

Relyea, R. A. 2005. The lethal impacts of Roundup and predatory stress on six species of North American tadpoles. *Archives of Environmental Contamination and Toxicology* **48**: 351–357.

Relyea, R. A., and J. Hoverman. 2006. Assessing the ecology in ecotoxicology: A review and synthesis in freshwater ecosystems. *Ecology Letters* **9**: 1157–1171.

Remold, S. K., and R. E. Lenski. 2004. Pervasive joint influence of epistasis and plasticity on mutational effects in *Escherichia coli*. *Nature Genetics* **36**: 423–426.

Remy, J.-J. 2010. Stable inheritance of an acquired behavior in *Caenorhabditis elegans*. *Current Biology* **20**: R877–R878.

Ren, P., C.-S. Lin, R. Johnson, P. S. Albert, D. Pilgrim, and D. L. Riddle. 1996. Control of *C. elegans* larval development by neuronal expression of a TGF-β homolog. *Science* **274**: 1389–1391.

Renn, S. C., N. Aubin-Horth, and H. A. Hofmann. 2008. Fish and chips: Functional genomics of social plasticity in an African cichlid fish. *Journal of Experimental Biology* **211**: 3041–3056.

Resetarits, W. J., Jr. 1996. Oviposition site choice and life history evolution. *American Zoologist* **36**: 205–215.

Reznick, D. N. 2013. A critical look at reciprocity in ecology and evolution: Introduction to the symposium. *American Naturalist* **181**: S1–S8.

Reznick, D. N., and J. Travis. 1996. The empirical study of adaptation in natural populations. In M. R. Rose and G. V. Lauder, eds. *Adaptation*, pp. 243–289. Academic Press, San Diego.

Rice, K. J., and N. C. Emery. 2003. Managing microevolution: Restoration in the face of global change. *Frontiers in Ecology and the Environment* **1**: 469–478.

Richards, C. L., O. Bossdorf, N. Z. Muth, J. Gurevitch, and M. Pigliucci. 2006. Jack of all trades, master of some? On the role of phenotypic plasticity in plant invasions. *Ecology Letters* **9**: 981–993.

Richards, C. L., A. W. Schrey, and M. Pigliucci. 2012. Invasion of diverse habitats by few Japanese knotweed genotypes is correlated with epigenetic differentiation. *Ecology Letters* **15**: 1016–1025.

Richards, C. L., K. J. F. Verhoeven, and O. Bossdorf. 2012. Evolutionary significance of epigenetic variation. In J. F. Wendel, J. Greilhuber, J. Dolezel, and I. J. Leitch, eds. *Plant Genome Diversity Volume 1*, pp. 257–274. Springer-Verlag, Vienna.

Richards, E. J. 1997. DNA methylation and plant development. *Trends in Genetics* **13**: 319–323.

Richards, E. J. 2006. Inherited epigenetic variation: Revisiting "soft inheritance." *Nature Reviews Genetics* **7**: 395–401.

Richardson, D. M., N. Allsopp, C. M. D'Antonio, S. J. Milton, and M. Rejmanek. 2000. Plant invasions: The role of mutualisms. *Biological Reviews* **75**: 65–93.

Ridenour, W. M., and R. M. Callaway. 2001. The relative importance of allelopathy in interference: The effects of an invasive weed on a native bunchgrass. *Oecologia* **126**: 444–450.

Riek, A., and F. Geiser. 2012. Developmental phenotypic plasticity in a marsupial. *Journal of Experimental Biology* **215**: 1552–1558.

Riessen, H. P., and J. B. Trevett-Smith. 2009. Turning inducible defenses on and off: Adaptive responses of *Daphnia* to a gape-limited predator. *Ecology* **90**: 3455–3469.

Rifkin, S. A., D. Houle, J. Kim, and K. P. White. 2005. A mutation accumulation assay reveals a broad capacity for rapid evolution of gene expression. *Nature* **438**: 220–223.

Risch, S. J., and F. R. Rickson. 1981. Mutualism in which ants must be present before plants produce food bodies. *Nature* **291**: 149–150.

Roach, D. A., and R. D. Wulff. 1987. Maternal effects in plants. *Annual Review of Ecology and Systematics* **18**: 209–235.

Robinson, D. 1994. The responses of plants to non-uniform supplies of nutrients. *New Phytologist* **127**: 635–674.

Rodger, S., A. G. Bengough, B. S. Griffiths, V. Stubbs, and I. M. Young. 2003. Does the presence of detached root border cells of *Zea mays* alter the activity of the pathogenic nematode *Meloidogyne incognita*? *Phytopathology* **93**: 1111–1114.

Roelofs, D., M. G. M. Aarts, H. Schat, and N. M. van Straalen. 2008. Functional ecological genomics to demonstrate general and specific responses to abiotic stress. *Functional Ecology* **22**: 8–18.

Rogers, S. M., D. A. Cullen, M. L. Anstey, et al. 2014. Rapid behavioural gregarization in the desert locust, *Schistocerca gregaria* entails synchronous changes in both activity and attraction to conspecifics. *Journal of Insect Physiology* **65**: 9–26.

Rohlfs M., M. Albert, N. P. Keller, and F. Kempken. 2007. Secondary chemicals protect mould from fungivory. *Biology Letters* **3**: 523–525.

Rose, C. S. 2004. Thyroid hormone-mediated development in vertebrates: What makes frogs unique. In B. K. Hall, R. D. Pearson, and G. B. Müller, eds. *Environment, Development and Evolution: Toward a Synthesis*, pp. 197–237. MIT Press, Cambridge.

Rosengaus, R. B., M. R. Guldin, and J. F. A. Traniello. 1998. Inhibitory effect of termite fecal pellets on fungal spore germination. *Journal of Chemical Ecology* **24**: 1697–1706.

Rosengaus, R. B., K. Mead, W. S. Du Comb, R. W. Benson, and V. G. Godoy. 2013. Nest sanitation through defecation: Antifungal properties of wood cockroach feces. *Naturwissenschaften* **100**: 1051–1059.

Roy, B. A. 1993. Floral mimicry by a plant pathogen. *Nature* **362**: 56–58.

Roy, B. A. 1994. The effects of pathogen-induced pseudo-flowers and buttercups on each other's insect visitation. *Ecology* **75**: 352–358.

Royle, N. J., P. T. Smiseth, and M. Kölliker, eds. 2012. *The Evolution of Parental Care*. Oxford University Press, Oxford.

Rumpho, M. E., K. N. Pelletreau, A. Moustafa, and D. Bhattacharya. 2011. The making of a photosynthetic animal. *Journal of Experimental Biology* **214**: 303–311.

Rumpho, M. E., E. J. Summer, and J. R. Manhart. 2000. Solar-powered sea slugs: Mollusc/algal chloroplast symbiosis. *Plant Physiology* **123**: 29–38.

Rumpho, M. E., J. M. Worful, J. Lee, et al. 2008. Horizontal gene transfer of the algal nuclear gene *psbO* to the photosynthetic sea slug *Elysia chlorotica*. *Proceedings of the National Acadamy of Sciences USA* **105**: 17867–17871.

Ryser, P., and L. Eek. 2000. Consequences of phenotypic plasticity vs. interspecific differences in leaf and root traits for acquisition of aboveground and belowground resources. *American Journal of Botany* **87**: 402–411.

Saastamoinen, P., D. van der Sterren, N. Vastenhout, B. J. Zwaan, and P. M. Brakefield. 2010. Predictive adaptive responses: Condition-dependent impact of adult nutrition and flight in the tropical butterfly *Bicyclus anynana*. *American Naturalist* **176**: 686–698.

Salinas, S., and S. B. Munch. 2012. Thermal legacies: Transgenerational effects of temperature on growth in a vertebrate. *Ecology Letters* **15**: 159–163.

Salter, M. G., K. A. Franklin, and G. C. Whitelam. 2003. Gating of the rapid shade avoidance response by the circadian clock in plants. *Nature* **426**: 680–683.

Saltz, J. B., and S. V. Nuzhdin. 2014. Genetic variation in niche construction: Implications for development and evolutionary genetics. *Trends in Ecology and Evolution* **29**: 8–14.

Sanford, E., B. Gaylord, A. Hettinger, E. A. Lenz, K. Meyer, and T. M. Hill. 2014. Ocean acidification increases the vulnerability of native oysters to predation by invasive snails. *Proceedings of the Royal Society B* **281**: 20132681.

Sarkar, S. 2004. From the reaktionsnorm to the evolution of adaptive plasticity: A historical sketch, 1909–1999. In T. J. DeWitt and S. M. Scheiner, eds. *Phenotypic Plasticity: Functional and Conceptual Approaches*, pp. 10–30. Oxford University Press, Oxford.

Sarkar, S. 2006. From genes as determinants to DNA as resource. In E. M. Neumann-Held and C. Rehmann-Sutter, eds. *Genes in Development: Re-reading the Molecular Paradigm*, pp. 77–95. Duke University Press, Durham.

Sarre, S. D., A. Georges, and A. Quinn. 2004. The ends of a continuum: Genetic and temperature-dependent sex determination in reptiles. *BioEssays* **26**: 639–645.

Sassi, P. L., C. E. Borghi, and F. Bozinovic. 2007. Spatial and seasonal plasticity in digestive morphology of cavies

(*Microcavia australis*) inhabiting habitats with different plant qualities. *Journal of Mammalogy* **88**: 165–172.

Schaum, C. E., and S. Collins. 2014. Plasticity predicts evolution in a marine alga. *Proceedings of the Royal Society B* **281**: 20141486.

Scheiner, S. M. 1993. Genetics and evolution of phenotypic plasticity. *Annual Review of Ecology and Systematics* **24**: 35–68.

Scheiner, S. M. 2002. Selection experiments and the study of phenotypic plasticity. *Journal of Evolutionary Biology* **15**: 889–898.

Scheiner, S. M. 2006. Genotype–environment interactions and evolution. In C. W. Fox and C. M. Wolf, eds. *Evolutionary Genetics: Concepts and Case Studies*, pp. 326–338. Oxford University Press, London.

Scheiner, S. M. 2013. The genetics of phenotypic plasticity. XII. Temporal and spatial heterogeneity. *Ecology and Evolution* **3**: 4596–4609.

Scheiner, S. M. 2014. The genetics of phenotypic plasticity. XIII. Interactions with developmental instability. *Ecology and Evolution* **4**: 1347–1360.

Scheiner, S. M., M. Barfield, and R. D. Holt. 2012. The genetics of phenotypic plasticity. XI. Joint evolution of plasticity and dispersal rate. *Ecology and Evolution* **2**: 2027–2039.

Scheiner, S. M., K. Donohue, L. A. Dorn, S. J. Mazer, and L. M. Wolfe. 2002. Reducing environmental bias when measuring natural selection. *Evolution* **56**: 2156–2167.

Scheiner, S. M., and C. J. Goodnight. 1984. The comparison of phenotypic plasticity and genetic variation in populations of the grass *Danthonia spicata*. *Evolution* **38**: 845–855.

Scheiner, S. M., and R. D. Holt. 2012. The genetics of phenotypic plasticity. X. Variation versus uncertainty. *Ecology and Evolution* **2**: 751–767.

Scheiner S. M., R. J. Mitchell, and H. S. Callahan. 2000. Using path analysis to measure natural selection. *Journal of Evolutionary Biology* **13**: 423–433.

Schenk, H. J. 2006. Root competition: Beyond resource depletion. *Journal of Ecology* **94**: 725–739.

Schenk, P. M., K. Kazan, I. Wilson, et al. 2000. Coordinated plant defense responses in *Arabidopsis* revealed by microarray analysis. *Proceedings of the National Academy of Sciences USA* **97**: 11655–11660.

Schiestl, F. P., and S. D. Johnson. 2013. Pollinator-mediated evolution of floral signals. *Trends in Ecology and Evolution* **28**: 307–315.

Schlichting, C. D. 1986. The evolution of phenotypic plasticity in plants. *Annual Review of Ecology and Systematics* **17**: 667–693.

Schlichting C. D., and M. Pigliucci. 1998. *Phenotypic Evolution: A Reaction Norm Perspective*. Sinauer Associates, Sunderland.

Schlichting, C. D., and H. Smith. 2002. Phenotypic plasticity: Linking molecular mechanisms with evolutionary outcomes. *Evolutionary Ecology* **16**: 189–211.

Schluter, D. 2000a. *The Ecology of Adaptive Radiation*. Oxford University Press, New York.

Schluter, D. 2000b. Ecological character displacement in adaptive radiation. *American Naturalist* **156**: S4–S16.

Schmalhausen, I. I. 1949. *Factors of Evolution*, transl. I. Dordick. Blakiston Press, Philadelphia.

Schmid-Hempel, P. 2009. Immune defence, parasite evasion strategies and their relevance for macroscopic phenomena such as virulence. *Philosophical Transactions of the Royal Society B* **364**: 85–98.

Schmitt, J., J. Niles, and R. Wulff. 1992. Norms of reaction of seed traits to maternal environments in *Plantago lanceolata*. *American Naturalist* **139**: 451–466.

Schmitt J., J. R. Stinchcombe, M. S. Heschel, and H. Huber. 2003. The adaptive evolution of plasticity: Phytochrome-mediated shade avoidance responses. *Integrative Comparative Biology* **43**: 459–469.

Schmitz, O. J. 2004. From mesocosms to the field: The role and value of cage experiments in understanding top–down effects in ecosystems. In W. W. Weisser and E. Siemann, eds. *Insects and Ecosystem Function*, pp. 277–302. Springer Series in Ecological Studies. Springer-Verlag, Berlin.

Schmitz, O. J. 2006. Predators have large effects on ecosystem properties by changing plant diversity not biomass. *Ecology* **87**: 1432–1437.

Schmitz, O. J., J. H. Grabowski, B. L. Peckarsky, E. L. Preisser, G. C. Trussel, and J. R. Vonesh. 2008. From individuals to ecosystem function: Toward an integration of evolutionary and ecosystem ecology. *Ecology* **89**: 2436–2445.

Schmitz, O. J., O. Ovadia, and V. Krivan. 2004. Trophic cascades: The primacy of trait-mediated interactions. *Ecology Letters* **7**: 153–163.

Schmitz, O. J., and K. B. Suttle. 2001. Effects of top predator species on the nature of indirect effects in an old field food web. *Ecology* **82**: 2072–2081.

Schmitz, R. J., M. D. Schultz, M. G. Lewsey, et al. 2011. Transgenerational epigenetic instability is a source of novel methylation variants. *Science* **334**: 369–373.

Schnekenburger, M., M. Dicato, and M. Diederich. 2014. Plant-derived epigenetic modulators for cancer treatment and prevention. *Biotechnology Advances* **32**: 1123–1132.

Schoener, T. W. 2011. The newest synthesis: Understanding the interplay of evolutionary and ecological dynamics. *Science* **331**: 426–429.

Schoeps, A., A. Rudolph, P. Seibold, et al. 2014. Identification of new genetic susceptibility loci for breast cancer through consideration of gene-environment interactions. *Genetic Epidemiology* **38**: 84–93.

Schradin, C., M. Schubert, and N. Pillay. 2006. Winter huddling groups in the striped mouse. *Canadian Journal of Zoology* **84**: 693–698.

Schultz, J. C. 2002. Shared signals and the potential for phylogenetic espionage between plants and animals. *Integrative and Comparative Biology* **42**: 454–462.

Schultz, J. C., and H. M. Appel. 2004. Cross-kingdom cross-talk: Hormones shared by plants and their insect herbivores. *Ecology* **85**: 70–77.

Schutte, B. J., A. S. Davis, S. A. Peinado, and J. Ashigh. 2014. Seed-coat thickness data clarify seed size–seed-bank persistence trade-offs in *Abutilon theophrasti* (Malvaceae). *Seed Science Research* **24**: 119–131.

Schwilk, D. W. 2003. Flammability is a niche construction trait: Canopy architecture affects fire intensity. *American Naturalist* **162**: 725–733.

Schwilk, D. W., and D. D. Ackerly. 2001. Flammability and serotiny as strategies: Correlated evolution in pines. *Oikos* **94**: 326–336.

Schwilk, D. W., and A. C. Caprio. 2011. Scaling from leaf traits to fire behaviour: Community composition predicts fire severity in a temperate forest. *Journal of Ecology* **99**: 970–980.

Scott-Phillips, T. C., K. N. Laland, D. M. Shuker, T. E. Dickins, and S. A. West. 2013. The niche construction perspective: A critical appraisal. *Evolution* **68**: 12331–11243.

Scoville, A. G., L. L. Barnett, S. Bodbyl-Roels, J. K. Kelly, and L. C. Hileman. 2011. Differential regulation of a MYB transcription factor is correlated with transgenerational epigenetic inheritance of trichome density in *Mimulus guttatus*. *New Phytologist* **191**: 251–263.

Seebacher, F., and Franklin, C. E. 2005. Physiological mechanisms of thermoregulation in reptiles: A review. *Journal of Comparative Physiology* B **175**: 533–541.

Seely, M. K., and W. J. Hamilton III. 1976. Fog catchment sand trenches constructed by tenebrionid beetles, *Lepidochora*, from the Namib desert. *Science* **193**: 484–486.

Seino, Y., Y. Suzuki, and K. Sogawa. 1996. An ovicidal substance produced by rice plants in response to oviposition by the whitebacked planthopper, *Sogatella furcifera* (Homoptera: Delphacidae). *Applied Entomology and Zoology* **31**: 467–473.

Sell, A. F. 2000. Morphological defenses induced in situ by the invertebrate predator *Chaoborus*: Comparison of responses between *Daphnia pulex* and *D. rosea*. *Oecologia* **125**: 150–160.

Semchenko, M., E. A. John, and M. J. Hutchings. 2007. Effects of physical connection and genetic identity of neighbouring ramets on root-placement patterns in two clonal species. *New Phytologist* **176**: 644–654.

Seong, K.-H., D. Li, H. Shimizu, R. Nakamura, and S. Ishii. 2011. Inheritance of stress-induced, ATF-2-dependent epigenetic change. *Cell* **145**: 1049–1061.

Seopher. 2006. *Experiment: The Best Working Environment Is . . . ?* <www.seopher.com/articles/experiment_the_best_working_environment>, accessed February 2, 2015.

Seymour, J. R., R. Simó, T. Ahmed, and R. Stocker. 2010. Chemoattraction to dimethylsulfoniopropionate throughout the marine microbial food web. *Science* **329**: 342–345.

Seymour, R. S. 2001. Biophysics and physiology of temperature regulation in thermogenic flowers. *Bioscience Reports* **21**: 223–235.

Shane, M. W., M. E. McCully, M. J. Canny, J. S. Pate, and H. Lambers. 2011. Development and persistence of sandsheaths of *Lyginia barbata* (Restionaceae): Relation to root structural development and longevity. *Annals of Botany* **108**: 1307–1322.

Shavit, A., and J. Griesemer. 2009. There and back again, or, the problem of locality in biodiversity surveys. *Philosophy of Science* **76**: 273–294.

Shea, N., I. Pen, and T. Uller. 2011. Three epigenetic information channels and their different roles in evolution. *Journal of Evolutionary Biology* **24**: 1178–1187.

Sheard, L. B., X. Tan, H. Mao, et al. 2010. Jasmonate perception by inositol-phosphate-potentiated COI1-JAZ co-receptor. *Nature* **468**: 400–405.

Sheldon, C. C., E. J. Finnegan, D. T. Rouse, et al. 2000. The control of flowering by vernalization. *Current Opinion in Plant Biology* **3**: 418–422.

Shi, Y. B. 1994. Molecular biology of amphibian metamorphosis. A new approach to an old problem. *Trends in Endocrinology and Metabolism* **5**: 14–20.

Shimizu, K. K., and M. D. Purugganan. 2005. Evolutionary and ecological genomics of *Arabidopsis*. *Plant Physiology* **138**: 578–584.

Shukla, S., S. M. Meeran, and S. K. Katiyar. 2014. Epigenetic regulation by selected dietary phytochemicals in cancer chemoprevention. *Cancer Letters* **355**: 9–17.

Silver, M., and E. Di Paolo. 2006. Spatial effects favour the evolution of niche construction. *Theoretical Population Biology* **70**: 387–400.

Simola, D. F., C. Ye, N. S. Mutti, et al. 2013. A chromatin link to caste identity in the carpenter ant *Camponotus floridanus*. *Genome Research* **23**: 486–496.

Simpson, G. G. 1953. *Major Features of Evolution*. Simon and Schuster, New York.

Simpson, S. J., G. A. Sword, and N. Lo. 2011. Polyphenism in insects. *Current Biology* **21**: R738–R749.

Slatkin, M. 2009. Epigenetic inheritance and the missing heritability problem. *Genetics* **182**: 845–850.

Slobodkin, L. B. 1968. Toward a predictive theory of evolution. In R. C. Lewontin, ed. *Population Biology and Evolution*, pp. 187–205. Syracuse University Press, Syracuse, NY.

Smeekens, J., and H. A. Hellmann. 2014. Sugar sensing and signaling in plants. *Frontiers in Plant Science* **5**: 113.

Smeyne, R. J., K. Schilling, L. Robertson, et al. 1992. *Fos-lacZ transgenic mice: Mapping sites of gene induction in the central nervous system. Neuron* **8**: 13–23.

Smirnova, A., H. Li, H. Weingart, et al. 2001. Thermoregulated expression of virulence factors in plant-associated bacteria. *Archives of Microbiology* **176**: 393–399.

Smith, C. R., K. E. Anderson, C. V. Tillberg, J. Gadau, and A. V. Suarez. 2008. Caste determination in a polymorphic social insect: Nutritional, social, and genetic factors. *American Naturalist* **172**: 497–507.

Smith, E. N., and L. Kruglyak. 2008. Gene–environment interaction in yeast gene expression. *Public Library of Science Biology* **6**:e83.

Smith, H. 1982. Light quality, photoperception, and plant strategy. *Annual Review of Plant Physiology* **33**: 481–518.

Smith, H. 1995. Physiological and ecological function within the phytochrome family. *Annual Review of Plant Physiology* **46**: 289–315.

Smith, H. 2000. Phytochromes and light signal perception by plants: An emerging synthesis. *Nature* **407**: 585–591.

Smith, H., and Whitelam G. C. 1997. The shade avoidance syndrome: Multiple responses mediated by multiple phytochromes. *Plant, Cell and Environment* **20**: 840–844.

Smith, T. B., and S. Skúlason. 1996. Evolutionary significance of resource polymorphisms in fishes, amphibians, and birds. *Annual Review of Ecology and Systematics* **27**: 111–133.

Smith, V. H., and D. W. Schindler. 2009. Eutrophification science: Where do we go from here? *Trends in Ecology and Evolution* **24**: 201–207.

Snell-Rood, E. C., J. D. Van Dyken, T. Cruickshank, M. J. Wade, and A. P. Moczek. 2010. Toward a population genetic framework of developmental evolution: The costs, limits, and consequences of phenotypic plasticity. *BioEssays* **32**: 71–81.

Sollid, J., and G. E. Nilsson. 2006. Plasticity of respiratory structures: Adaptive remodeling of fish gills induced by ambient oxygen and temperature. *Respiratory Physiology and Neurobiology* **154**: 241–251.

Sommer, R. J., and A. Ogawa. 2011. Hormone signaling and phenotypic plasticity in nematode development and evolution. *Current Biology* **21**: R757–R766.

Song, X., D. N. Kristie, and E. G. Reekie. 2009. Why does elevated CO_2 affect time of flowering? An exploratory study using the photoperiodic flowering mutants of *Arabidopsis thaliana*. *New Phytologist* **181**: 339–346.

Soranno, P. A., K. S. Cheruvelil, E. G. Bissell, et al. 2014. Cross-scale interactions: Quantifying multi-scaled cause-effect relationships in macrosystems. *Frontiers in Ecology and the Environment* **12**: 65–73.

Soubry, A., C. Hoyo, R. L. Jirtle, and S. K. Murphy. 2014. A paternal environmental legacy: Evidence for epigenetic inheritance through the male germ line. *BioEssays* **36**: 359–371.

Spitz, M. R., C. M. Duphorne, M. A. Detry, et al. 2000. Dietary intake of isothionates: Evidence of a joint effect with glutathione *S*-transferase polymorphisms in lung cancer risk. *Cancer Epidemiology, Biomarkers and Prevention* **9**: 1017–1020.

Sprent, J. I. 2001. *Nodulation in Legumes*. Royal Botanic Gardens, Kew.

Sprent, J. I. 2007. Evolving ideas of legume evolution and diversity: A taxonomic perspective on the occurrence of nodulation. *New Phytologist* **174**: 11–25.

Stanton, M. L., and C. Galen. 1997. Life on the edge: Adaptation versus environmentally mediated gene flow in the snow buttercup, *Ranunculus adoneus. The American Naturalist* **150**: 143–178.

Stanton-Geddes, J., and C. G. Anderson. 2011. Does a facultative mutualism limit species range expansion? *Oecologia* **167**: 149–155.

Stappenbeck, T. S., L. V. Hooper, and J. I. Gordon. 2002. Developmental regulation of intestinal angiogenesis by indigenous microbes via Paneth cells. *Proceedings of the National Academy of Sciences USA* **99**: 15451–15455.

Star, B. R., R. J. Stoffels, and H. G. Spencer. 2007a. Single-locus polymorphism in a heterogeneous two-deme model. *Genetics* **176**: 1625–1633.

Star, B. R., R. J. Stoffels, and H. G. Spencer. 2007b. Evolution of fitnesses and allele frequencies in a population with spatially heterogeneous selection pressures. *Genetics* **177**: 1743–1751.

Stearns, S. C. 1982. The role of development in the evolution of life-histories. In J. T. Bonner, ed. *The Role of Development in Evolution*, pp. 237–258. Springer-Verlag, New York.

Stearns, S. C. 1989. The evolutionary significance of phenotypic plasticity. *BioScience* **39**: 436–445.

Stearns, S. C. 2003. Safeguards and spurs. *Nature* **424**: 501–504.

Stearns, S. C. 2009. Endless forms most beautiful and wonderful. *Nature Genetics* **41**: 507–507.

Stearns, S. C., and R. F. Hoekstra. 2000. *Evolution: An Introduction*. Oxford University Press, Oxford.

Stebbins, G. L. 1980. Botany and the synthetic theory of evolution. In E. Mayr and W. B. Provine, eds. *The Evolutionary Synthesis*, pp. 139–152. Harvard University Press, Cambridge.

Steinbrenner, A. D., S. Gómez, S. Osorio, A. R. Fernie, and C. M. Orians. 2011. Herbivore-induced changes in tomato (*Solanum lycopersicum*) primary metabolism: A whole plant perspective. *Journal of Chemical Ecology* **37**: 1294–1303.

Stenseth, N. C., and A. Mysterud. 2002. Climate, changing phenology, and other life history traits: Nonlinearity and match-mismatch to the environment. *Proceedings of the National Academy of Sciences USA* **99**: 13379–13381.

Stern, D. L., and V. Orgogozo. 2009. Is genetic evolution predictable? *Science* **323**: 746–751.

Steufer, J. F., H. J. During, and H. de Kroon. 1994. High benefits of clonal integration in two stoloniferous species in response to heterogeneous light environments. *Journal of Ecology* **82**: 511–518.

Stief, P., and F. Hölker. 2006. Trait-mediated indirect effects of predatory fish on microbial mineralization in aquatic sediments. *Ecology* **87**: 284–307.

Stilling, R. M., T. G. Dinan, and J. F. Cryan. 2014. Microbial genes, brain and behaviour: epigenetic regulation of the gut–brain axis. *Genes, Brain and Behavior* **13**: 69–86.

Stinchcombe, J. R., R. Izem, M. S. Heschel, B. V. McGoey, and J. Schmitt. 2010. Across-environment genetic correlations and the frequency of selective environments shape the evolutionary dynamics of growth rate in *Impatiens capensis*. *Evolution* **64**: 2887–2903.

Stinchcombe, J. R., M. T. Rutter, D. S. Burdick, P. Tiffin, M. D. Rausher, and R. Mauricio. 2002. Testing for environmentally induced bias in phenotypic estimates of natural selection: Theory and practice. *The American Naturalist* **160**: 11–523.

Stinchcombe, J. R., and J. Schmitt. 2006. Ecosystem engineers as selective agents: The effects of leaf litter on emergence time and early growth in *Impatiens capensis*. *Ecology Letters* **9**: 258–270.

Stireman, J.O., III, and D. Cipollini. 2008. Stealth tactics of galling parasites and their potential indirect effects. *New Phytologist* **178**: 462–465.

Stocker. R., J. R. Seymour, A. Samadani, D. E. Hunt, and M. F. Polz. 2008. Rapid chemotactic response enables marine bacteria to exploit ephemeral microscale nutrient patches. *Proceedings of the National Academy of Sciences USA* **105**: 4209–4214.

Stork, W., C. Diezel, R. Halischke, I. Gális, and I. T. Baldwin. 2009. An ecological analysis of the herbivory-elicited JA burst and its metabolism: Plant memory processes and predictions of the moving target model. *Public Library of Science ONE* **4**: e4697.

Stratton, D. A. 1994. Genotype-by-environment interactions for fitness of *Erigeron annuus* show fine-scale selective heterogeneity. *Evolution* **48**: 1607–1618.

Strogatz, S. H. 2001. Exploring complex networks. *Nature* **410**: 268–276.

Suez, J., T. Korem, D. Zeevi, et al. 2014. Artificial sweeteners induce glucose intolerance by altering the gut microbiota. *Nature* **514**: 181–186.

Sultan, S. E. 1987. Evolutionary implications of phenotypic plasticity in plants. *Evolutionary Biology* **21**: 127–178.

Sultan, S. E. 1992. Phenotypic plasticity and the neo-Darwinian legacy. *Evolutionary Trends in Plants* **6**: 61–71.

Sultan, S. E. 1995. Phenotypic plasticity and plant adaptation. *Acta Botanica Neerlandica* **44**: 1–21.

Sultan, S. E. 1996. Phenotypic plasticity for offspring traits in *Polygonum persicaria*. *Ecology* **7**: 1791–1807.

Sultan, S. E. 2000. Phenotypic plasticity for plant development, function and life history. *Trends in Plant Science* **5**: 537–542.

Sultan, S. E. 2001. Phenotypic plasticity for fitness components in *Polygonum* species of contrasting ecological breadth. *Ecology* **82**: 328–343.

Sultan, S. E. 2003a. The promise of ecological developmental biology. *Journal of Experimental Zoology* **296**B: 1–7.

Sultan, S. E. 2003b. Phenotypic plasticity in plants: A case study in ecological development. *Evolution and Development* **5**: 25–33.

Sultan, S. E. 2004. Promising directions in plant phenotypic plasticity. *Perspectives in Plant Ecology, Evolution and Systematics* **6**: 227–233.

Sultan, S. E. 2007. Development in context: The timely emergence of eco-devo. *Trends in Ecology and Evolution* **22**: 575–582.

Sultan, S. E. 2010. Plant developmental responses to the environment: Eco-devo insights. *Current Opinion in Plant Biology* **13**: 96–101.

Sultan, S. E. 2011. Evolutionary implications of individual plasticity. In E. Jablonka and S. Gissis, eds. *From Subtle Fluids to Molecular Biology: Transformations of Lamarckism*, pp. 193–204. MIT Press, Cambridge.

Sultan, S. E., K. Barton, and A. M. Wilczek. 2009. Contrasting patterns of transgenerational plasticity in ecologically distinct congeners. *Ecology* **90**: 1831–1839.

Sultan, S. E., and F. A. Bazzaz. 1993a. Phenotypic plasticity in *Polygonum persicaria*. I. Diversity and uniformity in genotypic norms of reaction to light. *Evolution* **47**: 1009–1031.

Sultan, S. E., and F. A. Bazzaz. 1993b. Phenotypic plasticity in *Polygonum persicaria* II. Norms of reaction to soil moisture and the maintenance of genetic diversity. *Evolution* **47**: 1032–1049.

Sultan, S. E., and F. A. Bazzaz. 1993c. The evolution of ecological breadth for nutrient environment. *Evolution* **47**: 1050–1071.

Sultan, S. E., T. Horgan-Kobelski, L. M. Nichols, C. E. Riggs, and R. K. Waples. 2013. A resurrection study reveals rapid adaptive evolution within populations of an invasive plant. *Evolutionary Applications* **6**: 266–278.

Sultan, S. E., and H. G. Spencer. 2002. Metapopulation structure favors plasticity over local adaptation. *American Naturalist* **160**: 271–283.

Sultan, S. E., and S. C. Stearns. 2005. Environmentally contingent variation: Phenotypic plasticity and norms of reaction. In B. Hall and B. Hallgrímsson, eds. *Variation: A Hierarchical Examination of a Central Concept in Biology*, pp. 303–332. Elsevier Academic Press, New York.

Sultan, S. E., A. M. Wilczek, D. L. Bell, and G. Hand. 1998. Physiological response to complex environments in annual *Polygonum* species of contrasting ecological breadth. *Oecologia* **115**: 564–578.

Suomi, S. J. 2004. How gene–environment interactions influence emotional development in rhesus monkeys. In C. G. Coll, E. L. Bearer, and R. M. Lerner, eds. *Nature and Nurture: The Complex Interplay of Genetic and Environmental Influences on Human Behavior and Development*, pp. 35–51. Lawrence Erlbaum Associates, London.

Syers, J. K., and J. A. Springett. 1984. Earthworms and soil fertility. *Plant and Soil* **76**: 93–104.

Szathmary, E., F. Jordan, and C. Pal. 2001. Can genes explain biological complexity? *Science* **292**: 1315.

Tanaka, S., and Y. Nishide. 2012. Do desert locust hoppers develop gregarious characteristics by watching a video? *Journal of Insect Physiology* **58**: 1060–1071.

Tate, R. L. 1995. *Soil Microbiology*, 3rd ed. John Wiley and Sons, New York.

Taylor, C. T., and J. C. McElwain 2010. Ancient atmospheres and the evolution of oxygen sensing via the hypoxia-inducing factor in metazoans. *Physiology* **25**: 272–279.

Teh, A. L., H. Pan, L. Chen, et al. 2014. The effect of genotype and *in utero* environment on interindividual variation in neonate DNA methylomes. *Genome Research* **24**: 1064–1074.

Terrien, J., M. Perret, and F. Aujard. 2011. Behavioral thermoregulation in mammals: A review. *Frontiers in Bioscience* **16**: 1428–1444.

The Genome Institute at Washington University. 2014. *Genomes*. <http://genome.wustl.edu/genomes>, accessed April 2, 2015.

Thelen, G. C., J. M. Vivanco, B. Newingham, et al. 2005. Insect herbivory stimulates allelopathic exudation by an invasive plant and the suppression of natives. *Ecology Letters* **8**: 209–217.

Theoharides, K. A., and J. S. Dukes. 2007. Plant invasion across space and time: Factors affecting nonindigenous species success during four stages of invasion. *New Phytologist* **176**: 356–273.

Thibert-Plante, X., and A. P. Hendry. 2011. The consequences of phenotypic plasticity for ecological speciation. *Journal of Evolutionary Biology* **24**: 326–342.

Thomas, D. B., and R. E. Fordyce. 2012. Biological plasticity in penguin heat-retention structures. *The Anatomical Record* **295**: 249–256.

Thompson, D. B. 1992. Consumption rates and the evolution of diet-induced plasticity in the head morphology of *Melanoplus femurrubrum* (Orthoptera: Acrididae). *Oecologia* **89**: 204–213.

Tilman, D., J. Hill, and C. Lehman. 2006. Carbon-negative biofuels from low-input high-diversity grassland biomass. *Science* **314**: 1598–1600.

Tilman, D., P. B. Reich, J. Knops, D. Wedin, T. Mielke, and C. Lehman. 2001. Diversity and productivity in a long-term grassland experiment. *Science* **294**: 843–845.

Tishkoff, S. A., R. Varkonyi, N. Cahinhinan, et al. 2001. Haplotype diversity and linkage disequilibrium at human *G6PD*: Recent origin of alleles that confer malarial resistance. *Science* **293**: 455–462.

Thoday, J. M. 1953. Components of fitness. *Symposia of the Society for Experimental Biology* **7**: 96–112.

Tollrian, R., and C. D. Harvell. 1999. *The Ecology and Evolution of Inducible Defenses*. Princeton University Press, Princeton.

Toma, D. P., K. P. White, J. Hirsch, and R. J. Greenspan. 2002. Identification of genes involved in *Drosophila melanogaster* geotaxis, a complex behavioral trait. *Nature Genetics* **31**: 349–353.

Tomlinson, S., P. C. Withers, and S. K. Maloney. 2012. Flexibility in thermoregulatory physiology of two dunnarts, *Sminthopsis macroura* and *S. ooldea* (Marsupialia; Dasyuridae). *Journal of Experimental Biology* **215**: 2236–2246.

Tooker, J. F., J. R. Rohr, W. G. Abrahamson, and C. M. De Moraes. 2008. Gall insects can avoid and alter indirect plant defenses. *New Phytologist* **178**: 657–671.

Travis, J. 1994. Ecological genetics and life-history traits. In L. A. Real, ed. *Ecological Genetics*, pp. 171–204. Princeton University Press, Princeton.

Travis, J., J. Leips, and F. H. Rodd. 2013. Evolution in population parameters: Density-dependent selection or density-dependent fitness? *American Naturalist* **181**: S9–S20.

Travis, J., D. Reznick, R. D. Bassar, A. López-Sepulcre, R. Ferriere, and T. Coulson. 2014. Do eco-evo feedbacks help us understand nature? Answers from studies of the Trinidadian Guppy. *Eco-Evolutionary Dynamics* **50**: 1–40.

Trewavas, A. 2005. Green plants as intelligent organisms. *Trends in Plant Sciences* **10**: 413–419.

Trewavas, A. 2014. *Plant Behaviour and Intelligence*. Oxford University Press, London.

Trussell, G. C. 1996. Phenotypic plasticity in an intertidal snail: The role of a common crab predator. *Evolution* **50**: 448–454.

Tufto, J. 2000. The evolution of plasticity and nonplastic spatial and temporal adaptations in the presence of imperfect environmental cues. *American Naturalist* **156**: 121–130.

Turck, F., and G. Coupland. 2013. Natural variation in epigenetic gene regulation and its effects on plant developmental traits. *Evolution* **68**: 620–631.

Turelli, M., and N. H. Barton. 2004. Polygenic variation maintained by balancing selection: Pleiotropy, sex-dependent allelic effects and $G \times E$ interactions. *Genetics* **166**: 1053–1079.

Turnbaugh, P. J., R. E. Ley, M. Hamady, C. M. Fraser-Liggett, R. Knight, and J. I. Gordon. 2007. The human microbiome project. *Nature* **449**: 804–810.

Turner, A. M., and G. G. Mittelbach. 1990. Predator avoidance and community structure: Interactions among piscivores, planktivores, and plankton. *Ecology* **71**: 2241–2254.

Turner., J. S. 2000. *The Extended Organism: The Physiology of Animal-Built Structures*. Harvard University Press, Cambridge.

Underwood, W., M. Melotto, and S. Y. He. 2007. Role of plant stomata in bacterial invasion. *Cellular Microbiology* **9**: 1621–1629.

Urano, K., Y. Kurihara, M. Seki, and K. Shinozaki. 2010. "Omics" analyses of regulatory networks in plant abiotic stress. *Current Opinion in Plant Biology* **13**: 132–138.

Valladares, F., and Ü. Ninimets. 2007. The architecture of plant crowns: From design rules to light capture and performance. In F. Pugnaire and F. Valladares, eds. *Functional Plant Ecology*, pp. 101–151. Taylor and Francis, New York.

Valladares, F., and Ü. Niinemets. 2008. Shade tolerance, a key plant feature of complex nature and consequences. *Annual Review of Ecology, Evolution and Systematics* **39**: 237–257.

Valladares, F., and R. Pearcy. 1997. Interactions between water stress, sun-shade acclimation, heat tolerance and photoinhibition in the sclerophyll *Heteromeles arbutifolia*. *Plant, Cell and Environment* **20**: 25–36.

Van Beest, F. M., A. Mysterud, L. E. Loe, and J. M. Milner. 2010. Forage quantity, quality and depletion as scale-dependent mechanisms driving habitat selection of a large browsing herbivore. *Journal of Animal Ecology* **79**: 910–922.

Van Buskirk, J., and U. Steiner. 2009. The fitness costs of developmental canalization and plasticity. *Journal of Evolutionary Biology* **22**: 852–860.

Van den Hout, P. J., T. H. Piersma, A. Dekinga, S. K. Lubbe, and G. H. Visser. 2006. Ruddy turnstones *Arenaria interpres* rapidly build pectoral muscle after raptor scares. *Journal of Avian Biology* **37**: 425–430.

Van der Putten, W. H., M. Macel, and M. E. Visser. 2010. Predicting species distribution and abundance responses to climate change: Why it is essential to include biotic interactions across trophic levels. *Philosophical Transactions of the Royal Society B* **365**: 2025–2034.

Van Dijk, P. L. M., G. Staaks, and I. Hardewig. 2002. The effect of fasting and re-feeding on temperature preferance, activity and growth of *Rutilus rutilus*. *Oecologia* **130**: 496–504.

Van Dyken, J. D., and M. J. Wade. 2010. The genetic signature of conditional expression. *Genetics* **184**: 557–570.

Van Rheede van Oudtshoorn, K., and M. W. van Rooyen. 1999. *Dispersal Biology of Desert Plants*. Springer-Verlag, Berlin.

Van Zandt, P. A., and A. A. Agrawal. 2004. Community-wide impacts of herbivore-induced plant responses in milkweed (*Asclepias syriaca*). *Ecology* **85**: 2616–2629.

Varley, K. E., J. Gertz, K. M. Bowling, et al. 2013. Dynamic DNA methylation across diverse human cell lines and tissues. *Genome Research* **23**: 555–567.

Vazquez, D. M., R. Eskandari, A. Phelka, and J. F. Lopez. 2003. Impact of maternal deprivation on brain corticotropin-releasing hormone circuits. *Neuropsychopharmacology* **28**: 898–909.

Vedder, O., S. Bouwhuis, and B. C. Sheldon. 2013. Quantitative assessment of the importance of phenotypic plasticity in adaptation to climate change in wild bird populations. *Public Library of Science Biology* **11**: 1001605.

Venkatesh, B., E. F. Kirkness,Y.-H. Loh, et al. 2006. Ancient noncoding elements conserved in the human genome. *Science* **314**: 1892–1893.

Venn, A. A., J. E. Loram, and A. E. Douglas. 2008. Photosynthetic symbioses in animals. *Journal of Experimental Biology* **59**: 1069–1080.

Vercken, E., M. de Fraipont, A. M. Dufty Jr., and J. Clobert. 2007. Mother's timing and duration of corticosterone exposure modulate offspring size and natal dispersal in the common lizard *Lacerta vivipara*. *Hormones and Behavior* **51**: 379–386.

Vergeer, P., N. Wagemaker, and N. J. Ouborg. 2012. Evidence for an epigenetic role in inbreeding depression. *Biology Letters* **8**: 798–801.

Verhaage, A., I. Vlaardingerbroek, C. Raaymakers, et al. 2011. Rewiring of the jasmonate signaling pathway in *Arabidopsis* during insect herbivory. *Frontiers in Plant Science* 2: Article **47**.

Verheyen, E. M., and C. J. Gottardi. 2010. Regulation of Wnt/β-catenin signaling by protein kinases. *Developmental Dynamics* **239**: 34–44.

Verhoeven, K. J. F., J. J. Jansen, P. J. van Dijk, and A. Biere. 2010. Stress-induced DNA methylation changes and their heritability in asexual dandelions. *New Phytologist* **185**: 1108–1118.

Verhoeven, K. J. F., and V. Preite. 2014. Epigenetic variation in asexually reproducing organisms. *Evolution* **68**: 644–655.

Verhoeven, K. J. F., and T. P. van Gurp. 2012. Transgenerational effects of stress exposure on offspring phenotypes in apomictic dandelion. *Public Library of Science ONE* **7**: e38605.

Verschoor, A. M., M. Vos, and I. van der Stap. 2004. Inducible defences prevent strong population fluctuations in bi-and tritrophic food chains. *Ecology Letters* **7**: 1143–1148.

Via, S. 1987. Genetic constraints on the evolution of phenotypic plasticity. In V. Loeschcke, ed. *Genetic Constraints on Adaptive Evolution*, pp. 47–71. Springer, Berlin.

Via, S., R. Gomulkiewicz, G. De Jong, S. M. Scheiner, C. D. Schlichting, and P. H. Van Tienderen. 1995. Adaptive phenotypic plasticity: Consensus and controversy. *Trends in Ecology and Evolution* **10**: 212–217.

Via, S., and R. Lande. 1985. Genotype–environment interaction and the evolution of phenotypic plasticity. *Evolution* **39**: 505–522.

Via, S., and R. Lande. 1987. Evolution of genetic variability in a spatially heterogeneous environment: Effects of genotype–environment interaction. *Genetical Research* **49**: 147–156.

Vickers, M. H., B. H. Breier, W. S. Cutfield, et al. 2000. Origins of hyperphagia, obesity and hypertension and its postnatal amplification by hypercaloric nutrition. *American Journal of Physiology/Endocrinology and Metabolism* **279**: E83–E87.

Vieira, C., E. G. Pasyukova, Z.-B. Zeng, J. K. B. Hackett, R. F. Lyman, and T. F. C. Mackay. 2000. Genotype–environment interaction for quantitative trait loci affecting lifespan in *Drosophila melanogaster*. *Genetics* **154**: 213–227.

Visser, M. E. 2008. Keeping up with a warming world: Assessing the rate of adaptation to climate change. *Proceedings of the Royal Society B* **275**: 649–659.

Visser, M. E., and C. Both. 2005. Shifts in phenology due to global climate change: The need for a yardstick. *Proceedings of the Royal Society B* **272**: 2561–2569.

Visser, M. E., C. Both, and M. M. Lambrechts. 2004. Global climate change leads to mistimed avian reproduction. *Advances in Ecological Research* **35**: 89–110.

Vitousek, P., and L. R. Walker. 1989. Biological invasion by *Myrica faya* in Hawai'i: Plant demography, nitrogen fixation, ecosystem effects. *Ecological Monographs* **59**: 247–265.

Vitzthum, V. J. 2003. A number no greater than the sum of its parts. The use and abuse of heritability. *Human Biology* **75**: 539–558.

Vivanco, J. M., H. P. Bais, F. R. Stermitz, G. C. Thelen, and R. M. Callaway. 2004. Biogeographical variation in community response to root allelochemistry: Novel weapons and exotic invasion. *Ecology Letters* **7**: 285–292.

Voesenek, L. A. C. J., T. D. Colmer, R. Pierik, F. F. Millenaar, and A. J. M. Peeters. 2006. How plants cope with complete submergence. *New Phytologist* **170**: 213–226.

Voesenek L. A. C. J., J. H. G. M. Rijnders, A. J. M. Peeters, H. M. van de Steeg, and H. de Kroon. 2004. Plant hormones regulate fast shoot elongation under water: From genes to communities. *Ecology* **85**: 16–27.

Vorholt, J. A. 2012. Microbial life in the phyllosphere. *Nature Reviews Microbiology* **10**: 828–840.

Vos, J. G., E. Dybing, H. A. Greim, O. Ladefoged, C. Lambre, J. V. Tarazona, I. Brandt, and A. D. Vethaak. 2000. Health effects of endocrine-disrupting chemicals on wildlife, with special reference to the European situation. *Critical Reviews of Toxicology* **30**: 71–133.

Vreeburg, R. A. M., J. J. Benschop, A. J. M. Peeters, et al. 2005. Ethylene regulates fast apoplastic acidification and expansin A transcription during submergence-induced petiole elongation in *Rumex palustris*. *The Plant Journal* **43**: 597–610.

Wada, M., T. Kagawa, and Y. Sato. 2003. Chloroplast movement. *Annual Review of Plant Biology* **54**: 455–468.

Waddington, C. H. 1957. *The Strategy of the Genes*. Allen and Unwin, London.

Waddington, C. H. 1960. Experiments on canalizing selection. *Genetical Research* **1**: 140–150.

Wade, M. J. 2007. The co-evolutionary genetics of ecological communities. *Nature Reviews Genetics* **8**: 185–195.

Wade, M. J., and S. Kalisz. 1990. The causes of natural selection. *Evolution* **44**: 1947–1955.

Wade, M. J., R. G. Winther, A. F. Agrawal, and C. J. Goodnight. 2001. Alternative definitions of epistasis: Dependence and interaction. *Trends in Ecology and Evolution* **16**: 498–504.

Wagner, A. 1996. Does evolutionary plasticity evolve? *Evolution* 50: 1008–1023.

Wagner, G. P. 2000. What is the promise of developmental evolution? Part I: Why is developmental biology necessary to explain evolutionary innovations? *Journal of Experimental Zoology* **288**B: 95–98.

Wagner, G. P., and J. Zhang. 2011. The pleiotropic structure of the genotype–phenotype map: The evolvability of complex organisms. *Nature Review Genetics* **12**: 204–213.

Wagner, M. R., D. S. Lundberg, D. Coleman-Derr, et al. 2014. Natural soil microbes alter flowering phenology and the intensity of selection on flowering time in a wild *Arabidopsis* relative. *Ecology Letters* **17**: 717–726.

Walbot, V. 1996. Sources and consequences of phenotypic and genotypic plasticity in flowering plants. *Trends in Plant Science* **1**: 27–32.

Walker, T. S., H. P. Bais, E. Grotewold, and J. M. Vivanco. 2003. Root exudation and rhizosphere biology. *Plant Physiology* **132**: 44–51.

Wang, E. T., R. Sandberg, S. Luo, et al. 2008. Alternative isoform regulation in human tissue transcriptomes. *Nature* **456**: 470–475.

Wang, W. B. Vinocur, O. Shoseyov, and A. Altman. 2004. Role of plant heat-shock proteins and molecular chaperones in the abiotic stress response. *Trends in Plant Science* **9**: 245–252.

Wang, Z., and J. Zhang. 2011. Impact of gene expression noise on organismal fitness and the efficacy of natural selection. *Proceedings of the National Academy of Sciences USA* **108**: E67–E76.

Wang, Z., X. Pang, Y. Lu, et al. 2013. A dynamic framework for quantifying the genetic architecture of phenotypic plasticity. *Briefings in Bioinformatics* **14**: 82–95.

Waser, N. M. 2006. Specialization and generalization in plant–pollinator interactions: A historical perspective. In N. M. Waser and J. Ollerton, eds. *Plant–Pollinator Interactions: From Specialization to Generalization*, pp. 3–18. University of Chicago Press, Chicago.

Watt, M., M. E. McCully, and M. J. Canny. 1994. Formation and stabilization of rhizosheaths of *Zea mays* L. *Plant Physiology* **106**: 179–186.

Weiner, J., S. B. Andersen, W. K.-M. Wille, H. W. Griepentrog, and J. M. Olsen. 2010. Evolutionary agroecology: The potential for cooperative, high density, weed-suppressing cereals. *Evolutionary Applications* **3**: 473–479.

Weinig, C., L. A. Dorn, N. C. Kane, et al. 2003. Heterogeneous selection at specific loci in natural environments in *Arabidopsis thaliana*. *Genetics* **165**: 321–329.

Weinig, C., and J. Schmitt. 2004. Environmental effects on the expression of quantitative trait loci and implications for phenotypic evolution. *BioScience* **54**: 627–635.

Werner, D., and P. Müller. 2003. Communication and efficiency in the symbiotic signal exchange. In G. Heldmaier and D. Werner, eds. *Environmental Signal Processing and Adaptation*, pp. 9–34. Springer-Verlag, Berlin.

Werner, E. E., and S. D. Peacor. 2003. A review of trait-mediated indirect interactions in ecological communities. *Ecology* **84**: 1083–1100.

West, A. C., and R. W. Johnstone. 2014. New and emerging HDAC inhibitors for cancer treatment. *Journal of Clinical Investigation* **124**: 30–39.

West-Eberhard, M. J. 2003. *Developmental Plasticity and Evolution*. Oxford University Press, New York.

West-Eberhard, M. J. 2005a. Developmental plasticity and the origin of species differences. *Proceedings of the National Academy of Sciences USA* **102**: 6543–6549.

West-Eberhard, M. J. 2005b. Phenotypic accomodation: Adaptive innovation due to developmental plasticity. *Journal of Experimental Zoology* **304**B: 610–618.

Wickremasinghe, M. G. V., and H. F. V. Emden. 1992. Reactions of adult female parasitoids, particularly *Aphidius rhopalosiphi*, to volatile chemical cues from the host plants of their aphid prey. *Physiological Entomology* **17**: 297–304.

Whitham, T. G., J. K. Bailey, J. A. Schweitzer, et al. 2006. A framework for community and ecosystem genetics: From genes to ecosystems. *Nature Reviews Genetics* **7**: 510–523.

Whitlock, M. C. 1996. The red queen beats the jack-of-all-trades: The limitations on the evolution of phenotypic plasticity and niche breadth. *American Naturalist* **148**: S65–S77.

Whitman, D. W., and A. A. Agrawal. 2009. What is phenotypic plasticity and why is it important? In D. W. Whitman and T. N. Ananthakrishna, eds. *Phenotypic Plasticity of Insects: Mechanisms and Consequences*, pp. 1–63. Science Publishers, CRC Press, Enfield, NH.

Whittle, C. A., S. P. Otto, M. O. Johnston, and J. E. Krochko. 2009. Adaptive epigenetic memory of ancestral temperature regime in *Arabidopsis thaliana*. *Botany* **87**: 650–657.

Wied, A., and C. Galen. 1998. Plant parental care: Conspecific nurse effects in *Frasera speciosa* and *Cirsium scopulorum*. *Ecology* **79**: 1657–1668.

Wikelski, M., and C. Thom. 2000. Marine iguanas shrink to survive El Niño. *Nature* **403**: 37–38.

Wilczek, A. M., L. T. Burghardt, A. R. Cobb, M. D. Cooper, S. M. Welch, and J. Schmitt. 2010. Genetic and physiological bases for phenological responses to current and predicted climates. *Philosophical Transactions of the Royal Society B* **365**: 3129–3147.

Wilczek, A. M., J. L. Roe, M. C. Knapp, et al. 2009. Effects of genetic perturbation on seasonal life history plasticity. *Science* **323**: 930–934.

Wilks, M. 2007. Bacteria and early human development. *Early Human Development* **83**: 165–170.

Williams, C. T., J. C. Gorrell, J. E. Lane, A. G. McAdam, M. M. Humphries, and S. Boutin. 2013. Communal nesting in an "asocial" mammal: Social thermoregulation among spatially dispersed kin. *Behavioral Ecology and Sociobiology* **67**: 757–763.

Williams, G. C. 1992. *Natural Selection*. Oxford University Press, Oxford.

Williams, W. E., H. L. Gorton, and S. M. Witiak. 2003. Chloroplast movements in the field. *Plant, Cell and Environment* **26**: 2005–2014.

Wilson, A. J., J. M. Pemberton, J. G. Pilkington, et al. 2006. Environmental coupling of selection and heritability limits evolution. *Public Library of Science Biology* **4**: e216.

Wimberger, P. H. 1992. Plasticity of fish body shape. The effects of diet, development, family and age in two species of *Geophagus*. *Biological Journal of the Linnean Society* **45**: 197–218.

Windig, J. J. 1994. Genetic correlations and reaction norms in wing patterns of the tropical butterfly *Bicyclus anynana*. *Heredity* **73**: 459–470.

Wohl, D. L., S. Arora, and J. R. Gladstone 2004. Functional redundancy supports biodiversity and ecosystem function in a closed and constant environment. *Ecology* **85**: 1534–1540.

Wolf, J. B., E. D. Brodie, and M. J. Wade, eds. 2000. *Epistasis and the Evolutionary Process*. Oxford University Press, New York.

Wolf, J. B., and M. J. Wade. 2009. What are maternal effects (and what are they not)? *Philosophical Transactions of the Royal Society B* **364**: 1107–1115.

Wolfe, B. E., and J. N. Klironomos. 2005. Breaking new ground: Soil communities and exotic plant invasion. *Bioscience* **55**: 477–487.

Wolkovich, E. M. 2010. Non-native grass litter enhances grazing arthropod assemblages by increasing native shrub growth. *Ecology* **91**: 756–766.

Wolkovich, E. M., B. I. Cook, J. M. Allen, et al. 2012. Warming experiments underpredict plant phenological responses to climate change. *Nature* **485**: 494–497.

Woltereck, R. 1909. Weitere experimentelle Untersuchungen über Arrveranderung, speziell über das Wesen quntitativer Artuntershiede bei Daphniden. *Verhandlungen der Deutschen Zoologischen Gesellschaft* **1909**: 110–172.

Wood, T. K. 1993. Diversity in the New World Membracidae. *Annual Reviews of Entomology* **38**: 409–435.

Wood, T. K., K. C. Olmstead, and S. I. Guttman. 1990. Insect phenology mediated by host-plant water relations. *Evolution* **44**: 629–636.

Woods, W. A., Jr., B. Heinrich, and R. D. Stevenson. 2005. Honeybee flight metabolic rate: Does it depend upon air temperature? *Journal of Experimental Biology* **208**: 1161–1173.

Wray, G. A. 2007. The evolutionary significance of *cis*-regulatory mutations. *Nature Reviews Genetics* **8**: 206–216.

Wray, G. A., H. E. Hoekstra, D. J. Futuyma, et al. 2014. Does evolutionary theory need a rethink? No, all is well. *Nature* **514**: 161–164.

Wright, J. P., and C. G. Jones. 2006. The concept of organisms as ecosystem engineers ten years on: Progress, limitations, and challenges. *Bioscience* **56**: 203–209.

Wright, J. P., C. G. Jones, and A. S. Flecker. 2002. An ecosystem engineer, the beaver, increases species richness at the landscape scale. *Oecologia* **132**: 96–101.

Wright, S. 1931. Evolution in Mendelian populations. *Genetics* **16**: 97–159.

Wu, C.-S., I. Gomez-Mestre, and Y.-C. Kam. 2012. Irreversibility of a bad start: Early exposure to osmotic stress limits growth and adaptive developmental plasticity. *Oecologia* **169**: 15–22.

Wu, C.-T., and J. R. Morris. 2001. Genes, genetics, and epigenetics: A correspondence. *Science* **293**: 1103–1105.

Wu, J., and I. T. Baldwin. 2009. Herbivory-induced signaling in plants: Perception and action. *Plant, Cell and Environment* **32**: 1161–1174.

Wu, J., and I. T. Baldwin. 2010. New insights into plant responses to the attack from insect herbivores. *Annual Review of Genetics* **44**: 1–24.

Yan, H., D. F. Simola, R. Bonasio, J. Liebig, S. L. Berger, and D. Reinberg. 2104. Eusocial insects as emerging models for behavioural epigenetics. *Nature Reviews Genetics* **15**: 677–688.

Yang, Y., and A. Joern. 1994. Gut size changes in relation to variable food quality and body size in grasshoppers. *Functional Ecology* **8**: 36–45.

Yao, M.-C. 2008. RNA rules. *Nature* **451**: 131–132.

Yoshida, T., L. E. Jones, S. P. Ellner, G. F. Fussmann, and N. G. Hairston. 2003. Rapid evolution drives ecological dynamics in a predator-prey system. *Nature* **424**: 303–306.

Young, T. P., M. L. Stanton, and C. E. Christian. 2003. Effects of natural and simulated herbivory on spine lengths of *Acacia drepanolobium* in Kenya. *Oikos* **101**: 171–179.

Yukilevich, R., J. Lachance, F. Aoki, and J. R. True. 2008. Long-term adaptation of epistatic genetic networks. *Evolution* **62**: 2215–2235.

Zahran H. H. 1999. Rhizobium-legume symbiosis and nitrogen fixation under severe conditions and in an arid climate. *Microbiology and Molecular Biology Reviews* **63**: 968–989.

Zarnetske, P. L., S. D. Hacker, E. W. Seabloom, et al. 2012. Biophysical feedback mediates effects of invasive grasses on coastal dune shape. *Ecology* **93**: 1439–1450.

Zeng, W. M. Melotto, and S. Y. He. 2010. Plant stomata: A checkpoint of host immunity and pathogen virulence. *Current Opinion in Biotechnology* **21**: 599–603.

Zhang, J., and M. Lechowicz. 1994. Correlation between time of flowering and phenotypic plasticity in *Arabidopsis thaliana*. *American Journal of Botany* **81**: 1336–1342.

Zhang, P.-J., S.-J. Zheng, J. J. A. van Loon, et al. 2009. Whiteflies interfere with indirect plant defense against spider mites in Lima bean. *Proceedings of the National Academy of Sciences USA* **106**: 21202–21207.

Zhang, T. Y., B. Labonté, X. L. Wen, G. Turecki, and M. J. Meaney. 2013. Epigenetic mechanisms for the early environmental regulation of hippocampal glucocorticoid receptor gene expression in rodents and humans. *Neuropsychopharmacology Reviews* **38**: 111–123.

Zhang, X. Y., O. Clarenz, S. Cokus, et al. 2007. Whole-genome analysis of histone H3 lysine 27 trimethylation in *Arabidopsis*. *Public Library of Science Biology* **5**: 1026–1035.

Zhang, Y., and B. Malmqvist. 1997. Phenotypic plasticity in a suspension-feeding insect, *Simulium lundstromi* (Diptera: Simuliidae), in response to current velocity. *Oikos* **78**: 503–510.

Zhang, Y.-Y., M. Fischer, V. Colot, and O. Bossdorf. 2013. Epigenetic variation creates potential for evolution of plant phenotypic plasticity. *New Phytologist* **197**: 314–322.

Zhou, H.-X. 2008. Protein folding in confined and crowded environments. *Archives of Biochemistry and Biophysics* **469**: 76–82.

Zhou, H.-X. 2013. Influence of crowded cellular environments on protein folding, binding, and oligomerization. *Federation of European Biochemical Societies Letters* **587**: 1053–1061.

Index